EAI/Springer Innovations in Communication and Computing

Series editor
Imrich Chlamtac, European Alliance for Innovation, Gent, Belgium

Editor's Note

The impact of information technologies is creating a new world yet not fully understood. The extent and speed of economic, life style and social changes already perceived in everyday life is hard to estimate without understanding the technological driving forces behind it. This series presents contributed volumes featuring the latest research and development in the various information engineering technologies that play a key role in this process.

The range of topics, focusing primarily on communications and computing engineering include, but are not limited to, wireless networks; mobile communication; design and learning; gaming; interaction; e-health and pervasive healthcare; energy management; smart grids; internet of things; cognitive radio networks; computation; cloud computing; ubiquitous connectivity, and in mode general smart living, smart cities, Internet of Things and more. The series publishes a combination of expanded papers selected from hosted and sponsored European Alliance for Innovation (EAI) conferences that present cutting edge, global research as well as provide new perspectives on traditional related engineering fields. This content, complemented with open calls for contribution of book titles and individual chapters, together maintain Springer's and EAI's high standards of academic excellence. The audience for the books consists of researchers, industry professionals, advanced level students as well as practitioners in related fields of activity include information and communication specialists, security experts, economists, urban planners, doctors, and in general representatives in all those walks of life affected ad contributing to the information revolution.

About EAI

EAI is a grassroots member organization initiated through cooperation between businesses, public, private and government organizations to address the global challenges of Europe's future competitiveness and link the European Research community with its counterparts around the globe. EAI reaches out to hundreds of thousands of individual subscribers on all continents and collaborates with an institutional member base including Fortune 500 companies, government organizations, and educational institutions, provide a free research and innovation platform.

Through its open free membership model EAI promotes a new research and innovation culture based on collaboration, connectivity and recognition of excellence by community.

More information about this series at http://www.springer.com/series/15427

Nishu Gupta • Sara Paiva
Editors

IoT and ICT for Healthcare Applications

Springer

EAI
RESEARCH MEETS INNOVATION

Editors
Nishu Gupta
Electronics and Communication
Engineering
Vaagdevi College of Engineering
Warangal, Telangana, India

Sara Paiva
Technology and Management School
Inst Politécnico de Viana do Castelo
Viana do Castelo, Portugal

ISSN 2522-8595 ISSN 2522-8609 (electronic)
EAI/Springer Innovations in Communication and Computing
ISBN 978-3-030-42933-1 ISBN 978-3-030-42934-8 (eBook)
https://doi.org/10.1007/978-3-030-42934-8

This Springer imprint is published by the registered company Springer Nature Switzerland AG
The registered company address is: Gewerbestrasse 11, 6330 Cham, Switzerland

Foreword

I am pleased to write this Foreword because there is a need for researchers as well as practitioners to constantly keep up with the evolution of technology and its impact, which this book enables. I also believe that educators at various stages of their career can enhance their teaching by incorporating leading patterns and practices such as the ones presented in this book. This book is comprised of various research explorations in the field of Internet of Things (IoT) and Information and Communication Technology (ICT) with special emphasis on Healthcare Applications. It also showcases many research groups that are involved in their study and research.

Although the focus of this work is on Internet of Things, it contains much more that will be of interest to those outside this field. In this book, I see that the editors and authors have conceptualized the intellectual foundations of Information and Communication Technology, elaborated its distinctive pedagogy in healthcare applications, and studied its patterns and impact on the common man with restricted access to medical facilities. This book will certainly help researchers and professional practitioners develop a shared vision and understanding of interpretive discussion. It provides a valuable window on information assurance and covers the necessary components from fundamentals of IoT to challenges in healthcare industry. Risk analysis is a critical process to define both the probability and impact of undesired events particularly in health safety and medical treatments. Its objective is not just to analyze the risk but to inform policies and methods by which risks could be managed.

We know the huge challenges faced in addressing societal problems such as providing universal, affordable access to healthcare. The challenges are both difficult and interesting. Researchers around the world are working on them with tenacity and dedication to develop new approaches that provide new solutions to keep up with the ever-changing potential threats. In this age of global interconnectivity and interdependence, it is necessary to provide practitioners, professionals, and students, with state-of-the-art knowledge on the frontiers in IoT and ICT. This book is a good step in that direction. The authors can be assured that because of their contributions, there will be many readers who will appreciate gaining a better understanding of

Internet of Things and its applications in healthcare industry. I hope that this book will become very useful for practitioners and researchers across the globe to learn and to get motivated about the tremendous capabilities of IoT and ICT in healthcare.

NUS School of Computing
National University of Singapore, Singapore
Mohan Kankanhalli

Preface

This book presents its readers to useful Internet of Things (IoT) along with Information and Communication Technology (ICT) applications and architectures that cater to their improved healthcare requirements. It provides an insight on the vast and fast-growing field of IoT along with ICT. The presented chapters cover the significant technological advancements that IoT and ICT solutions can have in taking care of people's health. Key features of this book are the inclusion and elaboration of recent and emerging developments in various specializations of curing health problems and their solutions by incorporating IoTs and ICTs. IoT solutions can directly serve the general public. Their benefits are enormous, and the range of applicability is also significant. The readability level of the book is highly diversified, and covers topics being taught at undergraduate engineering programs throughout the world. The contents of the book are also fully helpful to the students of graduate studies. The book showcases strong focus on practical implementations. The application domain shows a large diversity in which health, communication technology, medical devices, materials and devices, mobile communication, challenges, and other trendy topics are discussed. The presented solutions aim to help make everyday life easier throughout the world.

This book presents four main parts: fundamentals, data analysis, safety issues, and case studies. The first part talks about the basis of IoT and ICT with emphasis on their role in healthcare applications.

The second part focuses on real-time data processing and health data analytics and presents a detailed analysis of the implementation and computational level challenges in e-healthcare sector.

The third part highlights safety issues such as mimicking biometrics on smart devices, data privacy, security for Internet of medical wearable things, etc.

Finally, the fourth part presents various case studies and real-time experimental setups such as intelligent wearable IoT, continuous monitoring system for elderly based on deep learning algorithm, automated pill dispenser application based on IoT for the patient medication, etc.

This book attracted contributors from all over the world, and we would like to thank all the authors for submitting their works. We extend our appreciation to the reviewers for their review work and gratefully acknowledge all the authors and publishers of the books quoted in the references.

Warangal, India Nishu Gupta
Viana do Castelo, Portugal Sara Paiva

Acknowledgment

I, **Nishu Gupta**, acknowledge the inspiration and blessings of my mother, Smt. Rita Rani Gupta; father, Prof. K.M. Gupta; sister, Smt. Nidhi Gupta; brother-in-law, CA Ritesh Shankar Gupta; and other family members. I am full of gratitude to my wife, Smt. Anamika Gupta, and son, Master Ayaansh Gupta, for the patience shown and encouragement given to complete this venture. I am highly obliged to Prof. Sara Paiva, without whose help, guidance, and support, it would not have been possible to bring this book to fruition. She has been a major driving force toward this and many other such accomplishments. I extend my heartfelt gratitude to the Principal of Vaagdevi College of Engineering, Warangal, Dr. K. Prakash; Head of the ECE Department, Assoc. Prof. M. Shashidhar; and other colleagues and friends for their support and motivation in several ways.

I, **Sara Paiva**, acknowledge my family as main inspiration to my career achievements, my daughter and son, Diana and Leonardo, for being my greatest teachers in life and for providing me with my best moments ever; my husband, Rogério, for standing by me in all occasions and supporting all my decisions and options in life; and my parents, my rock and solid ground, for always being there, understanding what I don't even say. I am also blessed for every single person that come into my life as all of them make me learn something. A word of appreciation also to all my colleagues I work with for the last 15 years at Instituto Politécnico de Viana do Castelo and all researchers with whom I cooperate around the world, such as Prof. Nishu who has been a pleasure to work with and to share the editing of this book.

Last but not the least, **we both** express our heartfelt gratitude to Ms. Eliška Vlčková, Managing Editor, European Alliance for Innovation (EAI), for their continued support and cooperation in publishing this book.

Contents

Part I
Fundamentals

Emerging IoT Technologies in Smart Healthcare

Matthew N. O. Sadiku, Nishu Gupta, Yogita P. Akhare, and Sarhan M. Musa

1 Introduction

The integration of Internet-based technologies and medical services is much anticipated for the sustained growth of healthcare services and their ease of availability at the global level [1].

Technology has always been an integral part of healthcare delivery, enabling health practitioners to use various tools to detect, diagnose, treat, cure, and monitor patients. Typical examples of medical technologies include medicines, medical devices, and biotechnology products. The main goal is to enhance quality of life using these technologies [2]. This chapter deals with emerging technologies in healthcare. It begins by discussing the concept of emerging technology. Then, it covers several emerging technologies in healthcare and discusses some of the applications of the emerging technologies. The last section concludes with some comments.

M. N. O. Sadiku (✉) · Y. P. Akhare · S. M. Musa
Roy G. Perry College of Engineering, Prairie View A&M University, Prairie View, TX, USA
e-mail: sadiku@ieee.org; yakhare@student.pvamu.edu; smmusa@pvamu.edu

N. Gupta
Electronics and Communication Engineering, Vaagdevi College of Engineering, Warangal, Telangana, India
e-mail: nishugupta@ieee.org

N. Gupta, S. Paiva (eds.), *IoT and ICT for Healthcare Applications*, EAI/Springer Innovations in Communication and Computing,
https://doi.org/10.1007/978-3-030-42934-8_1

2 Concept of IoT-Based Healthcare Technology

IoT-based healthcare includes cognitive computing, microwave, voice search, voice recognition, chatbots, social media, blockchain, 3D telepresence technology, 3D printing, wireless technology, mobile technology, 3D ultrasound, biometrics, genetics and genomics, electronic health records, and magnetic resonance imaging (MRI). Among so many technologies, some of the prominent ones are as follows [3–7]:

- *Wireless and Mobile Technology:* This technology would allow medical practice from anywhere, any time, and from any device. It is touching virtually every aspect of our lives. Mobile devices include tablets and smartphone. The use of mobile devices in the healthcare is a recent update, and it is still in infancy. It has the potential for managing chronic illnesses of the aging population [8]. The rise of the Internet age and the proliferation of smart devices have brought profound changes for the practice of medicine.
- *Wearable Technology:* This technology allows wearing light-weight sensors unobtrusively using regular clothes. Wearable gadgets can screen person's physiological capacities 24 hours every day. 3D printing innovation is arriving at the improvement of wearable gadgets. The significant concern is that wearable gadgets present issues with client protection and security [9–11].
- *3D Printing:* 3D printing (3DP) is the method for creating three-dimensional strong items from an advanced model. It has been viewed as one of the mainstays of the third modern upset. From that point forward, it has been utilized in assembling, car, gadgets, avionics, aviation, buyer items, training, amusement, medication, concoction, and gems businesses. Advantages of 3DP in human services incorporate restorative items, medications, and hardware; cost adequacy; expanded profitability; the democratization of plan and producing; and improved coordinated effort. Emergency clinics might make things on request, and this would fundamentally adjust the healthcare inventory network [12].
- *Augmented/Virtual Reality*: Augmented reality (AR) is an exceptionally intelligent, PC-based media condition in which the client turns into the member in a PC-created world. For instance, careful understudies can utilize virtual overlays of the circulatory framework to help direct them during methods. Charging operators can utilize "shrewd glasses" to see understanding protection and charging data when they are away from their PCs. It can help lessen the measure of uneasiness a patient is feeling when medical procedure. It tends to be utilized to prepare specialists in a practical and generally safe mimicked condition. It offers remedial potential and restoration for intense agony and nervousness issue [13, 14].
- *Robotics:* Robots have been playing an increasingly important role in our daily life. They are imperative in numerous enterprises. Mechanical autonomy manages the structure, development, activity, and use of robots. Robots are turning into an indispensable piece of the human service toolbox. Robots assume a significant job in health services as they bring down the quantity of medicinal

blunders and improve human service conveyance. Robots guarantee to make another degree of value human service by giving specialists to quiet. A wide scope of robots is created to fill various needs inside the healthcare condition. This outcomes in different sorts of human service robots [15]. A typical robot for complex surgery is shown in Fig. 1 [16].

- *Cloud Computing:* Computing is changing the way human service suppliers (specialists, centers, and medical clinics) convey administrations to their patients. It might give versatile and financially savvy human service administrations. Healthcare suppliers are progressively confronting sharp challenge and are constrained support less. They are quickly going to the cloud to address the patient requirements [17].
- *Internet of Things* (IoT): IoT permits all devices to be associated with one another through wired or remote correspondence. It has been for some time anticipated that IoT-based healthcare services will upset the human service segment as far as social advantages, infiltration, available consideration, and cost proficiency [18]. A scenario of IoT-enabled healthcare system is depicted in Fig. 2 [9].
- *Blockchain*: This innovation comprises common or disseminated database to keep up developing rundown of exchanges called blocks. With blockchain (BC), exchange records are put away and disseminated overall system members instead of at a focal area. Blockchain in healthcare services will be in clinical preliminary records, administrative consistence, and therapeutic records. The innovation can enable therapeutic specialists to improve and increasingly precise judgments and endorse progressively powerful medications [19].

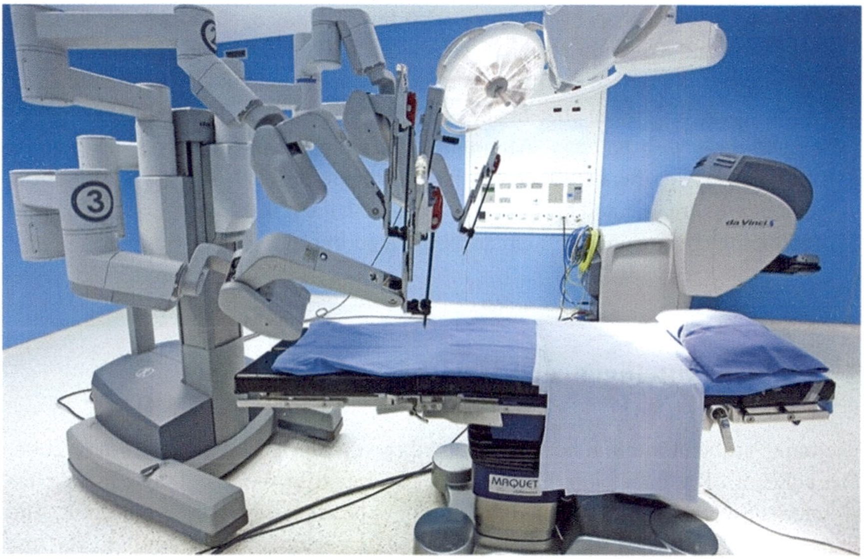

Fig. 1 A robot for complex surgery [16]

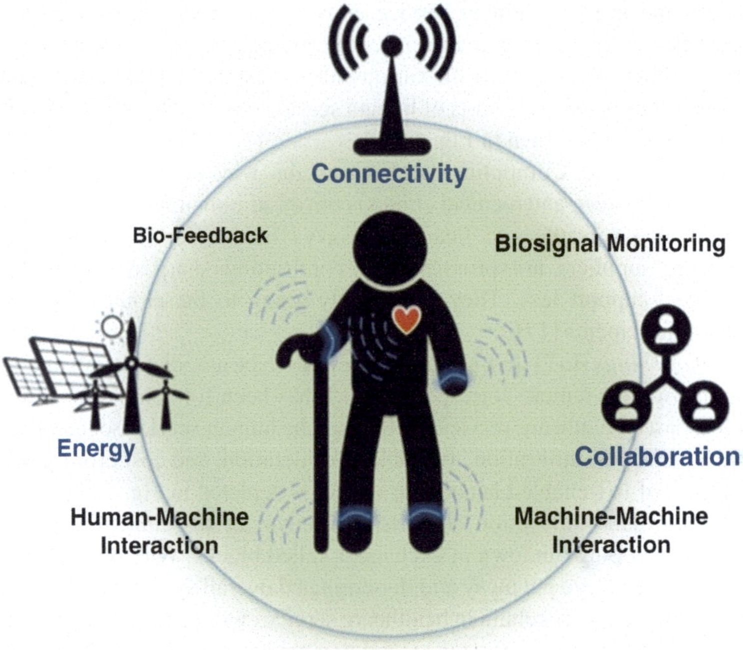

Fig. 2 A typical IoT-based healthcare system [9]

- *Social Media*: Advances in technology are impacting the future of healthcare, being more social than ever before. The Internet has engaged people to share health data and associate utilizing web-based social networking. Web-based life is fundamentally online instruments utilized for PC interceded correspondence. It is an incredible asset that healthcare industry experts can exploit to convey and associate with patients. Although Internet-based life is as yet developing, it has had a significant effect on the healthcare service industry [20]. These technologies are selected because they pose both the risk of disruption and reward of reducing costs. Some of them are illustrated in Fig. 3 [21].

3 Ambient Intelligence in Healthcare Technologies

The term "ambient intelligence" was coined by the European Commission in 2001. The concept of ambient intelligence was developed in the late 1990s in a series of internal workshops conducted by Philips (headquartered in Eindhoven, The Netherlands). The workshops were aimed at understanding the need to put the user in the center of product development. It soon became apparent that a single company could not possibly turn the broad concept of ambient intelligence into

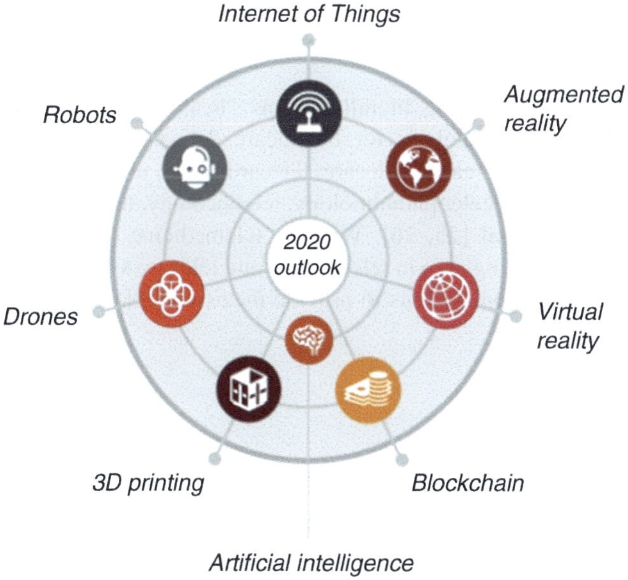

Fig. 3 Some emerging healthcare technologies [21]

reality. Although it was originally envisioned for consumer electronics, ambient intelligence has grown into other areas such as healthcare, public, and productivity domains.

Ambient intelligence is essentially a technology that combines Internet of things (IoT) and artificial intelligence (AI). It is intrinsically and thoroughly connected with AI. It is a technology-enriched environment which involves data acquisition systems, robotic systems, communication technology, wearable devices, decision support, analytics, machine actuators, machine learning, AI, human–computer interaction (HCI), and other capabilities. Figure 2 shows the confluence of different areas into ambient intelligence [22]. Developing advancements exhibited above can be applied to each part of medicinal services. Some famous territories incorporate the accompanying [23]:

- *Chronic Diseases:* A chronic disease is one that can hardly be completely cured and has a long-lasting effect on the human body. The aging population is growing rapidly, and emerging technologies are expected to meet their healthcare needs. Besides aging, another major cause of chronic diseases is unhealthy lifestyles that lead to obesity [24].
- *Electronic Health Records* (EHR): This is the most straightforward utilization of healthcare service innovation. The electronic health record (EHR) is a computerized detail of a patient. EHR comprises records from numerous sources, for example, emergency clinics, suppliers, centers, and general healthcare offices. Healthcare records keep on advancing because of innovation. Any adjustments in

documentation of care significantly affect nursing practice. EHR is accessible all day and every day and has worked in protections to guarantee persistent health data secrecy and security [6].
- *Telemedicine:* Telemedicine literally means "healing at a distance." This is an important core technology for healthcare delivery. Telemedicine's major applications span the areas of telecare, telecardiology, teleradiology, telepathology, teledermatology, teleophthalmology, teleoncology, telepsychiatry, or "tele-everything" in general [25, 26]. Wireless telemedicine (or "m-health") is an important and emerging area in telemedical and telecare systems. It uses current mobile communication systems to provide medical services with a high degree of mobility [27].

4 Benefits

Advancements in science and technology have brought healthcare services to virtually all corners of the world. Healthcare technologies can be used in a several beneficial ways. Emerging healthcare technologies offer several benefits mentioned below [28]

- Cheaper treatment.
- Enhanced monitoring and active care.
- Improved standard of living.
- Rise in life expectancy rate.
- Minimizing healthcare waste.
- Development of better drugs and facilities.

Significant numbers of healthcare service associations are grasping new innovations to build their productivity and adequacy. Receiving these advancements will help healthcare sector to give more patient consideration at a lower cost.

5 Challenges

Among various existing challenges for this upcoming field [29], other challenges include:

- Expenditure generated by healthcare technology will increase but may be expensive to the patients.
- Wireless devices are difficult to adopt because products from different vendors do not always work well together.
- Easy for unauthorized user to collect and analyze any personal or clinical data.
- Strict adherence to the regulatory authorities.

Much remains to be done before these technologies are incorporated into daily life of the elderly. Later on, there will stay a suffering differentiation among safety and security. The future patterns ought to incorporate fixing the way of life of innovation improvement. Notwithstanding their focal points and hindrances, developing advances truly make our lives simpler, particularly in the social insurance area.

6 Conclusion

Healthcare sector has no closure of issues: we as a whole need and anticipate better consideration, rising costs and execution are declining, we live longer with constant ailment, and so forth. In the event that we need healthcare services to improve later on, we should persistently get ready for it.

Today, rising innovations are blasting. Future developments (new medications, new medicines, new gadgets, and so on) will continue changing human service conveyance. Since innovation drives health services, the central issues of health and bliss will remain. We should know about the drivers, line up with them, and work with them to guarantee the best results for society in general.

The insights to healthcare applications shared in this chapter will have a great impact on delivering better and safer healthcare services. More information can be found in [27, 30–35].

References

1. S.M. Davet et al., A framework to manage the early value proposition of emerging healthcare technologies. Ir. J. Manag. **31**(1), 59–75 (2011)
2. M.N.O. Sadiku, Y.P. Akhare, S.M. Musa, Emerging technologies in healthcare: a tutorial. Int. J. Adv. Sci. Res. Eng. **5**(7), 199–204 (2019)
3. H. Thimbleby, Technology and the future of healthcare. J. Public Health Res. **2**(3), 160–167 (2013)
4. M. Halaweh, Emerging technology: what is it? J. Technol. Manag. Innov. **8**(3), 108–115 (2013)
5. 5 Technology trends impacting healthcare, March 2018, https://cbcommunity.comcast.com/browse-all/details/5-technology-trends-impacting-health-care
6. C. Huston, The impact of emerging technology on nursing care: warp speed ahead. Online J. Issues Nurs. **18**(2), 1–19 (2013)
7. M. Patel, J. Wan, Applications, challenges, and prospective in emerging body area networking technologies. *IEEE Wireless Communications*, February 2010, pp. 80–88
8. M. Bajwa, Emerging 21st century medical technologies. Pakistan J. Med. Sci. **30**(3), 649–655 (2014)
9. C.F. Pasluosta et al., An emerging era in the management of Parkinson's disease: wearable technologies and the Internet of things. IEEE J. Biomed. Health Inform. **19**(6), 1873–1881 (2015)
10. K. Zimmermann, Microwave as an emerging technology for the treatment of biohazardous waste: a mini-review. Waste Manag. Res. **35**(5), 471–479 (2017)

11. M.N.O. Sadiku, T.J. Ashaolu, S.M. Musa, Artificial intelligence in medicine: a primer. Int. J. Trend Res. Develop. **6**(1), 270–272 (2019)
12. M.N.O. Sadiku, J. Foreman, S.M. Musa, 3D printing in healthcare. Int. J. Sci. Eng. Technol. **7**(7), 65–67 (July 2018)
13. M.O. Onyesolu, F.U. Eze, Understanding virtual reality technology: advances and applications, https://www.semanticscholar.org/paper/Understanding-Virtual-Reality-Technology%3A-Advances-Onyesolu-Eze/12c4294ce620ec062e8dad1acb738b7cc9d211ba
14. M.N.O. Sadiku, T.J. Ashaolu, S.M. Musa, Nanomedicine: a primer. Int. J. Trend Res. Develop. **6**(1), 267–269 (2019)
15. M.N.O. Sadiku, Y. Wang, S. Cui, S.M. Musa, Healthcare robotics: a primer. Int. J. Adv. Res. Comput. Sci. Software Eng. **8**(2), 26–29 (2018)
16. The top 8 robotic healthcare breakthroughs: from gene mapping to robot nurses, https://www.telegraph.co.uk/education/stem-awards/innovation/robotic-healthcare-breakthroughs/
17. M.N.O. Sadiku, S.R. Nelatury, S.M. Musa, Cloud computing in healthcare. J. Sci. Eng. Res. **5**(9), 202–205 (2018)
18. M.N.O. Sadiku, S. Alam, S.M. Musa, IoT for healthcare. Int. J. Electron. Commun. Eng. **5**(11), 5–7 (2018)
19. M.N.O. Sadiku, K.G. Eze, S.M. Musa, Block chain technology in healthcare. Int. J. Adv. Sci. Res. Eng. **4**(5), 154–159 (2018)
20. M.N.O. Sadiku, N.K. Ampah, S.M. Musa, Social media in healthcare. Int. J. Trend Sci. Res. Develop. **2**(5), 665–668 (2018)
21. A. Datta, Top eight disruptive technologies and how they are relevant to geospatial, 2018, https://www.geospatialworld.net/blogs/top-disruptive-technologies-relevant-geospatial/
22. J.C. Augusto, Ambient intelligence: opportunities and consequences of its use in smart classrooms. Innovat. Teach. Learn. Informa. Comput. Sci. **8**(2), 53–63 (2009)
23. J. Goldsmith, Technology and the boundaries of the hospital: three emerging technologies. Health Aff. **23**(6), 149–156 (2004)
24. D. Gu et al., Discovering and visualizing knowledge evolution of chronic disease research driven by emerging technologies. IEEE Access **7**, 72994–73003 (2019)
25. M.N.O. Sadiku, M. Tembely, S.M. Musa, Telemedicine: a primer (Part 1). Int. J. Adv. Res. Comput. Sci. Software Eng. **9**(6), 43–46 (2019)
26. M.N.O. Sadiku, M. Tembely, S.M. Musa, Telemedicine: tele everything phenomena (Part 2). Int. J. Adv. Res. Comput. Sci. Software Eng. **9**(6), 35–38 (2019)
27. B. Fong, A.C.M. Fong, C.K. Li, *Telemedicine Technologies Information Technologies in Medicine and Telehealth* (Wiley, 2011)
28. B. Banova, The impact of technology in healthcare, June 2019, USA, https://www.aimseducation.edu/blog/the-impact-of-technology-on-healthcare/
29. R.S.H. Istepanian, J.C. Lacal, Emerging mobile communication technologies for health: Some imperative notes on m-health. *Proceedings of the 25the Annual International Conference of the IEEE EMBS*, Cancun, Mexico, September 2003, pp. 1414–1416
30. S. Jones, F. M. Groom (eds.), *Information and Communication Technologies in Healthcare* (CRC Press, Boca Raton, 2012)
31. J.M. Boyce, *Measuring Healthcare Worker Hand Hygiene Activity: Current Practices and Emerging Technologies* (Cambridge University Press, 2015)
32. J. A. Jacko (ed.), *The Human-Computer Interaction Handbook: Fundamentals, Evolving Technologies, and Emerging Applications*, 3rd edn. (CRC Press, Boca Raton, UK, 2012)
33. S.M. Ricchins, *Emerging Technologies in Healthcare* (CRC Press, Boca Raton, 2015)
34. S. Paiva, N. Gupta, Technologies and Systems to Improve Mobility of Visually Impaired People: A State of the Art. In '*Innovations in Communications and Computing*' by EAI/Springer, Springer International, pp. 105–123. https://doi.org/10.1007/978-3-030-16450-8
35. D. N. Le et al. (eds.), *Emerging Technologies for Health and Medicine: Virtual Reality, Augmented Reality, Artificial Intelligence, Internet of Things, Robotics, Industry 4.0* (Scrivener Publishing, USA, 2018)

Challenges of IoT in Healthcare

Sweta Anmulwar, Anil Kumar Gupta, and Mohammad Derawi

1 Introduction

Healthcare industry has actively started to incorporate Internet of Things (IoT) to improve the quality of healthcare services for the patient's overall health. For example, Open Artificial Pancreas System (OpenAPS) measures the amount of glucose in a patient's bloodstream and automatically delivers the required amount of insulin into the system [1]. Another example is connected inhalers; it has a sensor attached to the inhaler and is connected to the app on the mobile phone, and it aids patients to self-manage their health condition better [2]. It clearly shows that IoT is going to be an essential part of the patient's diagnosis, treatment, and recovery process [3]. IoT can significantly/effectively address the issue of the rising cost of healthcare through different applications like patient monitoring, clinical operations, drug/medicine development, fitness measurement, and preventive care. According to the Allied Market Research group, the worldwide market for IoT healthcare will reach $136.8 billion by 2021 [4]. It exhibits IoT's potential in the healthcare sector. There are several advantages of implementation of IoT in healthcare like fewer errors in data collection, faster diagnosis, efficient patient care, and better resource management in hospitals.

S. Anmulwar (✉)
Institute of Communication Systems University of Surrey, Guildford, UK
e-mail: s.anmulwar@surrey.ac.uk

A. K. Gupta
Centre for Development of Advanced Computing (C-DAC) R&D, Pune, Maharashtra, India
e-mail: anilg@cdac.in

M. Derawi
Department of Electronic Systems (IES), Norwegian University of Science and Technology, Trondheim, Norway

N. Gupta, S. Paiva (eds.), *IoT and ICT for Healthcare Applications*, EAI/Springer Innovations in Communication and Computing,
https://doi.org/10.1007/978-3-030-42934-8_2

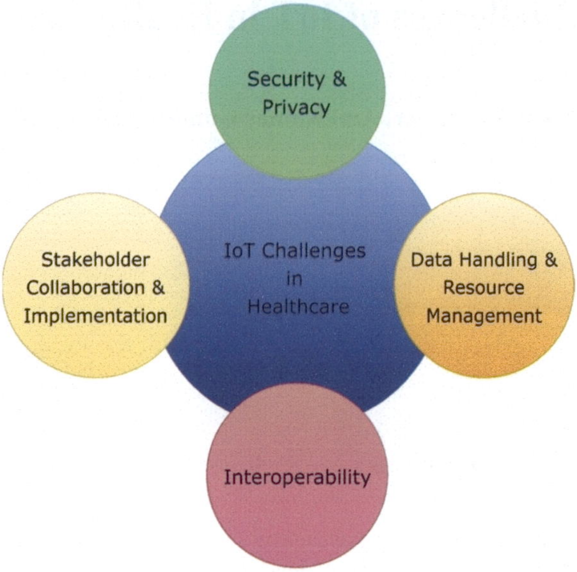

Fig. 1 IoT challenges in healthcare

Despite the advantages of the use of IoT in the healthcare sector, there are some challenges and are depicted in Fig. 1. Data are essential parts of the decision-making process for the patient's care. According to the International Data Corporation (IDC) by 2025, there will be 41.6 billion IoT devices, which will generate 79.4 ZB (Zettabytes) of data [5]. This enormous unstructured data need to be processed in a time-critical manner/real-time. Data collection, processing, and interpreting the results require huge resources in terms of storage, compute, and network. Critical data related to monitoring has to be synchronized and analyzed in time to take the appropriate and informed decisions by the treating physician.

Security is one of the major challenges to the IoT in general [6]. When it comes to healthcare, security becomes even more critical as the patient's sensitive data are flowing in the network without any encryption and can be used for medical identity theft, blackmailing, etc. Patient's data can be altered by the hacker, and these modified data could be used to take life and death decisions for the patient. IoT is on the verge to become pervasive in our lives, and it has already started to enter into our personal spaces like a home, car, and so on. There were incidences where the car brakes were hacked, doors were locked and opened wirelessly, which shows the lack of security and privacy into the existing IoT systems [7].

IoT devices are heterogeneous in nature. Earlier IoT systems were considered for small enterprises or home usage. Now, it has been envisioned that it will be used for large-scale projects like smart city, and almost everything will be connected to the internet through the IoT devices/sensors/systems, which makes the interoperability of systems a critical point. There is no open standard to follow [8]. Every manufacturer has different firmware, communication protocol, and frequencies on which the device operates. This could be the bottleneck for the successful implementation of the IoT.

Stakeholders will determine the success or failure of the IoT implementation in the healthcare industry. Various stakeholders, like healthcare service providers and businesses, have to work together to create valuable services for the patients, which makes stakeholder collaboration important. Unlike other sectors like industrial automation, in the healthcare sector, there is a huge part that is dependent on the human resource involved in diagnosis, educating the patient about the usage of the IoT devices/sensors and taking the critical decisions. It clearly indicates that actual implementation at the lower level is of extreme importance.

The remaining chapter is divided into four sections. It discusses the challenges of data handling and resource management; security and privacy; interoperability; and stakeholder collaboration and implementation for successful implementation of the IoT systems in the healthcare sector.

2 Data Handling and Resource Management

All the IoT applications like remote monitoring, clinical operations, etc. are data-centric which introduces several challenges as shown in Fig. 2. Data are collected, processed based on the application-specific algorithm, and finally, a decision is made. In the whole process, data play a vital role, and when it comes to healthcare, time-critical execution and synchronization are of extreme importance. There are different types of sensors in healthcare from which data are collected, like implantable sensors, wearable sensors, and others. There are 3.7 million devices in use today to monitor the various parts of the body, and this is going to exponentially increase with time [9]. Different applications have different time-criticality factor requirements as shown in Table 1.

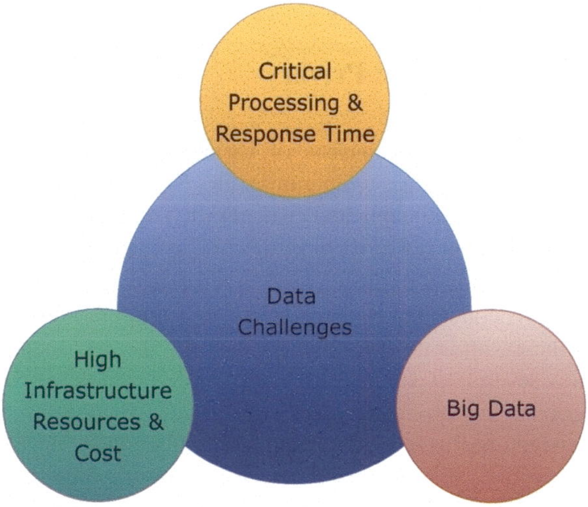

Fig. 2 Challenges related to the data in IoT healthcare systems

Table 1 Application and time-criticality factor

IoT application	Time-criticality factor
1. Remote patient monitoring	High
2. Inventory management at hospitals	Moderate
3. Workflow optimization	High
4. Fitness measurement	Low
5. Preventive care	Low

As mentioned in Table 1, remote patient monitoring has a high time-criticality factor. Processing and response time should be extremely low, e.g., monitoring after the major surgery, cardiac arrest, and so on. Workflow optimization is the IoT application in which the workflow, i.e., recording the patient's data (from patient's IoT device to the hospital's database), registering at the hospital, and diagnosis, is optimized. It significantly reduces the diagnosis time and can bypass all the paperwork related to healthcare. In this way, the quality of healthcare services can be improved. If there is any issue with the healthcare services, it can be detected and corrected in time. With the help of the sensors, medical supplies, measurement equipment, medication, and diagnostic reports will be tracked and hospital management will be notified in case of shortfall of the supply or expiry of the supplies and devices. Fitness measurement and preventive care are of relatively low priority compared to the above-mentioned IoT applications.

Data collected from different IoT devices are enormous and are of unstructured type which adds to the processing time further. These unstructured big data require a high-performance computing (HPC) systems to process the humongous amount of data, which will drastically increase the infrastructure cost. Initial implementation cost is going to be at higher end due to infrastructure investment for healthcare providers and business partners. Patients also have to invest in wireless setup at home, sensors, bands, etc. But it will be minimal due to the reduced cost of sensors.

3 Security and Privacy

IoT in healthcare consists of various types of body sensors to measure blood pressure, heart rate, temperature, etc. Security challenges related to IoT in general are directly applicable and critical to IoT in healthcare. Security of the IoT systems is the biggest reason behind the IoT systems not being adopted at a larger scale [10]. Mirai is a malware used in a large-scale network attack. It turns the Linux systems connected to the network into the remotely controlled bots. Mirai Botnet is the example of the exploitation of the security loopholes in the IoT systems and how severely it can affect the IoT systems and networks [11]. Security was never a focus in the initial development phase of IoT systems, and there was a competition between businesses to develop IoT applications. Heterogeneity of devices, lack of encryption, and several attacks through the physical and cyber medium have caused unauthorized access, data leakage, and hardware security issues.

It is evident that IoT devices will be ubiquitous. IoT is going to change our daily lives in a way that is unimaginable at this point of time. According to the Cisco reports, the number of IoT devices has already surpassed the world population [12]. IoT devices have already entered into our private spaces like home, car, and so on. This indicates that IoT devices/network will have personal information like illness information, medical reports, bank data, and social security numbers and raises the concern of the privacy.

There are several factors that cause security issues as shown in Fig. 3. They are discussed as follows:

- *Architecture:* Initially, the market focus was on IoT functionality rather than on the security as there was a competition between the businesses to launch the IoT products to catch up the market. Security was considered an add-on component rather than the important one.
- *Lack of encryption:* Data flowing in the IoT ecosystem are naked. Communication from the sensors to the gateway, from the gateway to the cloud, is not encrypted. Hacker can eavesdrop the communication or can hack the systems.
- *Access control:* Access to the IoT data is relatively easier because of the weak authentication schemes/vulnerable web portals. Access control is of extreme importance in the healthcare industry to protect the patient's sensitive data from malicious actors.

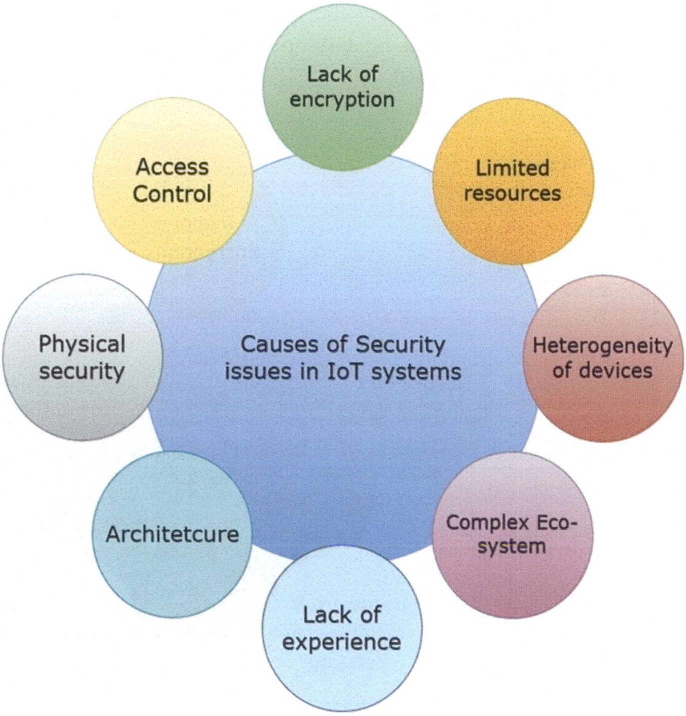

Fig. 3 Causes of security issues in IoT

- *Limited resources:* IoT devices have limited processing power, memory, and power (mostly, sensors are battery operated), making the implementation of security-related protocols difficult, which are compute-intensive.
- *Heterogeneity of devices:* IoT is a complex ecosystem, and it consists of various devices such as bracelets, cars, sensors implanted on people, home appliances, medical appliances, and many others. Each device sends and receives data in different formats, which presents an opportunity to the cybercriminals to enter into the system.
- *Lack of experience:* IoT security is a relatively new field. Expertise related to cybersecurity has to be applied to the IoT systems by taking into account the limited resources.
- *Physical security:* Not much attention has been paid to the actual hardware; hardware may also have backdoors. Safety of the IoT hardware is of extreme importance; intruders can just access the data or change the configuration from the device itself.
- *Complex ecosystem:* IoT consists of sensors, actuators, microcontroller, gateway, and cloud. These sensors can be incorporated with home appliances, cars, fitness bands, etc. It is difficult to secure the system having several components, which requires different types of expertise.

To successfully incorporate IoT in the healthcare industry, following measures as shown in Fig. 4 should be taken to address security and privacy of the IoT system:

- *Port communication*: Open ports are the readily available point of attack for cybercriminals. No ports should be left open, and after the communication, termination ports should be closed.

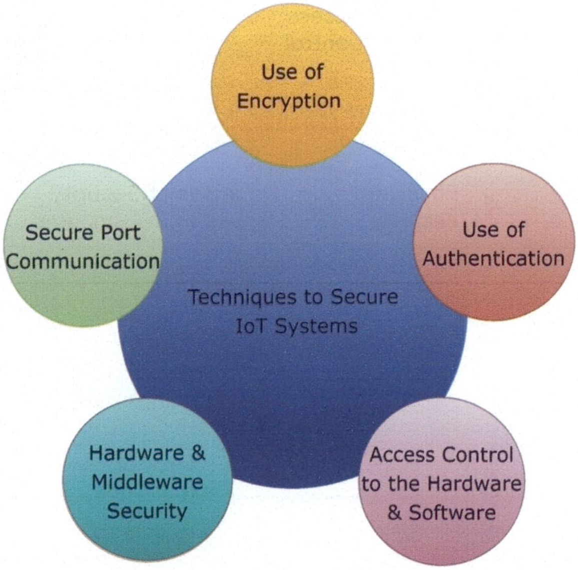

Fig. 4 Techniques to secure IoT systems

- *Use of encryption:* Data exchange in the IoT systems should be encrypted to ensure the security of the patient's sensitive data. It can help in preventing the unauthorized access of the patient's data, as well as eavesdropping of the information by the hacker.
- *Use of authentication:* IoT web portals should be protected with strong passwords and authentication protocols.
- *Access control to the hardware and software:* Role-based access should be designed to the IoT devices and data. So that only the authorized person should get the patient's sensitive data like treating physician pane, which is responsible for the diagnosis, medication, and monitoring of the patient.
- *Hardware and middleware security:* By ensuring the physical security of the device, various threats like tampering of the device, data theft through USB, or stealing of the storage media, etc. can be prevented. Middleware connects the applications to the things/sensors, and it should be secured.

4 Interoperability

Interoperability is the ability of systems to communicate with each other. Interoperability is the critical factor for the success of the IoT. According to McKinsey and Company predictions, IoT interoperability will create almost 40% of its value as a technology [13].

Healthcare industry makes use of various sensors to monitor patient's health remotely and in the hospital for medical supplies, etc. Wearable sensor devices used by patients can be from different vendors; there may be the case where a patient has to use more than one wearable sensor device. All these sensor devices are supposed to communicate with each other, but sensor devices from different vendors cannot communicate with each as they have different communication protocols. Because of this problem, sensors have to be used from the same vendor so that sensors could communicate with each other; it is a potential bottleneck for the large-scale implementation of IoT systems. It can cause vendor lock-in [14]. Interoperability issue even affects the development of the IoT applications exposing cross-platform.

IoT is an evolving field, and it is evident that new type of sensor devices in the different forms will be available in the market in the near future, and interoperability between already available devices and new devices is of extreme importance.

Interoperability has three significant concerns as shown in Fig. 5, i.e., heterogeneity of devices, nonuniformity of communication protocol, and no open standard to follow:

- *Heterogeneous devices:* IoT consists of devices starting from bracelets to home appliances, people, car, etc. It is a challenge to develop a capability to communicate with each other, and it has to be agreed upon collectively [15].
- *No open standard:* There is no open standard to follow for the manufacturers. Efforts have been made by the IEEE P2413 – Standard for an Architectural

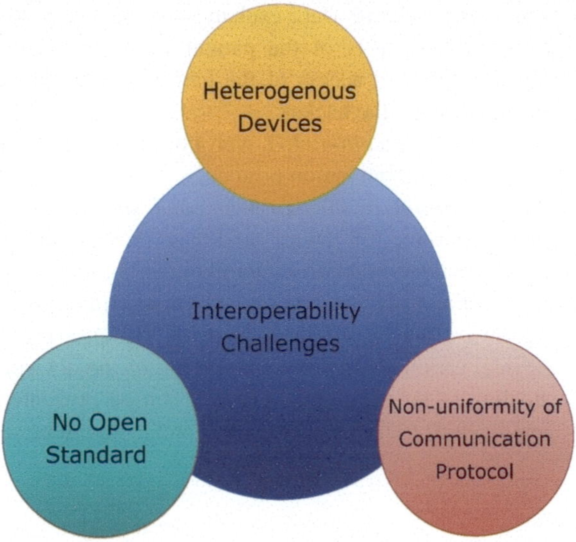

Fig. 5 IoT interoperability challenges

Infrastructure for the Internet of Things [16], Iot-A created to develop architectures that can be applied in different domains [17], IoTivity [18], and Industrial Internet Reference Architecture (IIRA) [19], but nothing has been formalized as open standard yet.

- *Nonuniformity of communication protocol:* Each vendor has different firmware and communication protocol. Only IoT devices from the same vendor can communicate with each other, and these devices cannot communicate with the IoT devices from different vendors, which is the barrier to the large-scale implementation of IoT [20].

5 Stakeholder Collaboration and Implementation

Although security is perceived as the biggest challenge for the successful implementation of IoT in the healthcare industry, stakeholder collaboration is of equal importance for the IoT implementation. Value creation is a collaborative process [21], and all the stakeholders must come together to improve the quality of healthcare by means of IoT. Figure 6 depicts the various challenges pertaining to stakeholder collaboration.

Healthcare providers, patients, healthcare payers, research laboratories, government authority, businesses, insurance agencies, and IoT application developers are the stakeholders for the IoT implementation in healthcare. Healthcare service providers and businesses have to work together to create value for all stakeholders through innovative IoT services. Understanding the needs of the patients, doctor, and hospital staff is of importance to create services. Data collection through interviews

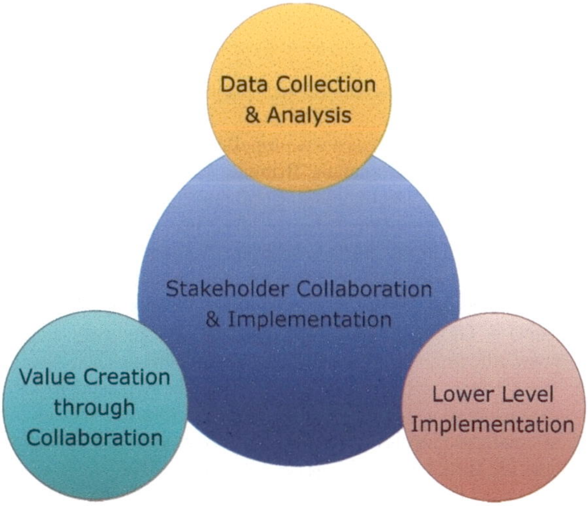

Fig. 6 Stakeholder collaboration challenges

and workshops has to be done. Further, these data have to be analyzed to get the insights/ideas for the creation of the IoT services. In this way, businesses around IoT and healthcare can be built.

Lower level implementation is crucial to the success of IoT in the healthcare industry [22]. Everything cannot be automated in the healthcare sector, and there is still a lot of human intervention in the healthcare industry. All the hospital staff has to be professionally trained to make use of the IoT devices and the ecosystem. Patients have to be educated for the use of IoT devices by the hospital staff. It is going to take some time to get used to the new IoT ecosystem.

6 Conclusion

IoT is going to be an integral part of the patient's diagnosis, treatment, and recovery process. IoT data collection mechanism coupled with predictive analytics will improve healthcare and reduce human errors. IoT in healthcare has several challenges such as data handling, resource management, security, privacy, interoperability, stakeholder collaboration, and actual implementation. At present, security is the biggest barrier to the success of IoT. Considering the enormous number of devices generating huge unstructured data, data handling and resource management are also going to be a challenge in the near future. Interoperability between IoT devices and creating values through stakeholder collaboration will decide the success or failure of the IoT systems in healthcare industry.

References

1. Open Artificial Pancreas System project, https://openaps.org/what-is-openaps/. Accessed 23 Sept 2019
2. Propeller health, https://www.propellerhealth.com/how-it-works/. Accessed 23 Sept 2019
3. C. Doukas, I. Maglogiannis, Bringing IoT and cloud computing towards pervasive healthcare, in 2012 Sixth International Conference on Innovative Mobile and Internet Services in Ubiquitous Computing, Palermo, 2012, pp. 922–926, https://doi.org/10.1109/IMIS.2012.26
4. Internet of Things (IoT) Healthcare Market by Component (Implantable Sensor Devices, Wearable Sensor Devices, System and Software), Application (Patient Monitoring, Clinical Operation and Workflow Optimization, Clinical Imaging, Fitness and Wellness Measurement) – Global Opportunity Analysis and Industry Forecast, 2014–2021 (2016), https://www.alliedmarketresearch.com/iot-healthcare-market. Accessed 23 Sept 2019
5. The Growth in Connected IoT Devices Is Expected to Generate 79.4ZB of Data in 2025, According to a New IDC Forecast (2019), https://www.telecomtv.com/content/iot/the-growth-in-connected-iot-devices-is-expected-to-generate-79-4zb-of-data-in-2025-according-to-a-new-idc-forecast-35522/. Accessed 23 Sept 2019
6. T. Xu, J.B. Wendt, M. Potkonjak, Security of IoT systems: Design challenges and opportunities, in *2014 IEEE/ACM International Conference on Computer-Aided Design (ICCAD), San Jose, CA*, (IEEE, Piscataway, 2014), pp. 417–423. https://doi.org/10.1109/ICCAD.2014.7001385
7. N. Lu, N. Cheng, N. Zhang, X. Shen, J.W. Mark, Connected vehicles: Solutions and challenges. IEEE Internet Things J. **1**(4), 289–299 (2014). https://doi.org/10.1109/JIOT.2014.2327587
8. Z. Song, A.A. Cardenas, R. Masuoka, Semantic middleware for the Internet of Things, in 2010 Internet of Things (IOT), Tokyo, 2010, pp. 1–8, https://doi.org/10.1109/IOT.2010.5678448
9. Internet of Things in Healthcare (2018), https://www.smartcity-iot-news.com/smart-living/internet-of-things-in-healthcare. Accessed 23 Sept 2019
10. Market Insight: Healthcare IoT in 2018 – Sell to CEOs and Set Realistic User Expectations (2018), https://www.gartner.com/en/documents/3857164/market-insight-healthcare-iot-in-2018-sell-to-ceos-and-s0 Accessed 23 Sept 2019
11. C. Kolias, G. Kambourakis, A. Stavrou, J. Voas, DDoS in the IoT: Mirai and other botnets. Computer **50**(7), 80–84 (2017). https://doi.org/10.1109/MC.2017.201
12. D. Evans, The Internet of Things-How the Next Evolution of the Internet Is Changing Everything, Cisco Internet Business Solutions Group (IBSG), April 2011
13. The Internet of Things: Mapping the Value Beyond the Hype (2015), https://tinyurl.com/y3xvw45q. Accessed 23 Sept 2019
14. S. Poudel, Internet of things: Underlying technologies, interoperability, and threats to privacy and security. Berkeley Technol. Law J. **31**(2016), 997 (2016)
15. S. Jabbar et al., Semantic interoperability in heterogeneous IoT infrastructure for healthcare. Wirel. Commun. Mob. Comput. **2017**, 1–10 (2017)
16. Standard for an Architectural Framework for the Internet of Things (IoT) IEEE (2016), http://grouper.ieee.org/groups/2413/Intro-to-IEEE-P2413.pdf. Accessed 23 Sept 2019
17. IOTA. https://www.iota.org/. Accessed 23 Sept 2019
18. IoTivity. https://iotivity.org/. Accessed 23 Sept 2019
19. Industrial Internet Reference Architecture. https://www.iiconsortium.org/IIRA.htm. Accessed 23 Sept 2019
20. M. Elkhodr, S. Shahrestani, H. Cheung, The internet of things: New interoperability, management and security challenges. Int. J. Netw. Secur. Appl. **8**(2), 1–18 (2016)
21. A.M. Mosadeghrad, Factors influencing healthcare service quality. Int. J. Health Policy Manage. **3**(2), 77 (2014)
22. A. Amato, A. Coronato, An IoT-aware architecture for smart healthcare coaching systems, in 2017 IEEE 31st International Conference on Advanced Information Networking and Applications (AINA), Taipei, 2017, pp. 1027–1034, https://doi.org/10.1109/AINA.2017.128

Internet of Things in Healthcare

Matthew N. O. Sadiku, Shumon Alam, and Sarhan M. Musa

1 Introduction to Internet of Things

IoT has been gaining popularity rapidly since its inception into the IT community and is being used in healthcare, education, gaming, finance, transportation, and several more. The healthcare industry is among the fastest to adopt the Internet of things. The primary goal of IoT in healthcare is to connect doctors with patients through a smart device. Healthcare providers are expecting the IoT to revolutionize the gathering of healthcare data and care delivery.

Internet of Things (IoT) is a worldwide network that connects devices to the Internet and to each other using wireless technology. IoT is expanding rapidly and it has been estimated that 50 billion devices will be connected to the Internet by 2020. These include smartphones, tablets, desktop computers, autonomous vehicles, refrigerators, toasters, thermostats, cameras, pet monitors, alarm systems, home appliances, insulin pumps, industrial machines, intelligent wheelchairs, wireless sensors, mobile robots, etc.

There are four main technologies that enable IoT [1]:

1. Radio-frequency identification (RFID) and near-field communication.
2. Optical tags and quick response codes: This is used for low-cost tagging.
3. Bluetooth low energy (BLE).
4. Wireless sensor network: They are usually connected as wireless sensor networks to monitor physical properties in specific environments.

Other related technologies are cloud computing, machine learning, and big data.

M. N. O. Sadiku (✉) · S. Alam · S. M. Musa
Roy G. Perry College of Engineering, Prairie View A&M University, Prairie View, TX, USA
e-mail: sadiku@ieee.org; shalam@pvamu.edu; smmusa@pvamu.edu

N. Gupta, S. Paiva (eds.), *IoT and ICT for Healthcare Applications*, EAI/Springer Innovations in Communication and Computing,
https://doi.org/10.1007/978-3-030-42934-8_3

The Internet of things (IoT) technology enables people and objects to interact with each other. It is employed in many areas such as smart transportation, smart cities, smart energy, emergency services, healthcare, data security, industrial control, logistics, retails, government, health, traffic congestion, manufacturing, industry, security, agriculture, environment, and waste management. Figure 1 shows the Internet of things and its application areas [2].

IoT supports many input-output devices such as camera, microphone, keyboard, speaker, displays, microcontrollers, and transceivers. It is the most promising trend in the healthcare industry. This rapidly proliferating collection of Internet-connected devices, including wearables, implants, skin sensors, smart scales, smart bandages, and home monitoring tools, has the potential to connect patients and their providers in a unique way.

Today, smartphone acts as the main driver of IoT. The smartphone is provided with healthcare applications.

IoT helps people and communities by making their systems smarter and their lives easier, more secure, and safer. IoT transforms ordinary products such as cars, buildings, and machines into smart, connected objects that can communicate with people and each other. These applications have given birth to smart everything, smart cars, smart homes, smart refrigerators, smart cities, smart parking, smart

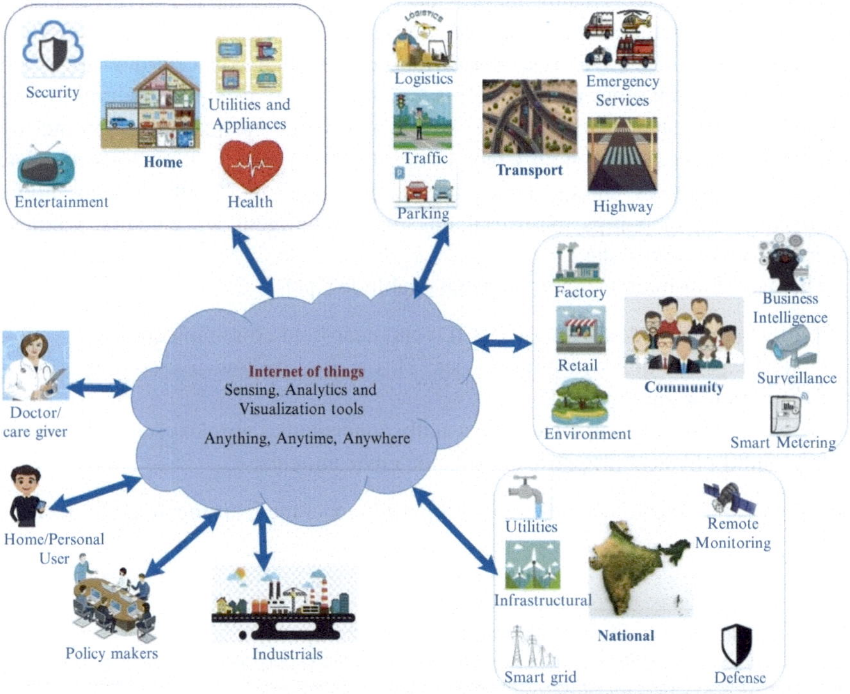

Fig. 1 The Internet of things and its application areas [2]

health, smart environment, smart transportation, smart shopping, smart agriculture, smart lighting, smart grid, smart bandages, and smart energy.

The narrowband version of IoT is known as narrowband IoT (NBIoT). This is an attractive technology for many sectors including healthcare because it has been standardized [3]. The main feature of NBIoT is that it can be easily deployed within the current cellular infrastructure with a software upgrade.

2 Healthcare

Healthcare is an essential part of the modern life. As they say, "Health is wealth." Healthcare is an essential service sector which has ubiquitous demand worldwide. It meets the basic need of every individual in the modern society. The healthcare system consists of patients, medical institutions, and healthcare resources to deliver healthcare services to meet human health needs. Unfortunately, the system is overwhelmed with problems such as expensive services, overworked doctors and nurses, illegitimate patient diagnoses, the growing rate of the aging population, increasing global demand for medical services, rise in the number of chronic diseases, and living environments with poor health [4]. Some of these problems are illustrated in Fig. 2 [5]. In addition, present approaches used for monitoring a patient in hospitals are time-consuming. The Internet of things (IoT) (also known as Future Internet) can resolve these issues quite well.

In healthcare system, the motivation of using modern technologies such as IoT is to offer promising solutions for efficiently delivering all kinds of medical healthcare services to patients at affordable cost. IoT could be a game changer for the healthcare services [6]. It makes it now possible to process data and remotely monitor a patient in real time.

The healthcare industry happens to be one of the fastest industry to adopt IoT. This is due to the fact that integrating IoT technologies into medical devices substantially improves the quality and effectiveness of service. The IoT enables

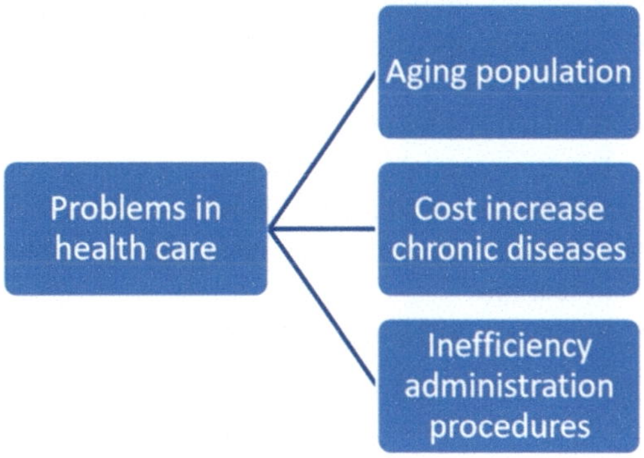

Fig. 2 Some problems in healthcare [5]

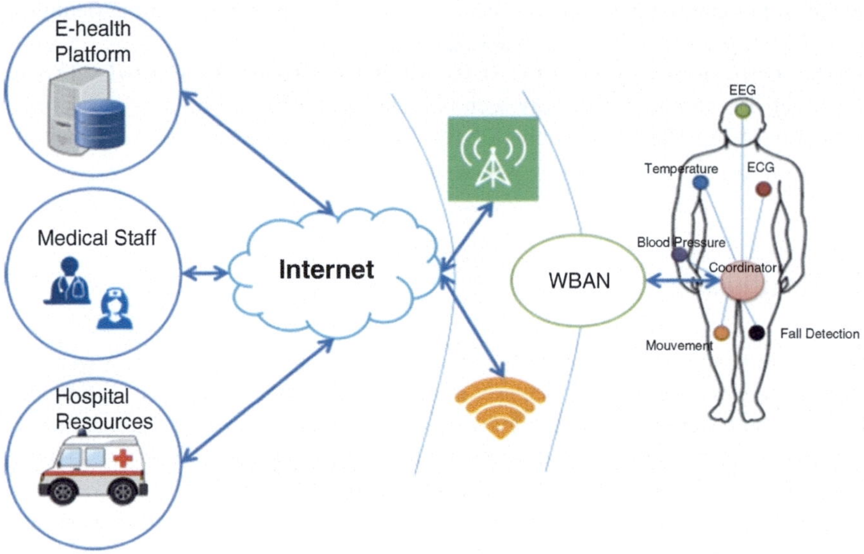

Fig. 3 A typical IoT in healthcare system [7]

practices in the area of healthcare for children, elderly, chronic care, real time monitoring of patients, operation theaters, and medicine dispenser. The application of IoT in healthcare can provide immediate treatment to the patient as well as monitor and keep track of health record for healthy person.

IoT has been identified as a technological solution to some medical challenges and a game changer for the healthcare services. Through the IoT, anything in the healthcare system can be identified and monitored anytime and anywhere. Monitoring the health parameters (such as blood pressure, heart rate, temperature, and humidity) of a patient remotely is achieved by IoT healthcare. A typical IoT in healthcare system is shown in Fig. 3 [7].

IoT in healthcare helps in [8]:

- Reducing emergency room wait time
- Tracking patients, staff, and inventory
- Keeping patients safe and healthy
- Enhancing drug management
- Ensuring availability of critical hardware
- Saving doctor's time and work
- Enabling nurses, doctors, and other team members to connect and communicate in real time
- Receiving critical information at the point of care without unnecessary alerts

3 Internet of Medical Things

The Internet of things in healthcare is variably referred to as IoT-MD, IoMT, Medical IoT, mIoT, and IoHT. Internet of medical things (IoMT), a healthcare application of the IoT technology, has emerged as a combination of advanced medical sensing system and computer communication technologies. The sensing systems include RFID, GPS, and wireless sensor networks. IoMT enables machine-to-machine interaction and real-time intervention solutions which are helping the healthcare industry increase its delivery, affordability, reliability, and productivity. When connected to the Internet, ordinary medical devices become smart and can collect more data, give insight into trends, enable remote care, and give patients more control. For example, IoT devices can be used for reminding patients about appointments, changes in blood pressure, calories burnt, and much more [9]. An illustration of IoMT is shown in Fig. 4 [10].

IoMT devices can sense real-time data for patient monitoring. Such devices are used to monitor parameters such as blood pressure, random blood sugar levels, and weight. IoMT will promote personalized care and high standard of living. Technologies used in IoMT can be divided into the three technical classes:

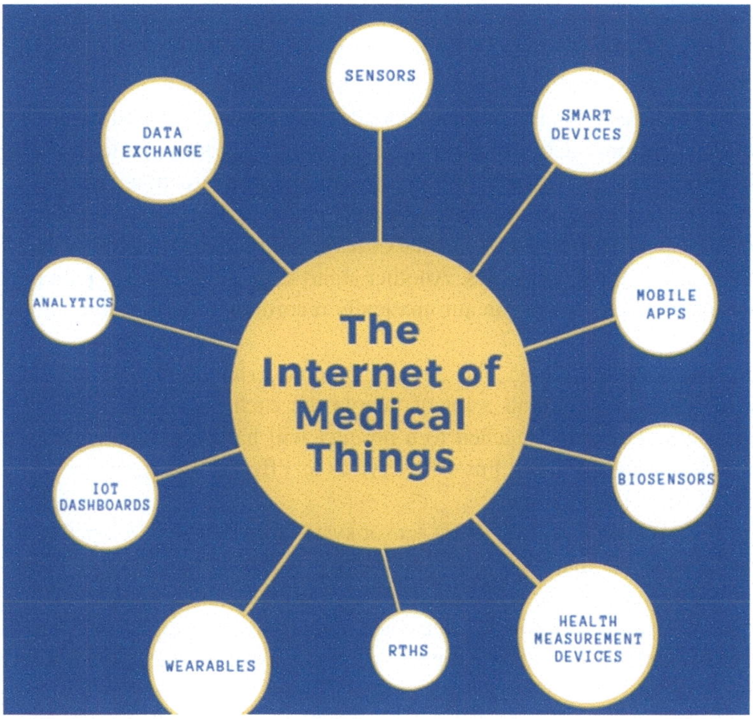

Fig. 4 The Internet of medical things [10]

local patient systems and controls, device connectivity and data management, and analytics solutions [11]. IoMT technology includes remote patient monitoring and medical system management. Smartphones are increasingly used as integral parts of IoMT. Various medical Internet of things platforms have been built for patient information management, telemedicine monitoring, and mobile medical [12].

4 Applications of IoT in Healthcare

Besides Internet of medical things, applications of IoT in healthcare are numerous, ranging from remote monitoring to smart sensors and medical device integration. The applications benefit patients, families, nurses, and physicians. IoT in healthcare is applicable in many medical instruments such as in ECG monitoring, glucose level sensing, and oxygen concentration detection. These various applications provide solutions for the patient and healthcare professionals. Some of the common applications are discussed below [11, 13]:

1. *Digital hospital*: Internet of Things has broad application prospects in the field of medical information management. Currently, the demand for medical information management in hospitals is in the form of identification, sample recognition, and medical record identification. Healthcare in hospitals is one way the medical segment is involved in IoT. With IoMT, hospital medical work is becoming increasingly intelligent, meticulous, and efficient.
2. *Cancer treatment*: Smart technology helps simplify care for both cancer patients and their care providers. By using smart monitoring system, patients experience less severe symptoms related to both the cancer and its treatment [11].
3. *Glucose monitoring:* Diabetes has been a fertile ground for developing smart devices. Such devices can help diabetics to continuously monitor their blood glucose levels for several days. Another smart device for diabetes patients is the smart insulin pen, which can automatically record the time, amount, and type of insulin [11].
4. *Drug anti-counterfeiting:* The amount of counterfeit medicines in the world has increased greatly, and a lot of people die each year as a result of wrong medication. The label attached to a product will have a unique identity that is very difficult to forge, and this will serve as an effective countermeasure against medical fraud [14].
5. *Elderly independent living*: RFID sensor systems are being developed to support older people so that they can safely stay independent. This application is important in view of an aging population. IoT applications can provide support for the elderly by detecting the activities of daily living using wearable devices.
6. *Remote monitoring*: Many patients continuously wear medical sensor-based devices to monitor their health statistics. Fitness, health electronics, and even smart watches have a role to play in monitoring and providing feedback and in some cases as a link to medical professionals. Remote monitoring translates into

a greater number of patients worldwide having access to adequate healthcare. Continuous patient monitoring provides the real-time tracking, collects patient data, and wirelessly transmits for ongoing display. This increases operational efficiency.

7. *Wearable devices*: Innovative devices, such as wearable devices, implantable chips, and embedded systems in biomedical devices, have been developed to continuously track continuous data on patient activity. Smart wearable devices allow the transfer of patient personal information between different devices. They support fitness, health education, symptom tracking, and disease management. They can be used to store health records especially for patients with diabetes, cancer, coronary heart disease, stroke, seizure disorders, and Alzheimer's disease [15].
8. *Body sensor network* (BSN): This technology is another IoT development in healthcare system, where a patient can be monitored using a collection of tiny-powered and lightweight wireless sensor nodes. It is essentially a collection of intelligent, miniaturized wireless sensor nodes used in monitoring the human body functions and surrounding environment. It opens the possibility for monitoring systems to operate wirelessly using low-cost wearable sensors [16, 17].

Other applications include people with disabilities, tracking and monitoring of objects and persons, identification and authentication, transport and data collection, clinical care, and continuous cardiac monitoring.

5 Benefits

Internet of things is a way of connecting devices to the Internet and to each other using wireless networks. It is injected into everything in healthcare, from X-ray machines to patient monitors. It creates new jobs and employment opportunities and bridges traditional engineering, computer sciences, and healthcare. It is transforming healthcare industry by increasing efficiency, lowering costs, and improving patient quality of care and safety. It ensures the personalization of healthcare services by providing digital identity for every patient. The doctors can break the limit of the geographical scope and provide medical education for medical personnel in remote areas. Figure 5 shows the most popular benefits of IoT in healthcare [18]. Besides these, other benefits include [19]:

- *Quality:* Integrating IoT features into healthcare devices greatly improves the quality and effectiveness of service. It enables a radical improvement of healthcare and quality of life. IoT healthcare principles are already being applied to improve access to care, increase the quality of care, reduce the cost of care, reduce medical errors, improve patient safety, and optimize the healthcare processes.
- *Connectivity and affordability*: Connectivity lies at the heart of Internet of things. It is the primary purpose of using IoT technology in healthcare, i.e., connect

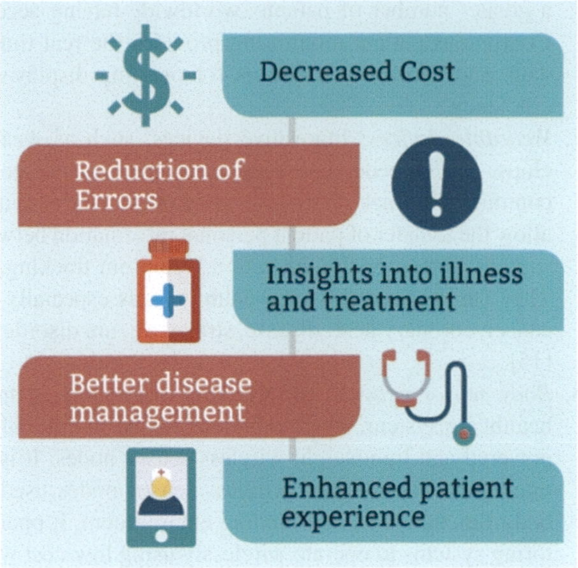

Fig. 5 The most popular benefits of IoT in healthcare [19]

doctors with patients through smart devices, without restrictions. The IoT links the medical devices with the virtual worlds, thereby enabling anytime and anyplace connectivity for anything and not only for anyone. IoT opens doors of opportunity for greater connectivity in healthcare. It enables interoperability, machine-to-machine communication, information exchange, and data movement that makes healthcare service delivery effective. It allows nurses, doctors, and other medical practitioners to connect and communicate instantly and receive information proactively in real time inside and outside the hospital. IoT in healthcare should provide better healthcare services to people at any time, from anywhere in a friendly manner. The IoT promises to make healthcare cheaper and better.

- *Monitoring:* Applications deliver care to people in remote locations and real-time monitoring systems that provide a stream of accurate data for better care decision-making [20]. IoT enables real-time monitoring of connected smart medical devices. Real-time monitoring can save lives in an event of a medical emergency like heart failure, diabetes, asthma attacks, etc. The IoT-connected devices can collect health data (such as blood pressure, oxygen and blood sugar levels, weight, and ECGs) and use smartphone to transfer the data to a doctor who may be several kilometers away. This makes healthcare service effective. The healthcare remote monitoring systems have contributed to the improvement of the elderly people's quality of life [21].
- *Tracking:* A healthcare facility needs to be able track all the devices and applications on the network continuously. IoT is used in tracking patients, staff, and inventory. It is difficult to maintain maximum security without the ability to track assets (patients, medical staff, and hardware) throughout the hospital. The

tracking may also include pharmaceutical inventory, helping elderly patients stay safe in their homes, and reminding patients when to take their medications. IoT and real-time location systems facilitate asset tracking. This is an inexpensive and effective method of monitoring and tracking day-to-day activities in a hospital setting. The ability to enable location tracking of assets using sensor-based technology has created a service which is known as location-as-a-service.

6 Challenges

Medical devices present some unique IoT challenges. These include the broad range of medical technologies, the diversity of network protocols, critical security and vulnerability considerations, regulatory compliance imperatives resulting from the handling of patient data, and stakeholders with varied interests. There is also an ambiguity about data ownership and a lack of EHR integration. This allows attackers/hackers to wreak havoc on the network. It is the responsibility of IT staff to bring more awareness to the health professionals about the challenges in supporting IoT devices. Besides these, other challenges include [8]:

- *Data security and privacy*: A significant challenge that IoT poses is of data security and privacy. The data that is being shared across the IoT devices are sensitive. Security and privacy of patients' medical data are crucial for wide acceptance and use of IoT in healthcare. Wearable sensors, for example, are prone to expose patient information and patient privacy. Medical security and privacy issues directly influence patient life and the healthcare system all over the world. Privacy issues may include misuse of medical information, leakage of prescriptions, and eavesdropping on medical data. An enemy may obtain your health status while you are busy exercising in a fitness center since medical sensors may be placed on your body [22]. Security solutions must be resource-efficient since medical sensors have limited processing power, memory, and communication bandwidth. Many countries prohibit privacy violations.
- *Integrating multiple devices*: Integrating complex medical devices is problematic due to lack of standards. Device manufacturers have not reached a consensus regarding communication protocols and standard.
- *Data overload*: The medical IoT generate massive data which can be utilized to gain insights and make smart decisions. The big data accumulated by IoT devices is a challenge for the IoT data processing. Handling the data is becoming very difficult for doctors, and this consequently affects the quality of their decision-making.
- *Workforce:* It is challenging to change the mentality of the current workforce. It can be difficult to convince those in the upper levels about the opportunities of IoT projects. Sometimes, there is not enough technical skill to gain valuable insights from the huge amount of data collected from IoT. Healthcare industry should hire experts with relevant IoT training.

- *High investment cost:* The high initial costs in IoT investments can scare some companies off. But IoT costs are declining rapidly. IoT project implementations with reasonable costs are recommended. Breakthroughs in the cost of sensors and processing power are enabling ubiquitous connections right now. The sensing devices such as RFID tags, sensors, actuator, etc. can be designed to minimize cost.

Other issues that negatively impact the adoption of IoT into healthcare include laws and policies, insurance coverage, standardization (lack of standards), data integrity, interoperability, compatibility, and cost [23]. These challenges may prevent healthcare from fully adopting the IoT technology.

7 Conclusion

The era of the Internet of things has already started and it will drastically transform our way of life. The central concept of the Internet of things is to connect anyone, anything, anytime, anyplace, any service, and any network. Healthcare is one of the major sectors where IoT can have the most relevant economic and social impact. IoT has enabled healthcare system to provide better healthcare services to people at any time and from anywhere in a friendly manner. It has opened up a world of possibilities in healthcare. From adherence to diagnosis, the applications are manifold. Due to these applications, the healthcare industry is changing at fast pace and is adopting the IoT rapidly. It has been long predicted that IoT in healthcare will revolutionize the healthcare sector in terms of social benefits, penetration, accessible care, and cost-efficiency. The IoT revolution is redesigning modern healthcare with extended benefits.

However, the rapid growth of IoT has presented some significant challenges. IoT's development has been restricted by the challenges. Security happens to be the most prominent challenge for physicians interested in IoT applications in medicine. More information about IoT in healthcare can be found in the book [24].

References

1. M.N.O. Sadiku, S.M. Musa, S.R. Nelatury, Internet of things: An introduction. Int. J. Eng. Res. Adv. Technol. **2**(3), 39–43 (2016)
2. A. Kulkarni, S. Sathe, Healthcare applications of the Internet of things: A review. Int. J. Comput. Sci. Inf. Technol. **5**(5), 6229–6232 (2014)
3. S. Anand, S.K. Routray, Issues and challenges in healthcare narrowband IoT, in International Conference on Inventive Communication and Computational Technologies, 2017, pp. 486–489
4. M.S.H. Talpur, The appliance pervasive of Internet of things in healthcare systems. Int. J. Comput. Sci. Issues **10**(1), 419–424 (2013)
5. J. Illegems, The Internet of things in health care, Master's Thesis, Universiteit Gent, 2017

6. A.S. Yeole, D.R. Kalbande, Use of Internet of things (IoT) in healthcare: A survey, in Proceedings of the ACM Symposium on Women in Research, March 2016
7. C.E.A. Zaouiat, A. Latif, Internet of things and machine learning convergence: The e-healthcare revolution, in Proceedings of the 2nd International Conference on Computing and Wireless Communication Systems, Larache, Morocco, November 2017
8. Peerbits, Internet of things in healthcare: Applications, benefits, and challenges, https://www.peerbits.com/blog/internet-of-things-healthcare-applications-benefits-and-challenges.html
9. G.J. Joyia et al., Internet of medical things (IOMT): Applications, benefits and future challenges in healthcare domain. J. Commun. **12**(4), 240–247 (2017)
10. D. Maltseva, IoT applications in healthcare. The Internet of medical things, October 2018., https://blak-it.com/blog/iot-applications-in-healthcare-internet-of-medical-things/
11. Econsultancy, 10 examples of the Internet of things in healthcare, February 2019., https://econsultancy.com/internet-of-things-healthcare/
12. C. Yaoa et al., A deep learning model for predicting chemical composition of gallstones with big data in medical Internet of things. Future Gener. Comput. Syst. **94**, 140–147 (2019)
13. M.N.O. Sadiku, S. Alam, S.M. Musa, IoT for healthcare. Int. J. Electron. Commun. Eng. **5**(11), 5–7 (2018)
14. L. Zhang, Applications of the Internet of things in the medical industry (Part 1): Digital hospitals, https://dzone.com/articles/applications-of-the-internet-of-things-in-the-medi-1
15. D. Bandyopadhyay, J. Sen, Internet of things: Applications and challenges in technology and standardization. Wirel. Pers. Commun. **58**(1), 49–69 (2011)
16. T. Wu et al., An autonomous wireless body area network implementation towards IoT connected healthcare applications. IEEE Access **5**, 11413–11422 (2017)
17. P. Gope, T. Hwan, BSN-care: A secure IoT-based modern healthcare system using body sensor network. IEEE Sens. J. **16**(5), 1368–1376 (March 2016)
18. MailMyStatements, 5 benefits of the convergence of IoT and healthcare, https://mailmystatements.com/2018/10/25/5-benefits-of-the-convergence-of-iot-and-healthcare/
19. A. Rghioui, A. Oumnad, Challenges and opportunities of Internet of things in healthcare. Int. J. Electr. Comput. Eng. **8**(5), 2753–2761 (2018)
20. D. Niewolny, How the Internet of things is revolutionizing healthcare, https://www.nxp.com/files-static/corporate/doc/white_paper/IOTREVHEALCARWP.pdf
21. S.F. Khan, Health care monitoring system in Internet of things (IoT) by using RFID, in Proceedings of the 6th International Conference on Industrial Technology and Management, 2017, pp. 198–204
22. P. Kumar, H.J. Lee, Security issues in healthcare applications using wireless medical sensor networks: A survey. Sensors **12**, 55–91 (2012)
23. R. Ajayi, Adoption of Internet of things into healthcare enterprise systems: A phenomenological study, Doctoral Dissertation, Colorado Technical University, September 2017
24. C. Bhatt, N. Dey, A. S. Ashour (eds.), *Internet of Things and Big Data Technologies for Next Generation Healthcare* (Springer, Cham, 2017)

Matthew N. O. Sadiku is a professor in the Department of Electrical and Computer Engineering at Prairie View A&M University, Prairie View, Texas. He is the author of several books and papers. His areas of research interest include computational electromagnetics and computer networks. He is a fellow of IEEE.

Shumon Alam is the director and researcher at the Center of Excellence of Communication Systems Technology Research (CECSTR) and SECURE Center of Excellence both at Prairie View A&M University. His research interests are in the areas of control, communication systems, networking, cybersecurity, and signal processing.

Sarhan M. Musa is a professor in the Department of Engineering Technology at Prairie View A&M University, Texas. He has been the director of Prairie View Networking Academy, Texas, since 2004. He is an LTD Sprint and Boeing Welliver Fellow. His research interests include computer networks and computational electromagnetics.

Internet of Things Applied to Mental Health: Concepts, Applications, and Perspectives

Ariel Teles, Flávio Barros, Ivan Rodrigues, Allan Barbosa, Francisco Silva, Luciano Coutinho, and Silmar Teixeira

1 Introduction

The World Health Organization (WHO) [79] conceptualises *mental health* as a well-being state in which individuals are able to use their abilities, recover themselves from the daily routine stress, be productive and contribute with the community. Such as highlighted by WHO [80], there are different terminologies used in mental health area (e.g. mental illness, mental disorder and mental disease) to imply the existence of a clinically recognisable set of symptoms or behaviour associated in most cases with distress and with personal dysfunction. Mental health disorder, or just mental disorder, is the general term chosen to be used here to designate problems related to the mental health. There are several types of mental and behavioural disorders, which can be classified using the 11th revision of the International Classification of Diseases (ICD-11) [80] into more than 100 such categories.

Here are listed some mental disorders: mood disorders (e.g. bipolar affective disorder and depressive episode), excessive anxiety and stress (e.g. phobic anxiety disorders and reaction to severe stress), disorders caused by psychoactive substance use (e.g. drug and alcohol addictions), schizophrenia and delusional disorders.

A. Teles (✉) · F. Barros
Federal Institute of Maranhão, São Luís, MA, Brazil
e-mail: ariel.teles@ifma.edu.br; ariel@lsdi.ufma.br; flavio.barros@ifma.edu.br

I. Rodrigues · A. Barbosa · F. Silva · L. Coutinho
Federal University of Maranhão, São Luís, MA, Brazil
e-mail: ivan.rodrigues@lsdi.ufma.br; afabio@lsdi.ufma.br; fssilva@lsdi.ufma.br; luciano.rc@lsdi.ufma.br

S. Teixeira
Federal University of Delta do Parnaíba, Parnaíba, PI, Brazil
e-mail: silmarteixeira@ufpi.edu.br

N. Gupta, S. Paiva (eds.), *IoT and ICT for Healthcare Applications*, EAI/Springer Innovations in Communication and Computing,
https://doi.org/10.1007/978-3-030-42934-8_4

These mental disorders have significantly reached a huge portion of the world population. To exemplify, some numbers about depression are worrisome [82]: it already affects more than 300 million people in the world (from 2005 to 2015, this number grew to 18.4%), and close to 800 thousand people die due to suicide every year, which is the second leading cause of death in young people aged between 15 and 29 years.

Concerned by that reason, the Information and Communication Technology (ICT) community has dedicated efforts to develop solutions applied to face such a worrying problem [65]. In particular, ubiquitous technologies associated with machine learning [47] provide promising tools for mental health professionals. For example, mobile applications already are widely proposed for patient monitoring, which take advantage of users' context data (e.g. physiological signals, heart rate, physical activity, social interactions) produced in real-world environments from mobile devices [28]. Context information can be useful to identify situations of interest, such as risk situations (e.g. suicide and depressive symptoms), which are meaningful inputs for a decision-making made by a mental health professional. Mobile applications can also provide ecological momentary assessments and interventions based on self-reports produced by the patients.

Internet of Things (IoT) technologies have also been applied to mental health [22] not only as innovative patient assistance and monitoring methods but also as user well-being measurement systems. IoT may be defined as the interaction of technologies from different areas, such as ubiquitous computing, context awareness, communication protocols and technologies and computing devices with embedded sensors and actuators, also known as smart objects [6]. The interconnection of thousands of heterogeneous addressable objects with network connectivity allows them to collect and share data about the environment where they are located [3, 26]. An IoT paradigm extension is the so-called Internet of Mobile Things (IoMT) [37, 48], which conceives situations where smart objects and IoT gateways can move or be moved with great flexibility. Wearable and mobile devices are some examples of moving objects. In IoMT scenarios, personal smartphones and tablets can act as a mobile edge gateway, promoting discovery and opportunistic connections with smart objects around them [63]. This chapter considers these definitions of IoT and IoMT, but it uses only the "IoT" abbreviation during the text to facilitate the understanding.

IoT solutions have been developed and proposed for monitoring the disorder evolution progress, performing interventions, by helping mental health professionals to manage medical treatment, specialised therapy and, when necessary, medicine. Such technologies should be sensitive, responsive, adaptive, transparent, ubiquitous and unobtrusive [25]. Wearable devices are an example of IoT technology that can bring those characteristics, which make them a natural choice for embedding new proposals for mental health and well-being monitoring systems. Those devices are worn by users during their daily life routines (e.g. smart watches, smart bands, fitness bracelets, actigraphy bands) and, as they are equipped with several sensors, can identify users' mental health and well-being evidences.

Mobile Mental Health (mMHealth) is the term used to represent the idea of the use of mobile and IoT technologies to support mental health service delivery [34]. Both mobile and IoT technologies are designed to give support to the users, mental health professionals and also mitigate limitations of traditional treatments, in which the remote and real-time monitoring does not happen. Therefore, they should be viewed as an additional support, and not a substitute, for the treatment and monitoring methods of mental health and well-being traditionally made. In addition, as solutions based on these technologies have become more and more popular and being used by different types of audiences, the adoption of these technologies has become very widespread. Moreover, mental health already is considered a focus of investment not just in terms of human development and dignity but mainly in terms of social and economic development [78]. As a consequence, regulatory boards have not only allowed its use but also motivated it. For example, in May 2018, the Brazilian Federal Council of Psychology has extended the possibilities of offering psychology services mediated by ICT [12]. From measures such as that, it is possible to verify an increasingly offer of mental health digital services.

This chapter intends to show the state of the art about the ubiquitous computing and IoT usage in mental healthcare and well-being monitoring, providing an analysis of proposed solutions. To achieve this objective, the remaining of this chapter presents a content organised as follows. Section 2 introduces the main concepts used in information and communications technology (ICT) applied to mental health. Section 3 details the clinical score systems, while Sect. 4 shows the importance of social relations for mental health and how IoT technologies can be used to identify relevant social situations. Section 5 discusses the influence of the sleep on well-being and mental status of an individual and presents IoT solutions to cope with sleep disorders. Section 6 presents the relation between mobility patterns and mental disorders and how studies have been trying to measure this relationship. Section 7 is responsible for drawing the available ICT solutions to prevent suicide. Section 8 discusses open issues of the research topic. In the end, Section 9 concludes the chapter.

2 Background

Ecological Momentary Assessment and Intervention (EMA/EMI) [33, 59], Digital Phenotyping [67] and Positive Computing [13] are some of the concepts and terms that have been coined to designate the application of ICT to well-being and mental health, developing new alternatives to deal with mental issues. These approaches have the potential to revolutionise traditional therapeutic processes and improve quality of life. Throughout this chapter, it is shown solutions with different types of proposals, which do contributions that can make all of these concepts possible to be used by patients and professionals in a near future.

2.1 Ecological Momentary Assessment and Intervention

Traditional methods of psychological evaluation are substantiated on clinical evidence and information self-reported by the patient. These approaches are typically based on retrospective reports of events experienced in patients' routine, in which the remembrance time can be days, weeks and even months. Therefore, these methods are exposed to a set of cognitive biases that may occur in the imprecision of the reported memory [4]. For example, patients can consciously or unconsciously shift the truth of the self-report to achieve a desirable outcome [72]. Another reason that limits the effectiveness in that type of methodology is that it occurs in clinical settings, where the therapy is coordinated by the mental health professional. Thus, there is no ecological validity in this method of monitoring such mental health behaviour, since the clinical environment is significantly different from the natural context of people [70]. Therefore, it is necessary to devise new methods able to automatically monitor behaviours of interest for mental health treatments.

From those requirements, EMA and EMI methods have emerged. EMA is a mechanism used to prompt individuals, at fixed or random times, to respond to questions about what they are doing (or have done) and/or are experiencing (or have experienced), repeatedly, throughout a period of time within their daily routine. Therefore, EMA is a set of methodologies used to collect behavioural information from the patient in their daily routine, where the time of data collection is defined based on situations of interest (e.g. after waking up, minutes after finishing physical activities). According to Shiffman [59], EMA aims "to assess the flow of experience and behaviour over time, capturing life as it is lived, moment to moment, hour to hour, day to day, as a way of faithfully characterising individuals and of capturing the dynamics of experience and behaviour over time and across settings".

Since EMA focuses on obtaining information from the patients as it is experienced by them, EMI "provides a framework for treatments characterised by the delivery of interventions to people as they go about their daily lives" [33]. EMI is a methodology to enable mental health professionals to perform interventions in the patients in their natural environment. Such remote interventions include clinical recommendations, encouraging messages, alerts based on different inputs, among others. Both methodologies (i.e. EMA/EMI) are ecological because they occur in the natural environment, and they are considered momentary by requesting, on time and in time, self-assessments and providing real-time support in the patient's everyday life.

Mobile devices have the appropriate characteristics for the implementation of these methodologies since they are technologies that are highly diffused in people's daily lives [28]. In addition, they have a pervasive and ubiquitous character that helps to reduce the intrusiveness and self-perception of monitoring, thus preventing the patient from consciously changing their natural behaviour. Thus, EMA/EMI methods have been implemented in mobile technologies by using phone calls, text messages and, more recently, mobile applications, which allow the reporting of experiences and mental status (e.g. mood, motivation, anxiety, sleep quality) in real-time, in real-world settings, over time and across contexts/situations. EMI mobile applications have been used to motivate patients, encourage engagements in practising physical activities and in using previously learned skills, aid in the development of new skills, notify or distract individuals when they are at risk of engaging in addictive behaviour and provide individuals with personalised summary data.

EMA/EMI triggers are made at predetermined times. Hence, requests for self-assessments and even interventions may not be made to the user at the most appropriate moment (e.g. when the user is sleeping or at workplace). *SituMan* [60, 64] is a proposed solution to face this limitation. It provides situation awareness to an EMA/EMI mobile application used for requesting self-assessments from patients in depression treatments. *SituMan* has a fuzzy inference engine developed to identify user daily routine situations (e.g. "studying", "working", "physical activity") by using context information obtained from mobile device sensors. Situations are specified jointly by the patient and mental health professional, and they represent the patient's daily routine. Taking advantage from the situation awareness provided by *SituMan*, the EMA/EMI application can request mental status self-assessments from patients at adequate moments.

2.2 *Digital Phenotyping*

EMA depends on the information self-reported by the patient, so presenting some limitations that may influence the efficiency of mental state monitoring. Therefore, it is necessary to develop monitoring approaches that can collect behavioural markers passively, removing the constraints presented by possible biased self-reports. The omnipresent adoption of IoT devices, including smartphones and tablets, can enable more efficient opportunities for tracking mental health status and disorders such as Digital Phenotyping, a term defined by Torous et al. [67]. It refers to the "moment-by-moment quantification of the individual-level human phenotype in situ using data from smartphones and other personal digital devices".

While EMA is based on self-reported static descriptions of emotions and behaviours, the objective of digital phenotyping is not only to implement features to request self-assessments but to collect and analyse large amounts of different types of social and behavioural information that represent experiences of the users and their interactions with people and places. Among others, digital phenotyping encompasses the collection of spatial trajectories, physical mobility patterns and audio samples [67].

The idea behind digital phenotyping have already been considered useful by exploring new dimensions of pathology largely inaccessible a few years before and enabling research discoveries and clinical advances [68]. In addition, results from an experimental study conducted with schizophrenic subjects that used a smartphone-based digital phenotyping tool have suggested that usage patterns of mobile applications may contain clinically relevant and potentially predictive information about future states of mental disorders [69].

2.3 *Positive Computing*

The term Positive Computing was created by Calvo and Peters [13]. It is a paradigm shift that defends the "design and development of technology to support psychological well-being and human potential" [13]. The notion of positive computing emerged from the necessity to face the negative effects of the burden of using some types of technology, which include, for example, the stress caused by excessive notifications and the feeling of privacy loss. Therefore, positive computing solutions are developed to promote mental well-being and help to make happen all human potentialities, respecting the psychological needs of individuals. In addition, they are conceived to improve the efficiency and effectiveness of knowledge workers.

Ubiquitous computing and IoT technologies have contributed to promote intelligent positive computing [40]. Some benefits of solutions that use such paradigm include facilitating new ways of detecting human behaviour changes that can indicate well-being problems or the beginning of mental disorders, provide timely therapeutic interventions and make it possible to track responses to evaluate the effectiveness of interventions [40].

3 Clinical Scores

Clinical scores have been used as part of modern clinical practice in recent decades. They are tools created to predict clinical outcomes, perform risk stratification, aid in clinical decision making, assess disease severity or assist diagnosis [1]. In the medical literature, there are many formally defined clinical scores, in which each one of them deals with a specific type of disease, especially those considered chronic. For example, in the mental health area, we have the *Clinical Dementia Rating Scale* [36] for dementia level detection, the *Hamilton Anxiety Scale* [31] for anxiety level rating and the *Major Depression Index* [9] for depression screening.

According to Graham Thompson [66], clinical scores are based on clinical prediction rules (CPRs), which are tools that use specific criteria to establish probabilities of outcomes or assist in management decisions. Gavin Falk and Tom Fahey [23] summarise the critical element of CPR as follows: "CPRs quantify the contribution of symptoms, clinical signs, and available diagnostic tests, and stratify patients according to the probability of having a target disorder. The outcome of interest can be diverse and be anywhere along the diagnostic, prognostic, and therapeutic spectrum."

In that context, some researchers have classified three types of CPRs and, consequently, three models of clinical scores: (1) *Diagnostic CPRs* that focus on

Table 1 AMT-4 Score Specification: Ask the patient to state each of the following

Variable	*Value*	*Score*
Age	Incorrect/correct	+1
Date of birth	Incorrect/correct	+1
Place	Incorrect/correct	+1
Year	Incorrect/correct	+1
Evaluation		*Result*
Normal cognition		4
Abnormal cognition		<4

factors related to the clinical diagnosis; (2) *Prognostic CPRs* that predict outcomes; (3) *Prescriptive CPRs* that provide recommendations for clinical intervention. The format of a CPR is variable and depends on the purpose for which it is intended, but it should include three or more variables obtained from patient history, physical examinations or necessary diagnostic tests [66]. The combination of these CPRs forms the clinical score specification skeleton.

In general, a clinical score specification consists of variables, rules for calculation of the score and evaluation. Variables are precisely the clinical predictors that influence the probability of a patient having a target disorder. Rules quantify the contribution of each specification variable in that probability through a specific punctuation. Evaluation describes the interpretation of a clinical score according to the result obtained by applying the rules for calculating the score. For example, the well-known *Abbreviated Mental Test 4* (AMT-4) Score [62], which assesses mental impairment in elderly patients, has variables, rules for calculating score and evaluations. Table 1 details the AMT-4 score specification, which shows declared variables, the set of rules associated with each specified variable and respective score and the two types of evaluation of this clinical score.

In recent years, there have been some initiatives involving mHealth and clinical scores. For example, in the area of respiratory diseases, Cook et al. [18] developed and evaluated a mobile application for increasing asthma control through the use of proactive actions and self-evaluation proposed by the *Asthma Control Test* (ACT). In turn, in the area of cancer, Pereira Azevedo et al. [52] designed and developed a mobile application for stratifying prostate cancer risk based on the *Rotterdam Prostate Cancer Risk Calculator* (RPCRC). In the area of metabolic procedures, Aminian et al. [5] developed a mobile application that offers quick access to three clinical scores – *Sleeve Gastrectomy Risk Calculator*, *Risk of Post-discharge Venous Thromboembolism after Bariatric Surgery* and *Individualised Metabolic Surgery Score* – for bariatric surgery decision-making.

Specifically in the field of mental health, mHealth solutions normally use digital biomarkers or propose a system to solve clinical score problems. Thus, Adams et al. [2] described mHealth approaches for acquisition of behavioural and psychological biomarkers aimed at psychiatric clinical research. The authors addressed clinical scores involving the sphere of psychiatry, such as smoking, alcoholism, drugs, stress, anxiety, autism and mood disorders. Stamate et al. [61] developed the *cloudUPDRS* application for clinical assessment of Parkinson's

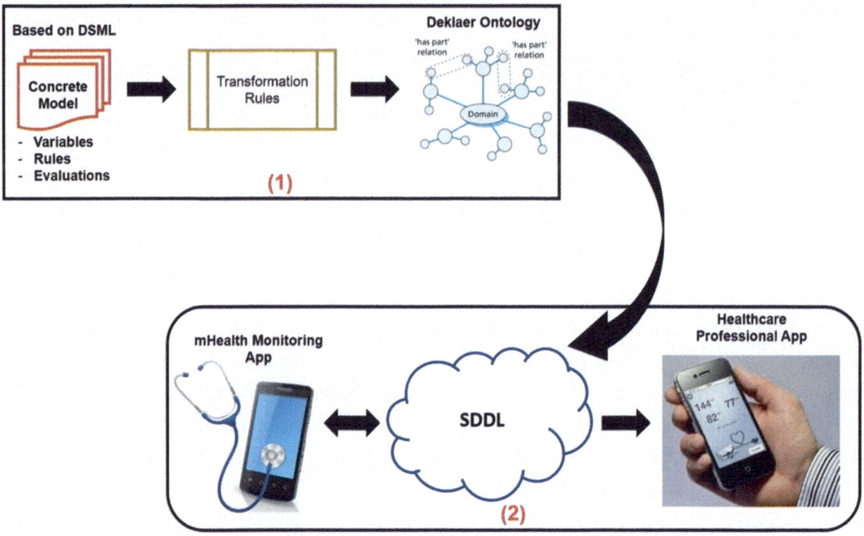

Fig. 1 Proposed approach of the *MDD4ClinicalScores* research project

disease motor symptoms, which is based on the *Unified Parkinson's Disease Rating Scale* (UPDRS). Those works and other ones described in the literature have the same characteristic: they developed a computational system for a specific clinical score.

An alternative approach to deal with the automation of clinical scores in mHealth is described in [8]. The authors employed Model Driven Development (MDD) concepts to solve different clinical scores (see Fig. 1). This study is part of the *MDD4ClinicalScores* research project that focuses on the software development to allow real-time remote patient monitoring using the processing of data streams generated from biomedical sensors. *MDD4ClinicalScores* proposed approach allows the specification of several clinical scores and the automatic generation of software components through the use of model transformation techniques.

The first part of Fig. 1 (identified by the label 1) corresponds to the clinical score specification. This step is based on a Domain-Specific Modelling Language (DSML) used to implement a high-level graphical authoring tool for supporting healthcare professionals in providing clinical score specifications [8]. Based on the provided specification (concrete model), *MDD4ClinicalScores* applies transformation rules for the generation of OWL ontology classes that are based on *Deklaer's Declarative Modeling Language* [53], an ontology used to describe IoT applications involving sensors. The generated ontology describes the rules used for patient monitoring according to the provided clinical score specification. The second part of Fig. 1 corresponds to the execution environment step of the proposed approach. This step involves the submission of the generated Deklaer's ontology to a middleware infrastructure called *Scalable Data Distribution Layer* (SDDL)

[21] for the semiautomatic generation of software components comprising the IoT application focusing on patient monitoring and the assessment of the specified clinical score. The process of an IoT application generation and the deployment of its software components are described in [53].

Clinical scores are useful tools that qualitatively enhance clinical practice within hospital environments, because they aid in the diagnosis, prevention and prognosis of various diseases, in addition to reducing costs with medical treatments. Therefore, research on the process of automating clinical scores in the mobile mental health field, with mobile and IoT support, should increase in the next years, with the emergence of new computational solutions for different medical specialties.

4 Social Relations

Social relations have important implications for mental health. The sociability of a person can directly affect their psychological state, by being able to play a protective role or contribute to the consequences of mental disorders. For example, social isolation presents associations with the symptoms of depression, anxiety and suicide [11]. There is also evidence that higher levels of social support represent a protective condition of depression and lower levels are associated with the presence, beginning or development of depression [27]. Therefore, the symptoms of mental disorders can be externalised through changes in social behaviours, hence characterising a situation of interest for the monitoring of mental health.

Sensors embedded in IoT devices (e.g. accelerometer, microphone, pedometer, location sensors) provide data capable of characterising social situations, thus allowing the development of systems capable of automating the process of monitoring social behaviour. These systems use data collected from the sensors to identify social situations, in which statistical techniques, data mining, machine learning and other methods of data analysis and processing are used. For example, log information of smartphone usage can be used to infer social ties (e.g. friends, family, co-worker) [46], while technologies such as Bluetooth, Near Field Communication (NFC) and Wi-Fi access points may indicate proximity and co-location [15, 45]. From the awareness of these situations, mental health professionals can make more effective interventions, such as advising the patient to socialise as soon as they recognise conditions of social isolation or propose activities that encourage social contact.

In the following subsections, we describe how these technologies can be used to identify social situations relevant to mental health monitoring.

4.1 Face-to-Face Encounters

Face-to-face encounters denote physical interactions, in which there is no presence of a mediating technology. These encounters are identified when involved individ-

Fig. 2 Detection of co-location and conversation between individuals

uals are physically co-present (i.e. over short distances) and interacting with each other. Devices that compose IoT (e.g. mobile and wearable devices) have sensors capable of characterising these situations, providing means of quantifying social relations. Most of these devices have microphones and wireless communication technologies (e.g. Wi-Fi, Bluetooth, NFC), which can be used to identify proximity and conversation [76]. Figure 2 presents the face-to-face interaction detection scenario between two individuals, where the discovery of devices using wireless technologies and microphone identifies co-location and conversation, respectively.

Next, we present methods capable of identifying face-to-face interactions using different technologies and their implications in the process of mental health monitoring.

4.1.1 Wireless Communication Technologies

Wireless communication interfaces are commonly used to identify the co-presence of people [15]. For this, researchers have used the ability of these interfaces to discover nearby devices to infer social interactions. So, the Bluetooth IDs (BTIDs) or Service Set Identifier (SSID) of all detected devices can be registered, acting as a proxy for the occurrence of social interactions [74]. For example, Wu et al.

[83] used features extracted from people's Bluetooth encounters to demonstrate correlations significant with stress levels. Miluzzo et al. [45] also used Bluetooth interfaces to retrieve MAC addresses of any detectable device, thus identifying the co-location.

The use of wireless communication interfaces as a proxy for social interaction has some limitations, which can result in poor performance. For example, there is a risk that a device discovered by the Bluetooth interface does not represent a person (e.g. it can be a smart TV or a printer), or the individual may carry more than one device. Another limiting factor is the detection range of these technologies, since Bluetooth radios have an estimated range of up to 10 m, while Wi-Fi signals cover an area average of 32 m in indoor environments [54]. Therefore, the interpersonal distance may not be accurately identified and may adulterate the number of people socially co-present. To mitigate this limitation, Carreras et al. [15] applied a machine learning algorithm to estimate the interpersonal distance, in which 20 samples of Wi-Fi Received Signal Strength Indicator (RSSI) were used to train the proposed model. This method was able to detect social interaction with an average error of 0.5 m, thus presenting a viable methodology for co-location detection.

In recent years, researchers have investigated the possibility of applying the method of detecting social interaction to monitor mental health. For example, Wang et al. [76] collected data from smartphone sensors from a set of college students to assess their mental health. Among the characteristics obtained, the Bluetooth encounters were used to identify the co-presence in between students. This study identified that the lower number of daily co-locations is related to higher levels of depressive symptoms.

4.1.2 Microphone

Microphones available in mobile devices can be used to identify when a person is engaged in a conversation, allowing to extract characteristics such as frequency and duration of communication. With the objective of identifying when a person participates in a conversation, studies have used machine learning algorithms to recognise the human voice in audio samples. For example, Lu et al. [41] used a combination of supervised and unsupervised machine learning techniques to identify general types of sound (e.g. voice, music) and distinguish specific events from each user. Wang et al. [76] segmented the audio stream into frames with 15 ms, in which the classifier extracted features to design a two-state Hidden Markov Model (HMM) capable of identifying natural language speech segments.

The use of microphone presents some factors that may limit the identification of social interactions, requesting specific care to preserve the efficiency of the proposed methods. Algorithms used to recognise conversations may have difficulty in distinguishing the human voice from other sounds, such as conversations derived from television and radio. Another concern that the studies must have is concerning the privacy of the individuals since methods able to process data of microphones can reveal the content of the speech, making it excessively invasive. Thus, the methods should only extract characteristics that do not allow the reconstruction of the speech content and make it impossible to identify the speaker, in order to preserve the users' privacy.

Studies have used the microphone embedded in IoT devices to identify situations of conversation and have investigated the relationship between those events and mental health. In a study involving college students, Wang et al. [76] found correlations between the frequency of conversations and their duration with depressive symptoms, indicating that students who socially interacted less were more likely to be depressed. Lane et al. [39] designed a mobile application capable of measuring the level of sociability through the identification of conversation, being able to provide feedback on the individual's social engagement. Gu et al. [30] developed a wearable computing platform capable of automatically extracting and analysing social cues and using a set of social characteristics to quantify the level of anxiety.

4.2 Device-Mediated Communication

Device-mediated communication is a social interaction that occurs through technology, such as mobile and wearable devices or online social networks. Social events that take place in these media let records with valuable information for monitoring social behaviour. For example, registrations of telephone calls and the social applications (e.g. SMS, email, WhatsApp, Facebook and Twitter) can identify characteristics such as social ties, social support, frequency of social interactions, among other aspects [46, 77]. Thus, computational approaches can use this rich source of behavioural data to identify characteristics of individuals' sociability, so representing a valuable tool for the process of mental health monitoring.

Some studies have provided promising results when using mobile device data to characterise social behaviour. Min et al. [46] used call logs and text messages as input to develop a machine learning algorithm capable of classifying social roles (e.g. family, co-worker and friends) with 90.5% of accuracy. Beiwinkel et al. [10] investigated the relationship between communication records (e.g. calls logs and text messages) and the symptoms of depressive and manic states, where the results of the analysis indicated that the increase in communication is related to manic symptoms, and its decrease is correlated with depressive symptoms.

Online social networks, such as Twitter, Facebook and Instagram, are increasingly indispensable for achieving social connection. Within these online platforms, people perform different types of activities, such as creating and sharing status, uploading photos and videos, establishing new friendships, exchanging messages with friends and other social events. From these actions, users express their feelings and thoughts, as well as expose information about their daily routine. In this way, the dynamics of users in online social networks produce a flow of behavioural data, which represents a profile of the social practice of this individual.

In recent years, researchers have recognised online social platforms as valuable sources of health data and behavioural indicators [77]. The public character and the high flow of social activities of these platforms present an adequate environment for the identification of significant social behaviours. In particular, the field of mental health care can take advantage of this behavioural information to identify possible indicators (e.g. type of status shared, publications with negative feelings, formation of friendships) of mental disorders [77]. Based on the indicators, it is possible to use a wide range of techniques, such as data mining, machine learning and complex event processing (CEP), to identify mental health problems. For example, techniques can be used to perform the analysis of feelings and identify meaningful actions and patterns of social activities within these platforms, acting as an indicator of behavioural change.

5 Sleep Quality

Sleep disorders are responsible for degrading the sleep quality of an individual, including insomnia, hypersomnia, excessive daytime sleepiness, among others [38]. Sleep is an aspect of life that directly influences the well-being of an individual, which can overwhelm a person's life. The relationship between sleep and mental state is bidirectional, in which the presence of sleep disorders is a substantial evidence of a degraded mental state, just as the reverse is also valid. For example, sleep disorders are often associated with depressive symptoms, since the sleep absence is a reason for caution in making a diagnosis of depression [49]. This link can provide a more efficient way of monitoring the state of mental health by tracking situations of sleep disorder in which the outcome of such situational awareness allows mental health professionals to make more effective interventions. Therefore, monitoring the presence of sleep disorders is an essential task in the field of mental health.

Currently, the gold standard for diagnosing sleep disorders is polysomnography, which consists of collecting a set of biomedical signals through sensors connected to the body, by using different types of equipment such as electroencephalogram (EEG) for neural signals, electrocardiogram (ECG) for heart rate and inertial sensors for body movement [51]. The experience lived by patients can last several nights to characterise their sleep, so polysomnography is not an ecological technique. This approach presents some peculiarities that may make it impossible to use, such as being a very invasive technique, high cost and by forcing the patient to frequently go to the clinic. Therefore, this technique is impractical to perform the monitoring of sleep in an ecological and long-term way. The solution for long-term sleep monitoring is wrist actigraphy, an accelerometer-based device that demonstrated satisfactory validity [44]. However, these devices have a high cost, which may limit their broad adoption. Thus, it is necessary to implement a mechanism capable of monitoring sleep in a non-invasive way, and that can be integrated into the natural environment of people.

Since the advent of IoT paradigm, new devices equipped with multimodal sensors have become part of people's daily lives, providing data that can identify the sleep quality ecologically and automatically. Among these devices, stand out the fitness bands, smartwatches and smartphones. Sensors embedded in these devices can provide personal and environmental information (e.g. body movement, heart rate, luminosity, ambient sound), allowing computational methods to process them in order to infer high-level situations, such as the sleep duration and the frequency of awakening.

Studies generally identify sleep quality by processing various contextual data sources such as body movement, ambient sound, luminosity and smartphone usage. For example, Daskalova et al. [20] developed a structure called *SleepCoacher*, which uses accelerometer motion data and microphone noise levels to identify the latency and frequency of awakening and sleep duration. *SleepCoacher* uses sleep quality awareness to make personalised recommendations to improve sleep, in which these interventions are modelled based on the knowledge of specialist physicians. Chen et al. [16] introduced an approach that combines context data to estimate the sleep time, in which information such as ambient light and sound, smartphone usage and device charging status were used to infer the total sleep time. Hao et al. [32] developed a system called *iSleep* that uses acoustic characteristics to design a decision tree algorithm able to measure events related to the sleep quality, which include body movement, coughing and snoring.

In conclusion, IoT devices have been used to identify the sleep quality in a non-invasive and ecological way, which allows the recognition of the influence of this situation on the mental state of individuals. The sleep monitoring techniques presented can serve as a viable solution to the limitations presented by polysomnography and actigraphy, enabling interventions from mental health professionals in the patient's environment, thus increasing the chances of achieving better results.

6 Mobility Patterns

This section presents studies focused on correlating the user's mobility with their mental health status. Mobility patterns are defined as a sequence of individuals' movements in relation to their location, speed and direction over a period of time. A pattern is a set of standardised rules and, in the context here presented, represents a sequence of individual geographical movements used together for a given function. Mobility patterns can be measured quantitatively via IoT devices such as smartphones and smartwatches by exploring location sensors such as GPS receivers and Wi-Fi records [57]. These sensors provide geographical coordinates that are used to characterise the mobility trajectory of users. The trait of a user's mobility is considered a sequence of stops and movements [14]. A stop represents a geographical location in which the user spends a certain amount of time.

Changes of mobility patterns can be a problem faced by individuals, which lead to limitations in their daily activities and may affect even more individuals with mental disorders [71]. Mobility can be measured by several aspects of the patient's location information [7], and it is represented by a set of statistical summaries [14].

Fig. 3 Location features related to mobility patterns

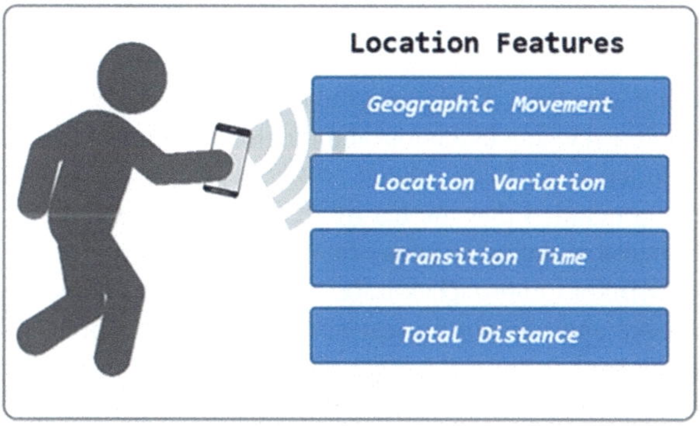

Figure 3 presents the main location features related to the mobility patterns of an individual.

Geographic movements are acquired from location information through the individual's location coordinates. With these coordinates, it is also possible to calculate the location variation, that is, the maximum travelled distance using geographic distance calculations. Already the transition time represents the duration in which an individual remained in an uninterrupted state, which is calculated from dividing the number of location samples in transition states by the total number of samples. The total distance is equivalent to the total sum measurement in meters or kilometres travelled by an individual.

Several studies have used the capabilities of IoT devices to attempt to verify the level of correlation between mobility patterns and mental disorders. These studies report that location data may be important for inferring and monitoring the mental health state of an individual. Localisation data attributes are extracted to develop machine learning models for the automation of this process. Machine learning models can be used to create tools that can support patients in their daily life and treatments, and the monitoring carried out by mental health professionals and caregivers.

Social anxiety is one of the major mental disorders and object of study. It is defined as extreme fear of being judged by other people in social or performance situations. This type of anxiety is one of the most common disorders faced by university students, and specifically in this population, it is associated with impairment in academic functioning. Huang et al. [35] performed a feasibility study using mobile sensors to passively assess social anxiety of university students. The study demonstrated the significance that the person's social anxiety risk correlates with his/her location trajectory. Participants also completed brief questionnaires about mood and affect at random times each day. Results of correlation and significance analysis indicated that visiting patterns at "home", "work" and "religious" places may be related to the social anxiety of college students.

The purpose of the research in [58] was to examine the ability of smartphone sensors to estimate semantic locations and to evaluate the relationship between patterns of semantic localisation and mental disorders such as depression and anxiety. The semantic location of an individual is important to enable suitable interventions such as medical interference and recommendations or even context-aware services. That knowledge is particularly useful in mental health care to monitor relevant behavioural indicators to improve the treatment delivery. The study included recordings related to the types of places that participants visited daily and concluded that the nature of the sites visited explains only a small part of the variation in the symptoms of anxiety and depression.

Different approaches have been proposed to investigate the use of smartphone sensors and interaction features, including location, SMS and call and application usage logs, to infer the depressive state of the users. In [43], the authors developed a solution to monitor depressive states by using smartphone multimodal sensors and metrics for measuring interaction with the smartphone. Examples of used metrics include the number of times that applications were launched and the phone was unlocked, the total duration time that applications were used and the phone was used. This study showed the coefficients that are calculated to evaluate the relationship between the PHQ-8 depression score and the metrics. Therefore, results showed that the depressive state of the users correlates moderately with the calculated metrics.

The work by Canzian and Musolesi [14] focused on the analysis and prediction of variations in depressive states. Application for automatic diagnosis and monitoring of depression that do not require user interaction were developed. To implement, mobility data extracted from metrics was used. *MoodTraces* is an Android application that automatically register samples of smartphone sensors to retrieve the user current location, which is represented by a timestamp, a longitude value and a latitude value. The study also proposes predictive mechanisms to predict depressive mood changes based on mobility data, demonstrating that it is possible to observe a significant correlation between mobility patterns and depressed mood using data collected through smartphones.

The research of Chow et al. [17] included undergraduate participants from a university, where symptoms of depression and social anxiety were assessed using self-reporting tools. An application called *Sensus* was developed and installed on participants' personal smartphones and collected location data within a time period of up to 2 weeks. It was identified that the greater social anxiety of participants was associated with more time spend at home.

Another study in [29] was performed with patients suffering from bipolar disorder, in which individuals may experience moments of manic and depressive states. The work showed that while the efficacy from rating scales has been proven in diagnosing bipolar disorder, they have their drawbacks, as they are a potential source of subjectivity in the diagnosis. A logging application used smartphone sensors and location traces to provide inputs for supporting the diagnosis of mental health professionals. The experiment was performed for a total period of more than 1000 days, and the application obtained a precision/recall of 96%/94% in detecting state changes and an accuracy of 80% in recognising states.

Beiwinkel et al. [10] developed a mobile system named *SIMBA* that monitors social information of patients with bipolar disorder. It tracks daily mood, physical activity and social communication of several patients. Patients used *SIMBA* for a time period of up to 12 months. Thus, the experiment investigated whether the tracked smartphone measurements could predict levels and changes of clinical symptoms. The study also used random-coefficient multilevel models that were computed to analyse the relationship between smartphone data and externally rated manic and depressive symptoms. Thus, overall symptom levels and change in clinical symptoms were related to smartphone measures. The conclusion of this study was that smartphones have the potential to monitor bipolar disorder symptoms in patients' daily life, but it is necessary to validate the monitoring tools in a larger sampling.

Yue et al. [84] developed an approach focused on combining smartphone location data collected from two sources (GPS and Wi-Fi association records) to track depression. An experiment was conducted for evaluating the performance of the proposed solution by using data collected from university students. The study demonstrated that the more complete data lead to features that are more strongly correlated with PHQ-9 questionnaire and lead to better depression screening. Therefore, results showed that smartphone location data are sufficient to achieve most of the performance gains in achieving accurate depression prediction.

7 Suicide Prevention

Mental disorders may trigger suicidal behaviour due to the significant degradation of mental state. For example, depression has a substantial impact on the lives of individuals, affecting social relationships, work performance, sleep quality, among other aspects that may lead to the attempt or even the execution of suicide. While the demand for immediate support from a trained professional is crucial, some factors still prevent adequate access to such services, such as the stigma associated with such a condition, the preference for the help of unprepared people (e.g. friends and acquaintances), relatively high costs, non-perception of need for help, among other factors. Given that suicide is a health problem that has a global scale, responsible for nearly 800 thousand deaths worldwide each year [81], it is possible to identify the urgency of new approaches that may improve the process of prevention and care of individuals at high risk of suicide.

The growing dissemination of IoT devices with embedded multimodal sensors and the use of online social networks present the potential of passive detection of a large amount of behavioural data, allowing the automatic detection of suicidal tendencies and associated risk situations. These smart devices are suitable for monitoring individuals who exhibit suicidal behaviour as they provide the following characteristics: (i) they allow immediate contact with specialised mental health professionals anywhere and at any time; (ii) they are able to recognise situations associated with suicidal behaviours, such as social isolation, negative mood, behavioural changes, among others; and (iii) they can be used to carry out interventions at the time when risk situations are identified.

Social isolation is one of the situations that can indicate the risk of suicidal behaviour [73], being able to be identified by relatives and friends. However, for people to recognise this abnormal behaviour, it usually takes considerable time, requiring more efficient means of monitoring. As shown in Sect. 4, IoT devices can be used to identify face-to-face and technology-mediated social interactions, so that the low frequency or non-identification of this situation over time can be considered a proxy for social isolation. Another possible approach is to explore location data to identify the coordinates or semantic place that represent the location of the individual's home, where, in combination with the temporal context, it is possible to identify signs of isolation social.

Another aspect that may reveal suicide risk situations is the extraction of behavioural clues from audio data [19]. The various cognitive, affective and physiological changes caused by the suffering condition experienced may reflect in the speech, allowing the identification of risk situations. By considering this opportunity, computing methods can use microphone data to identify the paralinguistic (i.e. non-verbal vocal signals) characteristics that represent the individual's crisis. The literature have presented a set of studies that associated prosodic features of speech with suicidal signs. The prosodic characteristics can have several components such as speech tone perception, speech rate, energy dynamics and sound intensity. Ozdas et al. [50] investigated the possibility of using two speech parameters based on excitation, vocal jitters and glottal flow spectra to identify short-term suicidal patients, major depression and high-risk non-suicidal patients. *StressSense* [42] used voice pitch (i.e. the rate of vibration of the vocal folds) and its derivatives, jitter, spectral centroid, high-frequency ratio, speaking rate and Mel frequency cepstral coefficient (MFCCs) to identify stress in the human voice.

In addition to the sensors present on the IoT devices, it is possible to use mobile social networks data to identify suicide risk situations [77]. In particular, these mobile online platforms are promising tools for performing this task, as individuals express thoughts and feelings in there, allowing the use of computing methods (e.g. machine learning and data mining) to identify risk situations such as changing patterns of interactions and emotions contained in comments, photos and videos. In the literature, Twitter, Facebook and Instagram are the most explored mobile social networks to recognise mental health information and evidence of suicidal thoughts [56]. Vioulès et al. [75] designed and developed behavioural features to identify an individual's suicide risk from the online activity pattern on Twitter, such as the number of social relationships and daytime activities. Reece et al. [55] developed a machine learning algorithm capable of identifying depressive symptoms from Instagram posts. This proposed solution analysed colour and metadata components, in addition to performing face detection.

The attempt or consummation of suicide is the result of the complex combination of several factors, such as sleep quality, social relations, physical activities, mobility patterns, among others. As described in this chapter, the advances in IoT devices allows the perception of such situations, and consequently, it becomes possible to perform adaptive interventions in response to the awareness of these risky situations [24]. In conclusion, the use of these intelligent devices provides significant indicators of risk situations, thus presenting potential use in the task of suicide prevention. Although this research area is new, the results have indicated that the use of ubiquitous devices can be exploited to achieve a practical, efficient and clinically validated solution.

8 Open Issues

Both mobile and IoT solutions proposed for mental health have limitations that may be seen from different points of view, either from a computational technical aspect or regarding their degree of maturity in the experiments performed for validation and evaluation. Such limitations should be explored, and they create perspectives for future research.

There are some challenges to be addressed within the domain involving clinical scores and ubiquitous computing. To computationally solve clinical score issues, the developer needs to deal with a huge variety of data from different sources. There is a variety of biomedical sensors of different models that can provide input data for a clinical score specification. In the literature, there are middleware proposals that act as intermediary for the communication between those devices and end applications. However, in a real scenario of a clinical score calculation, it is a hard task to predict which sensor models will be used to obtain patient measurements. This assumption makes it difficult to develop mobile applications for clinical scores.

Additionally, there are several proprietary solutions for electronic medical record systems. Electronic medical records are very important sources to provide useful data for calculating clinical scores. In this context, market offers several proprietary solutions whose systems usually do not communicate with each other. Hence, the developer may need to handle a large amount of data in different formats from closed-source systems. Moreover, manual data are very recurrent in the context of clinical scores, especially those related to mental health. They are usually obtained from interviews with patients and health professionals. Thus, it is a great challenge to computationally deal with this type of information, since it may be very subjective.

Data from physical and virtual sensors of IoT devices can be transformed into meaningful social behaviours through computational methods such as machine learning algorithms and data mining. Therefore, mobile applications can be developed to use this awareness of social situations in smarter services, such as providing feedback on social behaviour, performing personalised interventions and identifying conditions of risk near real time. Applying data from IoT devices to monitor social situations is a new and challenging research area. Wearable devices, such as smartwatches and smart bands, have not yet been fully explored to identify social interactions. Moreover, combining their data with smart environments will enable an increasingly accurate and detailed perception of social relationships of individuals.

Proposed solutions have used only one data source to identify social behaviour of patients, which may be a limiting factor in practical application. For example, some studies have explored wireless communication interfaces to detect co-locations [45, 74, 83], but they do not consider other social activities experienced by individuals, such as interactions on mobile social networks and communications through telephone calls. Thus, it is necessary to develop multimodal computational methods that explore various sources of social data to improve the recognition of patients' social behaviour. Another evident open issue is the need for conceiving solutions to explore the awareness of identified social situations to derive helpful high-level information for mental health professionals, such as recognising social behaviour patterns. The social routine recognition can be used to develop methods that identify abnormal behaviours in real time, in which there may be significant pieces of evidence of the presence, initiation or development of mental disorders.

Sleep quality detection is usually performed through multimodal sensing [16, 20, 32], which combines various context data to infer this information of interest. In this sense, a research opportunity is to explore and validate commercial wearable devices (e.g. Fitbit, Apple Watch, Mi Band) for sleep studies, including analysis and comparisons with actigraph devices. This will be an interesting scientific contribution since current wearable devices already are popular and may be integrated into the daily environment of patients. In the same line of research on sleep quality, another interesting study would be to extract high-level information from inferences and provide both patient and mental health professional with sleep patterns and their changes throughout daily life.

Studies that have investigated the feasibility of using ubiquitous technologies to detect suicide risk situations focus only on specific conditions, such as sentiment analysis in social network data, social isolation, among others. However, since suicidal behaviour is complex, which can be derived from the combination of different circumstances, further studies are needed to investigate the feasibility of developing tools able to detect risk situations in real time. These tools are required to provide easy-to-access communication channels for patients who are desperate, deliberately or not, to be attended by professionals capable of managing the emergency situation efficiently.

There is a significant correlation between mental disorders and mobility patterns detected by using location data collected from smartphone sensors. Smartphones and their embedded mobile sensors are very promising as assessment tools for measuring behaviour in daily life. Mobile devices have the opportunity to obtain additional data such as motion, light and sound to lead to a better estimation of the semantic location of an individual. The knowledge of the semantic location of an individual is important to provide medical interventions, recommendations and different types of context-aware services. That knowledge is particularly useful in mental health to monitor relevant behavioural indicators for improving treatment delivery. However, there are limitations in the current detection methods of correlation between mobility patterns and mental health made by ubiquitous devices, either due to hardware constraints or the demand for more efficient algorithms.

From a practical application perspective, although experimental researches with ubiquitous solutions for mental health have been performed with end-users (e.g. patients, professionals and caregivers), it is possible to cite different limitations regarding the methodological rigour and experiment restrictions that limit the pieces of evidence obtained in the studies, which are as follows: (i) first of all, they are still limited by involving a small number of participants (i.e. small sampling); (ii) most studies have not used randomised controlled trials; (iii) most of them do not consider control population; (iv) study duration times are relatively short, making it difficult to identify the long-term impacts and contributions of the proposed approaches and (v) there is a low prevalence of participants with clinical symptoms or patients diagnosed with mental disorder. As a result, all of these limitations create a huge window of opportunities to be explored by researchers. Therefore, despite recent advances in the research field, solutions are still far from being used in real clinical settings, requiring further studies to generate scientific evidence.

9 Conclusion

In this chapter, an overview of ubiquitous computing and IoT usage in the domain of mental health and well-being has been written. Different topics were covered during the text, namely clinical scores, social relations, sleep quality, mobility patterns and suicide prevention. By describing the main related concepts and current applications proposed in the literature, all of the topics were dedicated to display the state of the art of IoT applied to mental health. In addition, by analysing limitations of the proposed solutions, the provided content was able to show open issues and envisage research perspectives with possible problems to be addressed.

Notable advances in ubiquitous and IoT technologies, especially in terms of passive sensing, have allowed researchers to passively monitor user behaviour in real time and with a granularity that was not possible just a few years ago. Ubiquitous features of personal sensing technologies, such as mobile and wearable devices, allow the data collection able to infer situations of interest discreetly and ecologically. Collected data can be used to create machine learning models to be used for classifying or predicting contextual information about the patient, such as mood, physical activity level, stress, among other behavioural markers. That is, new metrics and patterns from data produced by IoT devices reveal promising opportunities. Therefore, improved models can create new possibilities for supporting mental health, potentially capable of aiding the decision-making process made by specialised mental health professionals.

Acknowledgements The authors would like to thank FAPEMA (State of Maranhão Research Funding Agency) for the financial support provided to their conducted research projects.

References

1. C. Aakre, M. Dziadzko, M. Keegan, V. Herasevich, Automating clinical score calculation within the electronic health record. Appl. Clin. Inform. **8**(2), 369–380 (2017)
2. Z. Adams, E. McClure, K. Gray, C. Danielson, F. Treiber, K. Ruggiero, Mobile devices for the remote acquisition of physiological and behavioral biomarkers in psychiatric clinical research. J. Psychiatr. Res. **85**, 1–14 (2017)
3. A. Al-Fuqaha, M. Guizani, M. Mohammadi, M. Aledhari, M. Ayyash, Internet of things: A survey on enabling technologies, protocols, and applications. IEEE Commun. Surv. Tutor. **17**(4), 2347–2376 (2015)
4. A. Althubaiti, Information bias in health research: Definition, pitfalls, and adjustment methods. J. Multidiscip. Healthc. **9**, 211 (2016)
5. A. Aminian, J.L. Alberts, S. Clemence, P. Schauer, Bariatric surgery decisionmaking calculator: A novel mobile app for evidence-based clinical practice. J. Surg. Obes. Relat. Dis. **13**(10), S147 (2017)
6. L. Atzori, A. Iera, G. Morabito, The internet of things: A survey. Comput. Netw. **54**(15), 2787–2805 (2010)
7. I. Avila-Palencian, L.I. Panis, A.D. Nazelle, T. Götschi, E. Raser, M. Gaupp-Berghausen, E. Stigell, F. Iacorossi, M. Laeremans, E.A. Boig, M. Nieuwenhuijsen, Active mobility and subjective general health: Roles of mental health, social support and physical activity. J. Transp. Health **5**, S76 (2017)
8. A. Barbosa, F. Silva, L. Coutinho, D. Santos, A. Teles, A domain-specific modeling language for specification of clinical scores in mobile health, in *14th International Conference on Evaluation of Novel Approaches to Software Engineering*, (SciTePress, Heraklion, 2019), pp. 104–113
9. P. Bech, N. Rasmussen, L. Olsen, V. Noerholm, W. Abildgaard, The sensitivity and specificity of the major depression inventory, using the present state examination as the index of diagnostic validity. J. Affect. Disord. **66**(2), 28–35 (2001)
10. T. Beiwinkel, S. Kindermann, A. Maier, C. Kerl, J. Moock, G. Barbian, W. Rössler, Using smartphones to monitor bipolar disorder symptoms: A pilot study. JMIR Ment. Health **3**(1), e2 (2016)
11. M.E. Beutel, E.M. Klein, E. Brähler, I. Reiner, C. Jünger, M. Michal, J. Wiltink, P.S. Wild, T. Münzel, K.J. Lackner, et al., Loneliness in the general population: Prevalence, determinants and relations to mental health. BMC Psychiatry **17**(1), 97 (2017)
12. Brazilian Federal Council of Psychology: Resolution number 11 published on May 11, 2018, https://site.cfp.org.br/wpcontent/uploads/2018/05/RESOLU%C3%87%C3%83O-N%C2%BA-11-DE11-DE-MAIO-DE-2018.pdf (only available in Portuguese). Online; Accessed 2 July 2019
13. R.A. Calvo, D. Peters, *Positive Computing: Technology for Well-Being and Human Potential* (The MIT Press, Cambridge, MA, 2014)
14. L. Canzian, M. Musolesi, Trajectories of depression: Unobtrusive monitoring of depressive states by means of smartphone mobility traces analysis, in *Proceedings of the 2015 ACM International Joint Conference on Pervasive and Ubiquitous Computing*, (ACM, New York, 2015), pp. 1293–1304
15. I. Carreras, A. Matic, P. Saar, V. Osmani, Comm2Sense: Detecting proximity through smartphones, in *2012 IEEE International Conference on Pervasive Computing and Communications Workshops*, (IEEE, Piscataway, 2012), pp. 253–258
16. Z. Chen, M. Lin, F. Chen, N.D. Lane, G. Cardone, R. Wang, T. Li, Y. Chen, T. Choudhury, A.T. Campbell, Unobtrusive sleep monitoring using smartphones, in *Proceedings of the 7th International Conference on Pervasive Computing Technologies for Healthcare*, (IEEE, Piscataway, 2013), pp. 145–152

17. P.I. Chow, K. Fua, Y. Huang, W. Bonelli, H. Xiong, L.E. Barnes, B.A. Teachman, Using mobile sensing to test clinical models of depression, social anxiety, state affect, and social isolation among college students. J. Med. Internet Res. **19**(3), e62 (2017)
18. K.A. Cook, R. Simon, B.D. Modena, Improvement in asthma control using a minimally burdensome and proactive smartphone application. J. Am. Acad. Allergy Asthma Immunol. **13**(10), 9 (2016)
19. N. Cummins, S. Scherer, J. Krajewski, S. Schnieder, J. Epps, T.F. Quatieri, A review of depression and suicide risk assessment using speech analysis. Speech Comm. **71**, 10–49 (2015)
20. N. Daskalova, D. Metaxa-Kakavouli, A. Tran, N. Nugent, J. Boergers, J. McGeary, J. Huang, Sleepcoacher: A personalized automated self-experimentation system for sleep recommendations, in *Proceedings of the 29th Annual Symposium on User Interface Software and Technology, UIST '16*, (ACM, New York, 2016), pp. 347–358
21. L. David, R. Vasconcelos, L. Alves, R. André, M. Endler, A DDS-based middleware for scalable tracking, communication and collaboration of mobile nodes. J. Internet Serv. Appl. **4**, 16 (2013)
22. I. de la Torre Díez, S.G. Alonso, S. Hamrioui, E.M. Cruz, L.M. Nozaleda, M.A. Franco, IoT-based services and applications for mental health in the literature. J. Med. Syst. **43**(1), 11 (2018)
23. G. Falk, T. Fahey, Clinical prediction rules. BMedJ **339**(2), b2899 (2009)
24. M.A. Franco-Martín, J.L. Muñoz-Sánchez, B. Sainz-de Abajo, G. Castillo-Sánchez, S. Hamrioui, I. de la Torre-Díez, A systematic literature review of technologies for suicidal behavior prevention. J. Med. Syst. **42**(4), 71 (2018)
25. E. Garcia-Ceja, M. Riegler, T. Nordgreen, P. Jakobsen, K.J. Oedegaard, J. Tørresen, Mental health monitoring with multimodal sensing and machine learning: A survey. Pervasive Mob. Comput. **51**, 1–26 (2018)
26. S.K. Goudos, P.I. Dallas, S. Chatziefthymiou, S. Kyriazakos, A survey of iot key enabling and future technologies: 5g, mobile iot, sematic web and applications. Wirel. Pers. Commun. **97**(2), 1645–1675 (2017)
27. S. Grav, O. Hellzèn, U. Romild, E. Stordal, Association between social support and depression in the general population: The hunt study, a cross-sectional survey. J. Clin. Nurs. **21**(1–2), 111–120 (2012)
28. F. Gravenhorst, A. Muaremi, J. Bardram, A. Grünerbl, O. Mayora, G. Wurzer, M. Frost, V. Osmani, B. Arnrich, P. Lukowicz, G. Tröster, Mobile phones as medical devices in mental disorder treatment: An overview. Pers. Ubiquitous Comput. **19**(2), 335–353 (2015)
29. A. Gruenerbl, V. Osmani, G. Bahle, J.C. Carrasco, S. Oehler, O. Mayora, C. Haring, P. Lukowicz, Using smart phone mobility traces for the diagnosis of depressive and manic episodes in bipolar patients, in *Proceedings of the 5th Augmented Human International Conference*, (ACM, New York, 2014), p. 38
30. J. Gu, B. Gao, Y. Chen, L. Jiang, Z. Gao, X. Ma, Y. Ma, W.L. Woo, J. Jin, Wearable social sensing: Content-based processing methodology and implementation. IEEE Sensors J. **17**(21), 7167–7176 (2017)
31. M. Hamilton, The assessment of anxiety states by rating. Br. J. Med. Psychol. **32**(1), 50–55 (1959)
32. T. Hao, G. Xing, G. Zhou, iSleep: Unobtrusive sleep quality monitoring using smartphones, in *Proceedings of the 11th ACM Conference on Embedded Networked Sensor Systems*, (ACM, New York, 2013), p. 4
33. K.E. Heron, J.M. Smyth, Ecological momentary interventions: Incorporating mobile technology into psychosocial and health behaviour treatments. Br. J. Health Psychol. **15**(1), 1–39 (2010)
34. D.M. Hilty, S. Chan, T. Hwang, A. Wong, A.M. Bauer, Advances in mobile mental health: Opportunities and implications for the spectrum of e-mental health services. mHealth **3**(8), 34 (2017)

35. Y. Huang, H. Xiong, K. Leach, Y. Zhang, P. Chow, K. Fua, B.A. Teachman, L.E. Barnes, Assessing social anxiety using GPS trajectories and point-of-interest data, in *Proceedings of the 2016 ACM International Joint Conference on Pervasive and Ubiquitous Computing*, (ACM, New York, 2016), pp. 898–903
36. C. Hughes, L. Berg, W. Danziger, L. Coben, R. Martin, A new clinical scale for the staging of dementia. Br. J. Psychiatry **140**(1), 566–572 (1982)
37. A. Kamilaris, A. Pitsillides, Mobile phone computing and the internet of things: A survey. IEEE Internet Things J. **3**(6), 885–898 (2016)
38. A.D. Krystal, Sleep and psychiatric disorders: Future directions. Psychiatr. Clin. **29**(4), 1115–1130 (2006)
39. N.D. Lane, M. Lin, M. Mohammod, X. Yang, H. Lu, G. Cardone, S. Ali, A. Doryab, E. Berke, A.T. Campbell, T. Choudhury, Bewell: Sensing sleep, physical activities and social interactions to promote wellbeing. Mob. Netw. Appl. **19**(3), 345–359 (2014)
40. U. Lee, K. Han, H. Cho, K.M. Chung, H. Hong, S.J. Lee, Y. Noh, S. Park, J.M. Carroll, Intelligent positive computing with mobile, wearable, and IoT devices: Literature review and research directions. Ad Hoc Netw. **83**, 8–24 (2019)
41. H. Lu, W. Pan, N.D. Lane, T. Choudhury, A.T. Campbell, SoundSense: Scalable sound sensing for people-centric applications on mobile phones, in *Proceedings of the 7th International Conference on Mobile Systems, Applications, and Services*, (ACM, New York, 2009), pp. 165–178
42. H. Lu, D. Frauendorfer, M. Rabbi, M.S. Mast, G.T. Chittaranjan, A.T. Campbell, D. Gatica-Perez, T. Choudhury, StressSense: Detecting stress in unconstrained acoustic environments using smartphones, in *Proceedings of the 2012 ACM Conference on Ubiquitous Computing, UbiComp '12*, (ACM, New York, 2012), pp. 351–360
43. A. Mehrotra, R. Hendley, M. Musolesi, Towards multi-modal anticipatory monitoring of depressive states through the analysis of human-smartphone interaction, in *Proceedings of the 2016 ACM International Joint Conference on Pervasive and Ubiquitous Computing*, (ACM, New York, 2016), pp. 1132–1138
44. L.J. Meltzer, L.S. Hiruma, K. Avis, H. Montgomery-Downs, J. Valentin, Comparison of a commercial accelerometer with polysomnography and actigraphy in children and adolescents. Sleep **38**(8), 1323–1330 (2015)
45. E. Miluzzo, N.D. Lane, K. Fodor, R. Peterson, H. Lu, M. Musolesi, S.B. Eisenman, X. Zheng, A.T. Campbell, Sensing meets mobile social networks: The design, implementation and evaluation of the CenceMe application, in *Proceedings of the 6th ACM Conference on Embedded Network Sensor Systems, SenSys '08*, (ACM, New York, 2008), pp. 337–350
46. J.K. Min, J. Wiese, J.I. Hong, J. Zimmerman, Mining smartphone data to classify life-facets of social relationships, in *Proceedings of the 2013 Conference on Computer Supported Cooperative Work, CSCW '13*, (ACM, New York, 2013), pp. 285–294
47. D.C. Mohr, M. Zhang, S.M. Schueller, Personal sensing: Understanding mental health using ubiquitous sensors and machine learning. Annu. Rev. Clin. Psychol. **13**(1), 23–47 (2017)
48. K. Nahrstedt, H. Li, P. Nguyen, S. Chang, L. Vu, Internet of mobile things: Mobility-driven challenges, designs and implementations, in *2016 IEEE First International Conference on Internet-of-Things Design and Implementation (IoTDI)*, (IEEE, Piscataway, 2016), pp. 25–36
49. D. Nutt, S. Wilson, L. Paterson, Sleep disorders as core symptoms of depression. Dialogues Clin. Neurosci. **10**(3), 329 (2008)
50. A. Ozdas, R.G. Shiavi, S.E. Silverman, M.K. Silverman, D.M. Wilkes, Investigation of vocal jitter and glottal flow spectrum as possible cues for depression and near-term suicidal risk. IEEE Trans. Biomed. Eng. **51**(9), 1530–1540 (2004)
51. T. Penzel, R. Conradt, Computer based sleep recording and analysis. Sleep Med. Rev. **4**(2), 131–148 (2000)
52. N. Pereira-Azevedo, A. Fraga, L. Osório, M.J. Roobol, Rotterdam prostate cancer risk calculator: Development and usability testing of the mobile phone app. JMIR Cancer **3**(1), e1 (2017)

53. V. Pinheiro, G. Neumann, M. Endler, F. Silva, Deklaer: An ontology-driven framework for generating IoT applications using contextnet, in *2018 IEEE Symposium on Computers and Communications (ISCC)*, (IEEE Computer Society, Natal, Brazil, 2018), pp. 608–617
54. L. Qiu, H. Rui, A. Whinston, When cellular capacity meets wifi hotspots: A smart auction system for mobile data offloading, in *2015 48th Hawaii International Conference on System Sciences*, (IEEE Computer Society, Los Alamitos, 2015), pp. 4898–4907
55. A.G. Reece, C.M. Danforth, Instagram photos reveal predictive markers of depression. EPJ Data Sci. **6**(1), 15 (2017)
56. J. Robinson, G. Cox, E. Bailey, S. Hetrick, M. Rodrigues, S. Fisher, H. Herrman, Social media and suicide prevention: A systematic review. Early Interv. Psychiatry **10**(2), 103–121 (2016)
57. S. Saeb, E.G. Lattie, S.M. Schueller, K.P. Kording, D.C. Mohr, The relationship between mobile phone location sensor data and depressive symptom severity. PeerJ **4**, e2537 (2016)
58. S. Saeb, E.G. Lattie, K.P. Kording, D.C. Mohr, Mobile phone detection of semantic location and its relationship to depression and anxiety. JMIR Mhealth Uhealth **5**(8), e112 (2017)
59. S. Shiffman, A.A. Stone, M.R. Hufford, Ecological momentary assessment. Annu. Rev. Clin. Psychol. **4**(1), 1–32 (2008)
60. A. Soares Teles, A. Rocha, F. João da Silva e Silva, J. Correia Lopes, D. O'Sullivan, P. Van de Ven, M. Endler, Enriching mental health mobile assessment and intervention with situation awareness. Sensors **17**(1), 127 (2017)
61. C. Stamate, G. Magoulas, S. Kueppers, E. Nomikou, I. Daskalopoulos, A. Jha, J. Pons, J. Rothwell, M. Luchini, T. Moussouri, M. Iannone, G. Roussos, The cloudUPDRS app: A medical device for the clinical assessment of Parkinson's disease. Pervasive Mob. Comput. **43**, 146–166 (2018)
62. D. Swain, P. Nightingale, Evaluation of a shortened version of the abbreviated mental test in a series of elderly patients. Clin. Rehabil. **11**, 243–248 (1997)
63. L.E. Talavera, M. Endler, I. Vasconcelos, R. Vasconcelos, M. Cunha, F. João da Silva e Silva, The mobile hub concept: Enabling applications for the internet of mobile things, in *2015 IEEE International Conference on Pervasive Computing and Communication Workshops (PerCom Workshops)*, (IEEE, Piscataway, 2015), pp. 123–128
64. A.S. Teles, A. Rocha, F.J. Silva, J.C. Lopes, D. O'Sullivan, P.V. de Ven, M. Endler, Towards situation-aware mobile applications in mental health, in *2016 IEEE 29th International Symposium on Computer-Based Medical Systems (CBMS)*, (IEEE Computer Society, Los Alamitos, 2016), pp. 349–354
65. A.S. Teles, I. Rodrigues, D. Viana, F. Silva, L. Coutinho, M. Endler, R. Rabêlo, Mobile mental health: A review of applications for depression assistance, in *IEEE 32th International Symposium on Computer-Based Medical Systems, CBMS'19*, (IEEE Press Books, Piscataway, 2019), pp. 708–713
66. G. Thompson, Appendicitis, Chap. 4, in A. Lander (ed.) *Clinical Scoring Systems in the Management of Suspected Appendicitis in Children*, (IntechOpen, 2012), https://doi.org/10.5772/25485
67. J. Torous, M.V. Kiang, J. Lorme, J.P. Onnela, New tools for new research in psychiatry: A scalable and customizable platform to empower data driven smartphone research. JMIR Ment. Health **3**(2), e16 (2016)
68. J. Torous, J.P. Onnela, M. Keshavan, New dimensions and new tools to realize the potential of RDoC: Digital phenotyping via smartphones and connected devices. Transl. Psychiatry **7**(3), e1053 (2017)
69. J. Torous, P. Staples, I. Barnett, L.R. Sandoval, M. Keshavan, J.P. Onnela, Characterizing the clinical relevance of digital phenotyping data quality with applications to a cohort with schizophrenia. NPJ Digit. Med. **1**(1), 15 (2018)
70. T.J. Trull, U. Ebner-Priemer, Ambulatory assessment. Annu. Rev. Clin. Psychol. **9**(1), 151–176 (2013)
71. F. Tseliou, A. Maguire, M. Donnelly, D. O'Reilly, The influence of mobility on mental health status in young people: The role of area-level deprivation. Health Place **42**, 96–103 (2016)

72. T.F. van de Mortel, Faking it: Social desirability response bias in self-report research. Aust. J. Adv. Nurs. **25**(4), 40 (2008)
73. K.A. Van Orden, T.K. Witte, K.C. Cukrowicz, S.R. Braithwaite, E.A. Selby, T.E. Joiner Jr., The interpersonal theory of suicide. Psychol. Rev. **117**(2), 575 (2010)
74. G. Vanderhulst, A. Mashhadi, M. Dashti, F. Kawsar, Detecting human encounters from wifi radio signals, in *Proceedings of the 14th International Conference on Mobile and Ubiquitous Multimedia, MUM '15*, (ACM, New York, 2015), pp. 97–108
75. M.J. Vioulès, B. Moulahi, J. Azé, S. Bringay, Detection of suicide-related posts in twitter data streams. IBM J. Res. Dev. **62**(1), 7:1–7:12 (2018)
76. R. Wang, F. Chen, Z. Chen, T. Li, G. Harari, S. Tignor, X. Zhou, D. Ben-Zeev, A.T. Campbell, StudentLife: Using smartphones to assess mental health and academic performance of college students, in *Mobile Health: Sensors, Analytic Methods, and Applications*, (Springer International Publishing, Cham, 2017), pp. 7–33
77. A. Wongkoblap, M.A. Vadillo, V. Curcin, Researching mental health disorders in the era of social media: Systematic review. J. Med. Internet Res. **19**(6), e228 (2017)
78. World Health Organization, Investing in mental health: Evidence for action. Technical Report. World Health Organization, WHO Document Production Services (2013). Online; Accessed 15 May 2019
79. World Health Organization, Mental health action plan 2013–2020. Technical Report. World Health Organization, WHO Document Production Services (2013). Online; Accessed 11 Jan 2019
80. World Health Organization, Mental, behavioural or neurodevelopmental disorders, in International Classification of Diseases 11th Revision. World Health Organization (2018), https://icd.who.int/en/
81. World Health Organization, National suicide prevention strategies: Progress, examples and indicators. Technical Report. World Health Organization (2018). Online; Accessed 2 July 2019
82. World Health Organization, Depression – key facts, https://www.who.int/newsroom/fact-sheets/detail/depression. Online; Accessed 11 Jan 2019
83. C. Wu, M. Boukhechba, L. Cai, L.E. Barnes, M.S. Gerber, Improving momentary stress measurement and prediction with bluetooth encounter networks. Smart Health **9–10**, 219–231 (2018). CHASE 2018 Special Issue
84. C. Yue, S. Ware, R. Morillo, J. Lu, C. Shang, J. Bi, J. Kamath, A. Russell, A. Bamis, B. Wang, Fusing location data for depression prediction, in *2017 IEEE SmartWorld, Ubiquitous Intelligence & Computing, Advanced & Trusted Computed, Scalable Computing & Communications, Cloud & Big Data Computing, Internet of People and Smart City Innovation*, (IEEE, Piscataway, 2017), pp. 1–8

Internet of Things and Communication Technology Synergy for Remote Services in Healthcare

Raluca Maria Aileni, George Suciu, Carlos Alberto Valderrama Sukuyama, and Sever Pasca

1 Introduction

Internet of Things (IoT) is an innovation now, and the usage in the health monitoring system is challenging due to the security problems of data can occur. Healthcare electronics and informatics are domains in continuous growth, and it develops new solutions nowadays. IoT has the leading role in empowering frameworks worldwide and in sharing data using advanced communication technologies and smart systems [1]. Currently, considering the latest technology developments, IoT solutions provide various advantages to the present computerized world. Healthcare represents one of the promising fields to observe the IoT advantages. A critical security problem in IoT applications is data sharing through communication technologies [2] and pervasive devices. Data sharing has the following benefits: improves the efficiency and the usefulness of the provided service; especially for continuous data sharing, reliability and secured correspondence are vital.

Regarding healthcare applications, in order to choose the appropriate technology, numerous factors are to be considered, such as node and network value, battery life, throughput, response time (latency), adaptability, dynamic range, coverage, and organizational model. The conclusion is the remote that no single innovation technology would almost certainly exceed all factors at the same time. Although many of the healthcare applications for remote monitoring are available on homes, hospital rooms, and wearable invasive or non-invasive devices and that for being

R. M. Aileni (✉) · G. Suciu · S. Pasca
Politehnica University of Bucharest, Faculty of Electronics, Telecommunication and Information Technology, Bucharest, Romania

C. A. Valderrama Sukuyama
Mons University, Faculty of Engineering, Department of Electronics and Microelectronics, Mons, Belgium

N. Gupta, S. Paiva (eds.), *IoT and ICT for Healthcare Applications*, EAI/Springer Innovations in Communication and Computing,
https://doi.org/10.1007/978-3-030-42934-8_5

necessary, a mix between the short- and long-range communication technologies will be necessary [1].

The synergy among IoT and medical devices guarantees to be useful for human healthcare, and they will represent the Internet of Medical Things (IoMT) [1–4]. IoMT is becoming nowadays more and more visible in the healthcare domain. Today, it is an increased interest in the development of smart medical systems capable of integrating technologies such as cloud computing, edge computing, and IoT [5]. According to the Grand View Research surveys, the healthcare IoT market is expected to increase to $330 billion by 2020 [6]. Numerous legislatures are willing to contribute from a financial point of view and promote the utilization of eHealthcare services.

By using remote monitoring systems, which are, in essence, IoMT systems, patients can be observed anywhere and anytime, so diseases are treated before they become unmanageable. In general, all systems for remote medical monitoring are interesting for care providers and families. Connected healthcare solutions and virtual infrastructure make treatment precise and better. The main advantage of using remote medical monitoring is that the patient can stay at home, not being necessary to be in hospitals or medical clinics. In addition, the implementation of those systems will help by reducing the time and financial resources allocated in hospitals for senior care programs, in this way, replacing the nurses with formal or informal caregivers help for a limited time per day. These apps for patients health remote monitoring can work through a smartphone using iOS or Android, cloud services for data storage, and edge computing for data processing. The biomedical sensors used by these applications will gather and transmit information regarding blood, heart rate, or heartbeat straightforwardly to the clients' devices. Emergency assistance can also be improved by using advanced automation and analysis of IoT.

Remote monitoring healthcare systems must include three layers:

- The first layer is composed of wearable devices containing apps capable of acquiring biomedical data: moisture, temperature, pulse, brain, or heart electrical signals.
- The second layer is based on edge computing services that allow data storage, signal processing by dedicated application, anonymization, and analysis in the cloud.
- The third layer is based on the users' (patients, medical staff that can access the information) interface accessible on a smartphone, or wristband, etc.

The wireless communication is usually done by Bluetooth, ZigBee, or Wi-Fi. For remote monitoring systems, smart sensors are used, usually wearable invasive or non-invasive, to collect all necessary biomedical signals and transmit real-time processed information to the doctors and the patients' dedicated interface.

In a smart system based on IoMT architecture, information is collected by wearable biomedical smart sensors. In a smart system, edge computing should also process and analyze the signals. In such a system, the analogical signal collected by the sensor is converted into a digital signal by two processes: sampling and quantization. After signal processing, the values for biomedical parameters are sent instantly to the users' interfaces (patients, doctors). In the edge computing layer, fog computing uses a large number of resources for data storage, processing, and backup [6, 7].

Fixed-line networks are there yet, for the most part, excessively restricted and costly to be executed. A very flexible, less expensive choice is the usage of wireless communication networks. However, the wireless communication technologies such as short-range communications (e.g., Bluetooth, ZigBee, Wi-Fi, infrared, or visible light communication) and long-range communications (e.g., 3G/4G cell or satellite communication) are typically considered [1, 7].

Because wearable devices are low-power devices, and the communication is wireless, the signal processing of analog signal to digital involves signal sampling in order to send aggregator only the numerical useful values, not redundant ones. Cybersecurity specialists appreciate that the information that is public from the wearable devices can be the primary target for hackers [8]. The accuracy in healthcare data minimizes errors and makes the precise medication.

The chapter is structured as follows: Sect. 1 presents an introduction to IoT and communication services for healthcare; Sect. 2 presents methods for signal processing, data mining, and machine learning, while Sect. 3 analyzes edge computing, security, and anonymization for the patient using 5G connected wearables. Furthermore, Sect. 4 presents the communication technology and architecture of the system based on low-power computing devices, and Sect. 5 investigates signal processing and methods to extract information data analyzing, while Sect. 6 deals with data anonymization using edge computing. Finally, Sect. 7 outlines future challenges regarding IoT connected devices for remote healthcare services.

2 Methods to Extract Valuable Data by Signal Processing, Data Mining, and Machine Learning

In many years, significant data used in many domains such as engineering sciences, social networks, bimolecular, and security are extracted by large data sets and analyzed. Despite these domains, data are processed fast and accurate, and the processing of these involves machine learning, training, and validation algorithms [9]. Machine learning provides the system capability to learn and improve from previous experiences such as biomedical parameters associated with already known diseases and is useful in predictive decision modeling from analysis of broad data sets [10].

Signal Processing
Analysis and processing of big data sets represent a significant challenge. In general, big data is from various domains such as engineering, social sciences, biochemistry research, commerce, and security. The big data analysis presents more challenges for digital signal processing, especially in signal filtering, noise-reducing, and spectral density analysis of large data sets [11–13].

The processing of signals on the graphs extends the classical theory of processing of the signal to the general graph. Some techniques used with the help of the data representations reduced based on Laplacian [11].

Considering the graph signal: s: $\nu \rightarrow C$, $\nu_{\mathrm{n}} \rightarrow s_{\mathrm{n}}$, where C is the set of complex numbers [12], the graph signal can be defined as a vector: $s = [s_0\ s_1\ \ldots\ s_{N-1}]$ $T \in CN$. For each value, s_{n} corresponds to a node ν_{n}. Some researches consider World Wide Web a direct graph, the vertices being the sites and edge the hyperlinks (e.g., the characteristics such as subject, view count, and relevance are graph signals indexed through graphs consisting of hyperlink references) [11, 12].

Data Mining

The data mining generates new information by using the signal patterns and mapping on large data sets [11, 13]. Data mining is linked with finding new connections among the information [10]. We consider this tool a multi-disciplinary one because it uses machine learning, statistics, and database. Other functionalities of data mining are knowledge discovery, knowledge extraction, data analysis, and information harvesting. The concept has been increasing nowadays from the commercial world, and the database clients were rapidly growing in size and number. Data are collected from large areas of modalities, including satellite imagery or high-altitude imagery, and then changes of urbanization and detect climate changes are monitored [14].

In data mining, the computer stores electronically and searches the data automatically. It is known that data can be automatically sought, validated, identified, and can also be used for prediction. We can refer to data mining as solving problems, analyzing data that are already in databases and summarizing them [15].

In terms of machine learning to data mining, explicit knowledge and structural description have extraordinary importance. Data mining is frequently used by people to gain knowledge.

Recently, machine-learning techniques have been active in massive numbers in fields such as medicine, biology, and astronomy, because these techniques give us proper solutions for discovering information that is hidden in the data.

3 Edge Devices and Computing for Patients Data Anonymization

Edge Computing

The edge computing replaces the previous cloud services due to the need for signal or data processing in an edge application in order to have quick response and reliability [1]. Additionally, bandwidth could be used efficiently if part of the data is handled on edge. The increase in the number of wearable IoT devices and the generalization of mobile services lead to transform the user from consumer in data provider [1, 4]. It is more efficient to process data at the edge of the network by

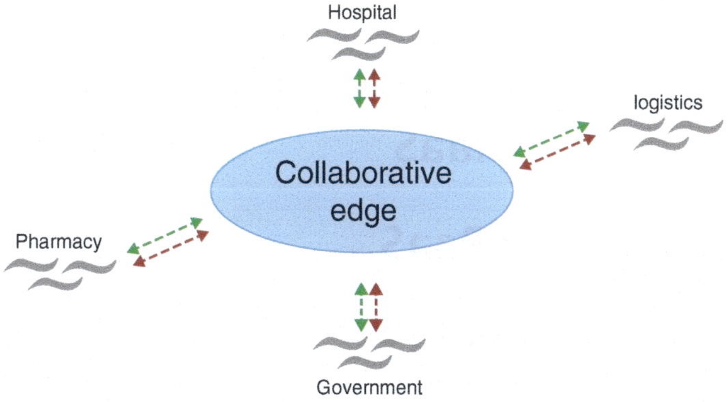

Fig. 1 Collaborative edge example: connected to health [19]

using dedicated apps in order to obtain the optimal computer performance in data processing. Edge computing played a vital role in data networks since the signal processing or data analyzing is generated at the edge of the network. In the past, the microdata center [16, 17], cloudlet [18], and fog computing [19] have been put in practice for the user communities because the cloud computing is not efficient in data processing when the data are at the edge of networks by sensor systems. Edge computing permits the processes at the edge of the network on downstream data (cloud services case) and upstream data (IoT case) (Fig. 1) [19, 20].

Collaborative Edge for Healthcare

The collaborative edge for healthcare involves connecting multiple healthcare edges, and this can create an advantage in data analyzing, statistics, stocks, and patient management [1].

To present the potential advantages of the edge computing, a case of healthcare connected will be presented.

For example, some hospitals analyze precisely and forward the information such as symptoms, costs, and population for a flu eruption. However, by collaborative edge computing, the pharmacy would be able to provide the purchasing record of a patient to the hospital, which considerably facilitates healthcare costs managements [1].

Data Anonymization

Data anonymization is the optimal safety possibility to use wearable biosensors for patient monitoring. This objective consists of minimizing the security risks by using the adequate methodology for working with personal sensitive data, anonymization, and data analytics in the cloud. The possibility of using the data is to filter data using a private cloud and to anonymize data that will be used for statistics. In this way, patient data security and privacy can be ensured by applying a different level of protection for different types of user accounts (medical staff and patients). The anonymization of the patients' data is relevant because data are used for a secondary

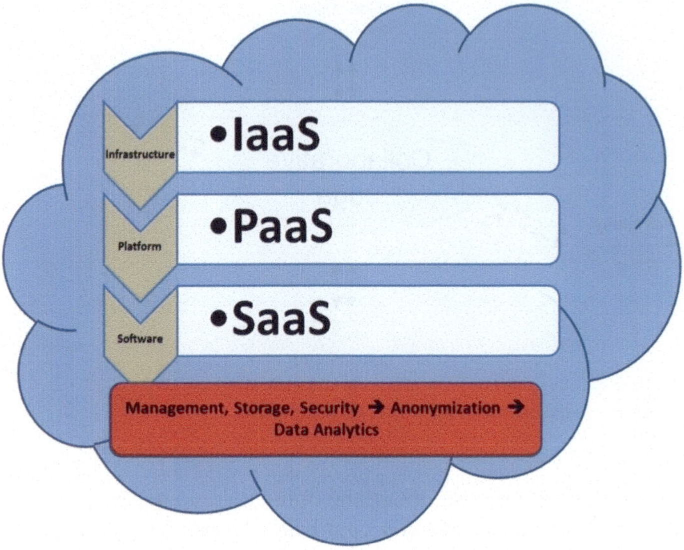

Fig. 2 System architecture based on the cloud for healthcare [9, 30]

purpose – predictive analysis modeling for understanding and anticipating the disease behaviors or for preventive medical actions [9].

Personal Data Management in Private Cloud Versus Public Cloud for Data Analytics

Cloud computing architecture for healthcare outlines several already known services:

- Applications, infrastructure as service (IaaS)
- Platforms as service (PaaS)
- Servers as a service (SaaS)

which allow data virtualization, storage, and management (Fig. 2).

The solution for managing personal data saved in a private-public hybrid cloud when we have high data volume is straightforward to use the data without having local servers for the database, but this requires data anonymization and data process with minimal concern that other user may capture the data in the public cloud [22–26].

Data anonymization in the cloud is a process that consists of removing sensitive information from data sets to anonymize the patients' data. For data analytics in the cloud, it is required to anonymize the data from the private cloud in order to analyze it in the public cloud. For patients' data security in the cloud, anonymization techniques such as k-anonymity [27–30], I-diversity [9], and t-closeness (Fig. 3) are used.

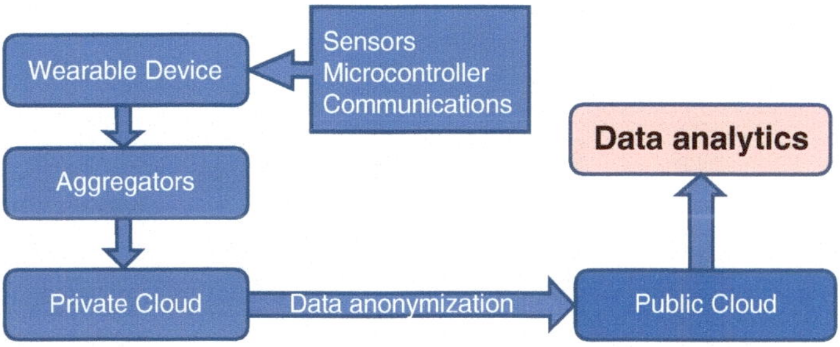

Fig. 3 Data management – private-public cloud [9, 30]

Fig. 4 Hybrid cloud PaaS [9, 31]

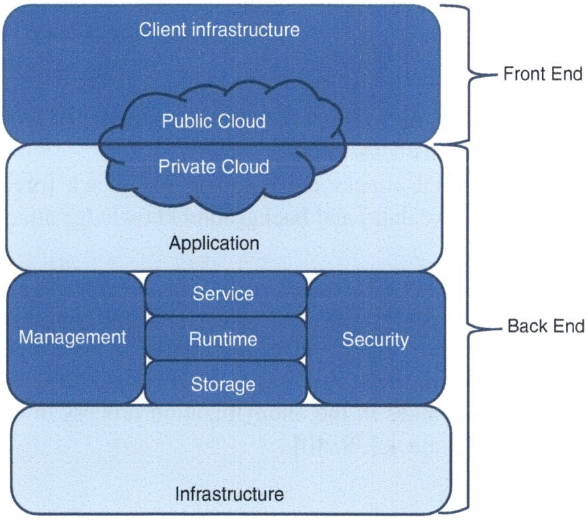

Figure 4 shows a hybrid cloud model based on PaaS architecture for medical wearable devices. Patients' data should be saved in a private cloud because wearable device involves the usage of wireless personal area network (WPAN) [33, 34].

Starting from the premise that a doctor must have access to sensitive, insensitive, quasi-identifying, or identifying attributes of patients' data, but patients should have access only to their own data, a software solution for data management must be architected by levels by setting the permissions rights. Acquisition of the biomedical signals by personal medical wearable devices can be used by doctors or for data analytics to develop predictive disorder models, but only after data anonymization (Fig. 5).

- *k-anonymity*

The solution of hiding data set against reidentification is by generalizing the attributes that might be utilized in a linkage attack (quasi-identifiers) [9, 36]. If the attack is targeting a single individual record, then all sensitive attributes can be obtained [9, 37]. The challenges in k-anonymity technique implementation are

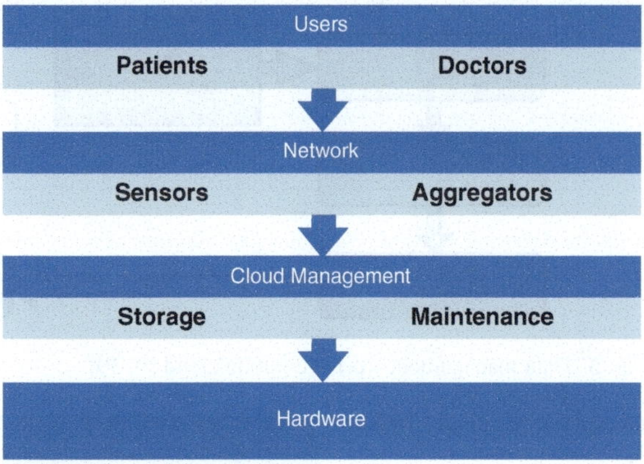

Fig. 5 Cloud service for healthcare [9, 31]

based on the computational complexity of the k-anonymization algorithm (O (ln k)), scalability, robustness, and data quality.

The potential attacks are homogeneity attack (prediction of raw values using similar sensitive data) and background knowledge attack [38].

- *L-diversity*

L-diversity method is an extension of the k-anonymity, which can allow data anonymization by reducing the granularity of data representation using techniques such as generalization and suppression [39]. The l-diversity model adds diversity for sensitive values in the anonymization process in order to minimize a potential homogeneity attack [39, 40].

- *T-closeness*

Being an extension of the method l-diversity, in this case the anonymization can be obtained by distribution of data values for a selected attribute [41].

The anonymization of the patients' data must be taken into account of the influence factors such as environment, medical providers, regulations, application type, data type, data semantics, and privacy and utility requirements. Data privacy by design can be released by anonymization of a few records from a multidimensional data table or by anonymities of a tuple in a record. The objective is to avoid sensitive data disclosure, and this can be achieved by random perturbation methods and group anonymization techniques, such as k-anonymity or I-diversity [9].

Methods for IoT Wearable Risk Reduction by Data Anonymization
Several methods from statistical disclosure control, such as k-anonymity, l-diversity, or t-closeness, are required to be used in healthcare data anonymization for respecting patient data privacy [9]. The risks of data anonymization for IoT analytics in the cloud are due to insufficient information missing or incorrect about patients due to l-diversity or l-closeness methods, or cohorts are incomplete. Scientific literature specified using two techniques such as masking and de-identification in order to remove the link between patient sensitive data and data related to diseases

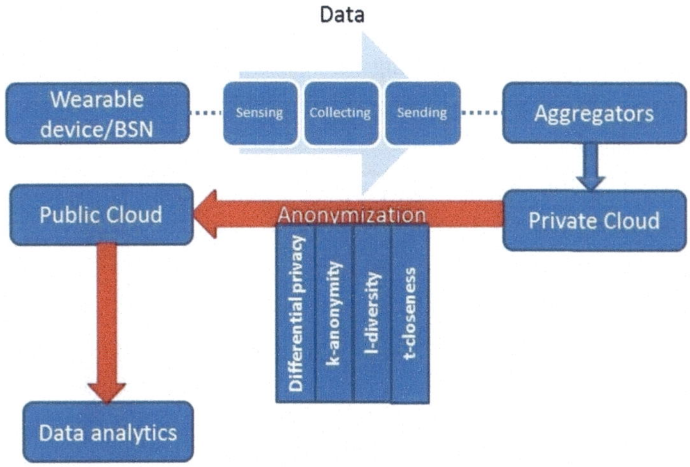

Fig. 6 Data flows from analog to digital data anonymization [9, 30]

behaviors (age, weight, geographic location, device used, parameters, and vital signs values).

For IoT wearable risk reduction, it is necessary to implement the next measures [28] such as wearable device authentication and control access and data and privacy protection by anonymization (Fig. 6).

Methods Used in Statistical Disclosure Control

The goal of statistical disclosure control (SDC) [37, 41, 42] is to present the relation and match percent of the data sets in order to obtain sensitive patients' data [43, 44]. For SDC, in general, perturbative and non-perturbative methods are used. Perturbative methods (noise adding, swapping, micro-aggregation, or PRAM) generate a distortion of the original data before publication. This method of masking data is to protect patients' privacy.

Non-perturbative methods (global recording, sampling, or local suppression) generate partial destruction or a reduction of the original data before publication. Perturbative and non perturbative methods can be used for continuous or discrete data values [9, 39, 45].

- *Perturbative Methods*
  - Noise addition is a perturbative method that can be used for continuous data [38, 45]. The masking is provided by the addition of correlated or uncorrelated noise and noise addition and linear or nonlinear transformation.
- *Swapping Method*
  - It can be applied for continuous or discrete data and suppose exchanging the values of sensitive data, and low-order frequency counts or marginal are maintained.

- *Micro-aggregation*
  - The method involves partitioning records, replacing the values from group-by-group centroid value [39] and lead to k-anonymity.
- *Post-randomization*
  - This method (PRAM) is used only for discrete data and is based on the modification of some discrete variables according to the known transition probability matrix [43, 44].
- *Non-perturbative Methods*

The generalization (global recording) method can be applied to continuous and discrete data. For a discrete variable, several categories are combined to obtain new categories.

Sampling applied for continuous data generates discrete values [39]. The sampling method can be used for sensors, for continuous signal readings, and for sampling by considerable time intervals. The time interval for sampling will affect the accuracy of the values recorded (e.g., temperature).

The local suppression method is applied to discrete data and assumes that the values of individual attributes are suppressed in order to increase the set data according to a combination of quasi-identifier values [45].

4 Technologies and System Architecture

Low-Power Technologies in Healthcare

The low-power wide area (LPWA) technologies can be used simultaneously with the smartphones and short-range wireless that allow the low-power computing and devices with low data transfer [1]. It is expected that the number of devices using LPWA technologies will increase from 59 million (in 2016) to 3 billion connections in 2025 [46, 47].

Sigfox

The Sigfox technology uses ultranarrow band (UNB) technique and operates in the sub-GHz ISM frequency band. UNB operation is achieved using channel bandwidths lower than 1 kHz [1]. Sigfox is based on a star network topology, and for data transmission, it uses Binary Phase Shift Keying (BPSK) modulation. However, the technology can support bidirectional communication with significant link asymmetry [1]. This technology is suitable for apps requiring the transfer of a small amount of data, such as alarm systems, pulse monitoring, or that required only one-way communication [1].

Lora

LoRa Alliance LPWAN solution operates in the sub-GHz ISM band [1] and is based on two components: LoRa and LoRaWAN. LoRa technology uses a chirp spread spectrum (CSS) scheme for modulation, which consists of spreading the narrowband signal into a wider channel bandwidth. To improve the robustness of the communication, LoRa includes a scheme for error correction. LoRa allows communication between multiple devices using a star topology. The LoRaWAN is optimized specifically for low power consumption devices. LoRaWAN end devices are classified into three classes, each being different in latency and power requirements. Class A supports devices that are highly delayed tolerant but require ultralow power consumption, class B supports devices with moderate latency requirements that need low power consumption, and class C supports devices that require the smallest possible latency. In addition, LoRa provides better data rates than Sigfox and is suitable for data transfer rates between 300 and 5000 bps [1]. One major drawback of using the LoRa LPWAN solution is that it requires a subscription from Semtech vendor [1].

Ingenu

Ingenu technology is operating in the 2.4 GHz ISM band, which provides high transmission power [1], and in comparison with other sub-GHz band technologies, Ingenu has an increased capacity [48]. On the other hand, Ingenu in comparison with Sigfox and LoRa provides a shorter range and is more vulnerable to interference [1].

Weightless is a set of open LPWAN standards, containing three standards such as the following:

- *Weightless-P* – provides a bidirectional communication, offering performance rate, network reliability, and security parameters according to 3GPP solutions. For this standard, the data rate is in the range between 0.2 and 100 Kbps and has lower costs compared to other LPWA technologies [1].
- *Weightless-N* – this standard operates in the sub-GHz band. Moreover, it uses a star network architecture, and it offers only one-way communication. Weightless-N devices have lifelong batteries and low network cost. To reduce interference, this standard uses frequency-hopping algorithms [1].
- *Weightless-W* – this standard operates in the TV white spaces spectrum while allowing a peak data rate between 1 Kbps and 10 Mbps [1].

In comparison with LoRa technology, all weightless standards employ symmetric key cryptography for security [1, 49].

Sensor Network Architecture

Sensors play an essential role in all sensor networks. Their quality directly correlates to industrial advancements in nanotechnology, signal conditioning, and microcontroller or microelectromechanical systems (MEMS). Biomedical sensors are used to measure the biomedical parameters such as blood pressure, temperature, pulse, blood oxygen, glucose level, and respiratory rate. Wireless Body Sensor Network (WBSN) amasses various physiological statistics based on vital signals recorded from blood pressure, heart rate, posture and motion, electrocardiography (ECG), pulse oximetry, etc. In the case of medical signals such as body temperature, they can vary very slowly or remain constant. Thus, biomedical sensors are tasked with converting the biosignals that are static in nature into dynamic signals. The

Table 1 Biomedical signal methods and measurement parameters [51, 52]

Measurement	Range of parameter	Frequency, Hz	Sensor or method
Blood flow	1–300 mL/s	0–20	Flow meter
Blood pressure	0–400 mmHg	0–50	Strain gauge or cuff
Electrocardiography	0.5–5 mV	0.05–150	Skin electrodes
Electroencephalography	5–300 μV	0.5–150	Scalp electrodes
Electromyography	0.1–5 mV	0–10,000	Needle electrode
pH	3–13	0–1	pH electrode
Respiratory rate	2–50 breaths/min	0–10	Impedance, fiber optic
Temperature	32–40 °C	0–0.1	Thermistor, thermocouple

biomedical signal analysis and smart system sensors used in measurement should be integrated in compact system in order to offer a real-time monitoring. Linear differential equations are used to describe the tools, and they relate the output signal to the input signal in the time domain [50]. Table 1 [51, 52] presents the main characteristics of the biomedical signals that can be measured by sensors in a predefined frequency.

GC Architecture

This architecture works in connection with the Meta Fog-Recognition framework providing services between Fog and Cloud computing [12], as well as protects and prevents intrusions and unauthorized access to big data. Various cloud data centers store big data that can be retrieved based on importance. AWS Cloud Trail is used in the process of logging the files to classify the data into different types. Amazon S3 bucket stores the log files, which are then forwarded to the Amazon EMR framework [12]. Apache HBase is used in connection with the Amazon S3 bucket in order to offer enough storage for the massive amount of log files distributed. Cloud Trail can establish connections with many apps through API and plug-in, collecting and storing the API call [12, 53].

MF-R Architecture

This architecture is made of three distinct phases: *Data Collection, Data Transfer, Big Data* [12].

Data acquisition consists of signal acquisition using medical devices by the patients [12]. For example, in the case of modifications of the pulse, respiratory rate, blood pressure, glucose levels, the body temperature of the individual, the devices contact the clinic, and the assigned doctor through alert messages using the wireless network and fog computing. A cloud is connected to fog computing in order to transfer and store the data collected from the sensors into a database [52].

Data Transfer consists of moving the biomedical data into an Amazon S3 bucket [12] using the "s3cmd utility" method. In order to store physiological data collected by the sensors, Healthcare organizations use the EC2 instance in Amazon's cloud [54]. Still, sensor data are stored only on the local disk of a running EC2 instance, and the data set are not available in Amazon EMR. This way, the Hadoop service provided by Amazon EMR in the cloud processes the big data. This issue is solved by using Apache Pig for moving the large volume of data from the Amazon S3 bucket to the Apache HBase [12].

In the *Big Data Storage*, Amazon EMR [12] grants several ways to get the big data onto a cluster. The usual approach is to synchronize the data to Amazon S3 [54] and then load it onto the HBase cluster by the built-in features of Amazon EMR [12]. HBase is a scalable and distributed database that can store a large number of rows and columns in a distributed manner, and it is available to all nodes in the cluster [12]. More precisely, it follows the columnar database storage model [12].

Multiagent Architecture
This hybrid architecture proposes to collect signals using reactive agents. This architecture uses reactive agents for learning and decision-making [47]. In this case, patient wears the biomedical in order to monitor the patient's health [47]. The predication based on biomedical data is used for implementation of the support decision for medical staff and extracting the main knowledge about parameter dynamics in particular diseases [55, 56].

Healthcare Data Gateway (HGD)
This HGD architecture allows the patient to decide and have control on his medical data quickly and securely based on block-chain technology and smartphones [55–58].

Establish Architecture
Establish is a project [59] that proposes a rehabilitation decision support system that combines parameters obtained from environmental sensor data with physiological sensor data to empower patients in a rehabilitation clinic with decision support tools for behavioral choices and treatment options. The goals of the platform are:

- Monitor health parameters to continually improve the health of the population through rehabilitation programs focused on physical recovery care, explicitly targeting the patient's functional aspect of integration in everyday life, environment, and work [59];
- Develop a decision support system and services based on the outdoors environment parameters and indoor location [59];
- Reduce operations costs and improve the quality of the services provided [59, 60].

In order to find links between the biological data and environmental conditions, the monitoring of the physical activities (monitoring heart rate, the burned calories, sleep patterns) during the recovery programs (recommended by physiotherapists, trainers, or physical education teacher) was performed [59]. For monitoring physical activity, Fitbit Charge 2 wristband was used. It is a vital hardware unit due to its heart rate monitoring function and other biological parameters like burned calories, steps, sleeping hours, and so on.

The pilot study group was included persons with various health problems such as for overweight, obesity, kyphosis, scoliosis, lombosacralgia, rheumatoid arthritis, and lymphedema [59].

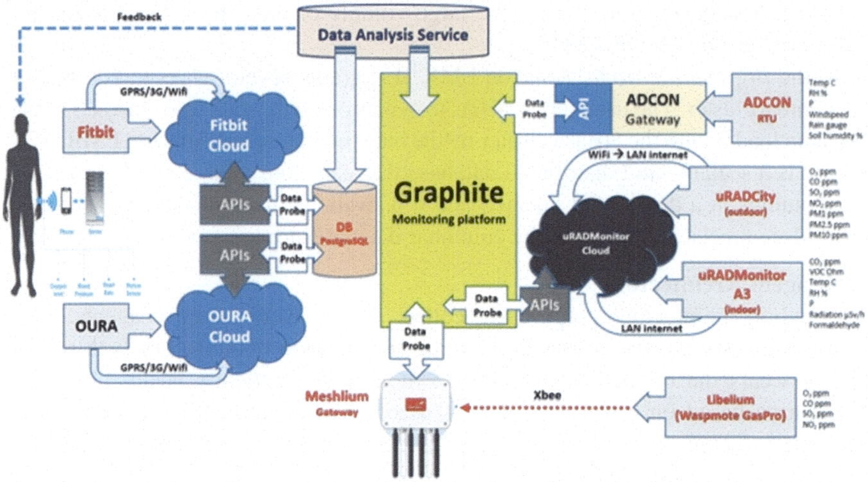

Fig. 7 Establish architecture [64]

Establish architecture for data collection, analysis, decision support, and visualization is presented in Fig. 7.

The principal goals of the Establish platform are the development of a personal recovery plan for different affections and consist of rehabilitation therapy to promote weight loss, amelioration of the column posture, increase of the column flexibility, increasing muscle strength, improving breathing, and increasing lung ventilation.

5 IoT-Connected Devices for Remote Healthcare Services: Use Cases for Patients with Higher Risk

The Internet of Things in the healthcare area allows generating the centralized network based on connected electronic devices (Things) that send/receive data using edge computing and cloud infrastructure. This means that patients can be monitored at hospitals or at home and can receive remote treatments. However, the most significant diseases that should be evaluated and remote treated using wearable electronic devices are diabetes, epilepsy, and cardiovascular management [34]. The patients at higher risks for severe illness have cancer, HIV, diabetes, cardiovascular diseases, and chronic kidney disease.

The main use cases based on IoT-connected devices for patients with higher risks are:

- Invasive continuous glucose monitor (CGM) EVERSENSE System (Senseonics, Incorporated) based on a fluorescent tiny sensor HPC and transmitter for wireless power of the sensor, with waterproof properties to be inserted under the skin, and the possibility to send vibe alerts for 90 days [62].

- Continuous monitoring Medtronic Guardian™ Connect, for continuous monitoring of the glucose levels [63].
- Continuous ambulatory cardiac monitoring MoMe® Kardia [64] using a platform available for real-time ECG data analysis.
- Invasive monitoring PillCam™ SB 3 system for sending endoscopic images to the PillCam™ SB 3 sensor array based on eight-lead electrodes mounted on the patient body [65].
- System for management of dialysis and nephrology (NEXADIA) [66].
- Wearable dialysis devices such as artificial kidney (AWAK, ViQWAK) [67–70].
- IoT for stroke rehabilitation [71–73] and gait monitoring [74, 75].
- IoT for cancer diagnosis and treatment [76, 77].

All these innovative systems have the facility, also named mHealth [78–79], to watch on real time the parameters on the mobile phone, tablet, or notebook.

In general, these are developed solutions for assisted living for the elderly because these patients have comorbidities, and it is necessary to live alone or in care homes. However, the studies show these elderly patients are not receptive to IoT wearable system. In antithesis, the young or adults are more responsive to a wearable device for monitoring using IoT systems.

6 Security Vulnerabilities of the IoT 5G-Connected Wearable Devices Using Low Power Consumption

In the medical industry, it has changed to the new wave of digitization of medical records. As a result, the industry of healthcare has seen an increase in data volume in terms of complexity, diversity, and opportunity. Data in this area are complex, used in calculations, stored, utilized, and, sometimes, communicated to conventional systems. All these are reflected in the cost of these systems and services. Scientists investigate all possible ways to reduce costs by care, system efficiency, and health management [1]; essential data appear to be a plausible solution, promising to turn the healthcare industry [1].

IoT systems represent a proactive way to manage healthcare and welfare as a result of the need to reduce the cost of standard medical services. Their costs have increased substantially both for services and for insurance. For example, real-time monitoring by wearable biomedical sensors invasive or non-invasive allows medical staff to be notified in case of an abnormality. In addition, digital and smart system integrated for healthcare is one of the next trends in health information technology (IT), with electronic records being a fundamental element of this vision [1].

Medical security is essential because data must be confidential and protected. Thus, a patient may be vulnerable by the medical data he inserts into a database. Therefore, healthcare IoT systems must ensure the safety and protection of personal data.

Furthermore, the collection of the confidential and private data of the patient must drive a strategy and a mechanism to secure this confidential data. Having all objects connected involve new security measurements; the system must answer for new exigencies about security, especially the system must be resilient against attacks. It must be resistant against the most straightforward attacks and failures. Moreover, after having recovered the information, the system must identify this information. Then, the data providers must do enforce and manage the access control of the information provided to permit control of information visibility.

However, in the case when the patient is at his home and needs a continuous observation, the necessity to use modern measuring devices (for assessing hypertension, respiratory diseases, heart failure) is high. Therefore, this involves the use of the healthcare system IoT and, besides this, a vulnerability of the data. When the health agents go to the patient's home, they can have access to the data. Moreover, they can control and correct access to avoid persons who are not authorized to read and extract the patient's data.

The equivocation about the regulation of the data ownership and the necessity of norms and data protocols of the IoT system involves that the IoT equipment is vulnerable. Furthermore, the multiplication of captors and connected devices expands the attack surface substantially on a network. As a result, the cybercriminal is a danger for this critical data in healthcare. The data of the patient can use adversely by the cybercriminal to take the identity of the patient. With that, they can defraud the health insurance, buy drugs and sell the drugs, or blackmail the patient about his data (for example, the cybercriminal can do this to a very important person – VIP). The IoT devices for health are particularly sensitive because today the production of the majority of these devices does not consider the device security or does not understand the current security requirements. Thereby, they do not secure the IoT equipment, or they secure them with easy to guess codes, such as 0000 or 1234. In addition, sometimes the patient cannot change the code, so he cannot create a first security layer with his code. Therefore, the IoT devices represent the weakness of cybersecurity of all hospitals and care establishments.

Over the last decade, the number of IT security breaches in healthcare systems has increased. Unfortunately, there have been countless cases in which patient data were not secured; for example, in 2013, Kaiser Permanente (USA) informed patients that health information was compromised by an unencrypted USB flash drive containing medical records of patients [77]. Hospitals are most vulnerable to security attacks; according to the study [25, 26], there is a 94% rate that patients' privacy and data security in hospitals are violated. These attacks have been made inside, very little outside, though there are countries where this is the reverse, for example, Romania [78]. The adoption of relevant healthcare data significantly increases patients' security and confidentiality issues, because the patient information is stored in data centers with different security levels [1].

Even if many data centers used are HIPAA (Health Insurance Portability and Accountability Act) certified, this certification still not guarantee the safety of patient records [79]. For this reason, HIPAA focuses more on security policies and procedures than on implementing them. Besides, the large data sets from various sources are an additional burden for storage, processing, and communication [1]. For IoT, the implementation of security in networks with limited resources is still a challenge and will continue to become more complicated because of the increased number of IoT devices [1].

For example, conventional symmetrical and asymmetric key distribution and revocation systems cannot be extended to a billion IoT devices. Therefore, the crucial scalable management solutions that lead to uninterrupted interoperability of different networks (for example, IoT and legacy Internet networks) are essential to integrate high-volume data into a cloud environment [1].

The healthcare industry is subject to processing large volumes of transformable data into accurate and real-time data. It is desirable to integrate healthcare systems with large volume of data, financial, genomic, social, and environmental data for real-time analysis and to understand the dynamic in health of the population in order to allow disease control and predictive analysis. This predictive analysis is useful to understanding health conditions and preventing the emergence of health problems [1, 80].

Lately, the scientific and industrial community had proposed and developed smart technologies in the medical field. This progress is due to development of the Internet and mobile communications [81, 82].

Massive low power connections and low data flow scenarios are specific to IoT services [83] and vertical industries where old systems cannot realize performance requirements. For 5G, these are extended scenarios. The low-power massive scenario focuses primarily on sensors and data usage cases such as smart cities, environmental monitoring, smart agriculture, healthcare, and forest fire prevention. Features of this scenario include small data packets, low power consumption, and reduced costs. Generally, these types of devices are numerous, with a wide geographic distribution. To support these applications, 5G has to ensure maintenance to 1 million connections per square kilometer and 100 billion connections [84] in an efficient way allowing the terminals to have low energy consumption and a final cost to reduce [59].

The main technical scenario and key challenges for 5G low-power massive connections are:

- Connection density: 106/km^2
- Low power consumption and low cost

In the field of 5G technology, the wireless technologies, massive multiple-input-multiple-output (MIMO), ultradense networking (UDN), and all-spectrum access are target for global ICT industry. The network architecture with software-defined networking (SDN) and network function virtualization (NFV) becomes the prevailing view worldwide.

Massive MIMO allows many more antennas than 4G systems to support dozens of independent data streams. This helps to enhance spectral efficiency of multiuser systems. Before massive MIMO can be deployed for 5G, some critical technical issues must be solved, including channel measurement and feedback, reference signal definition, antenna array design, and low-cost implementation.

Ultradense networking can theoretically increase the spectral reuse factor infinitely via deploying denser base stations. In practice, even limited by frequency interferences, site availability, and deployment cost, it can still achieve capacity improvement in hot-spot areas. Interference management and suppression, virtual cell, joint access, and feedback are important research areas in ultradense networking [85, 86].

The multiple accesses can improve spectral efficiency and access capability of 5G in many scenarios by super-positioning signals of numerous users in space/time/frequency/code domains [85, 86].

Moreover, grant-free-based multiple access schemes will significantly signal overhead, access latency, and power consumption of terminals [86, 87]. Currently, the potential systems proposed by the industry include sparse code multiple access (SCMA), multiuser shared access (MUSA), pattern division multiple access (PDMA), and non-orthogonal multiple access (NOMA) [87–89].

All-spectrum access can exploit a variety of spectrum resources for mobile communications, such as high and low, paired and unpaired, licensed and unlicensed, contiguous and non-contiguous frequency bands, for increasing data rates and system capacity. Besides, a large part of unused spectrum resources between 6 and 100 GHz can serve as supplementary bands of 5G [90].

Smartphones and tablets contain a set of sensors for environmental monitoring. Data collected from these sensors can be sent using a Wi-Fi network or a cellular network from mobile devices outdoors.

The significant amount of Wearable Devices include many options, such as smart watches, smart bracelets, wearable camera, smart glasses, smart clothes, smart rings, or smart bands. The sensors on these devices have the adequate functions based on specific apps. For example, a smartwatch is used as a phone replacement with similar features, so they have motion sensors, GPS, voice, and camera. In addition, smart bracelets are used as a personal management device for pulse or steps (motion). These wearable devices having a Bluetooth LE (low energy) interface communicate with other electronic devices. In this case, mobile devices are used as aggregators for processing and sending data from wearable devices.

The electronic devices that do not have phone connection have to obtain values and data from sensors that are not embedded in mobile phones (barometer, temperature, humidity sensors). These devices can be used in sport, fitness, environmental monitoring, and many more [88, 89, 91].

7 Future Challenges on IoT-Connected Devices for Remote Healthcare Services

Industry IoT presents many advantages for people, society, environment, and business.

IoT is very important for remote medical monitoring. IoT-based applications and systems have redefined the world and perception. Through IoT, the gap between doctors, patients, and healthcare services can be eliminated. IoT enables doctors and hospital staff to do their work more accurately and actively with minimum effort.

The future of IoT in healthcare is represented by 87% [89, 91, 92] of healthcare organizations that use the IoT technology in their utilities. Some statistical reports specify that 73% of IoT apps will be used in healthcare remote monitoring, 50% for cyber-surgery and control of healthy people, and 47% for location detection.

In addition, the wearable devices for patients can now send data to physicians and may allow real-time monitoring of the biomedical parameters such as heart rate, glucose, or patient location/position (fall detection). Because of the remote monitoring by wearable devices, patients can stay comfortable in their house while being monitored by medical staff or informal/non-informal caregivers.

Nowadays, smart sofa/beds are used and can provide detailed information about patients' positions and biomedical signals, helping to identify and prevent potential problems related to bedsores.

The applications of IoT in the healthcare [91–94] domain can help medical staff in healthcare management and remote monitoring of the patients.

In the appropriate future, the IoT will be more pervasive in hospitals/patient' homes, allowing real-time data processing and health diagnosis at home [93–96].

In the case of the wearable devices that request connectivity to a network and have the possibility to process the signals at the device level offline, they will not have the risks of data losses due to inexistent internet connection. For example, an insulin pump that operates independently of the internet will analyze glucose levels and release the right quantity of insulin and upload the data on cloud when accessing the internet [97–99].

The mixing of the technologies such as artificial intelligence and IoT in the healthcare area is to support devices can perform activities autonomously, such as devices that recognize the sick person and operate with them based on their treatment plan.

Acknowledgments Some of the results have been obtained in the PhD. thesis "Theoretical and experimental contributions to the monitoring of vital parameters by using intelligent control systems based on sensors integrated and Cloud Computing services", Aileni R. M., Doctoral School of the Faculty of Electronics, Telecommunications and Information Technology, UPB This work has been supported partially by UEFISCDI Romania and MCI through projects ESTABLISH, PARFAIT, funded in part by European Union's Horizon 2020 research and innovation program under grant agreement No. 777996 (SealedGRID project), No. 787002 (SAFECARE project) and No. 826452 (Arrowhead Tools).

References

1. M.M. Alam, H. Malik, M.I. Khan, T. Pardy, A. Kuusik, Y. Le Moullec, A survey on the roles of communication technologies in IoT-based personalized healthcare applications. IEEE Access **6**, 36611–36631 (2018)

2. M. Frustaci, P. Pace, G. Aloi, G. Fortino, Evaluating critical security issues of the IoT world: Present and future challenges. IEEE Internet Things J. **5**(4), 2483–2495 (2017)
3. Ericsson white paper: Cellular networks for massive IoT, https://www.ericsson.com/res/docs/whitepapers/wp_iot.pdf. Accessed 28 Mar 2017
4. K. Gautam, V. Puri, J.G. Tromp, C. Van Le, N.G. Nguyuen, Internet of things and healthcare technologies: A valuable synergy from design to implementation. Int. J. Mach. Learn. Netw. Collab. Eng. **02**(3), 128–142 (2018). ISSN: 2581-3242
5. F. Firouzi, A.M. Rahmani, K. Mankodiya, M. Badaroglu, G.V. Merrett, P. Wong, B. Farahani, Internet-of-Things and big data for smarter healthcare: From device to architecture, applications and analytics. Future Gener. Comput. Syst. **78**, 583–586 (2018). https://doi.org/10.1016/j.future.2017.09.016
6. Internet of Things in Healthcare Market Size, Share & Trends Analysis Report By End Use, By Component, By Connectivity Technology, By Application and Segment Forecasts, March 2019, By Grand View Research, ID: 978-1-68038-857-2
7. J.J. Rodrigues, D.B. Segundo, H.A. Junqueira, M.H. Sabino, R.M. Prince, J. Al-Muhtadi, V.H.C. de Albuquerque, Enabling technologies for the Internet of health things. IEEE Access **6**, 13129–13141 (2018). https://doi.org/10.1109/ACCESS.2017.2789329
8. S. Sneha, U. Varshney, A wireless ECG monitoring system for pervasive healthcare. Int. J. Electron. Healthc. **3**(1), 32–50 (2007)
9. R.M. Aileni, Theoretical and experimental contributions to the monitoring of vital parameters by using intelligent control systems based on sensors integrated and Cloud Computing service, PhD Thesis, 2015
10. Kirby, B. and Mosley, B. (2015). The Architecture of Wearable Technology, White Paper KS-WP-001, https://www.researchgate.net/publication/278327463_The_Architecture_of_Wearable_Technology?channel=doi&linkId=557f134c08aeea18b7795de5&showFulltext=true
11. J. Qiu, Q. Wu, G. Ding, Y. Xu, S. Feng, A survey of machine learning for big data processing. EURASIP J. Adv. Signal Process. **2016**(1), 67 (2016)
12. A. Sandryhaila, J.M. Moura, Big data analysis with signal processing on graphs: Representation and processing of massive data sets with irregular structure. IEEE Signal Process. Mag. **31**(5), 80–90 (2014)
13. G. Manogaran, R. Varantharajan, D. Lopez, P.M. Kumar, R. Sundarasekar, C. Thota, A new architecture of internet of things and big data ecosystem for secured smart healthcare monitoring and alerting system. Future Gener. Comput. Syst. **82**, 375–387 (2018)
14. W. Wei, Y. Jiong, Approximate patterns, in *Mining Sequential Patterns from Large Data Sets. Advances in Database Systems*, vol. 28, (Springer, Boston, 2005)
15. A. Sandryhaila, J.M.F. Moura, Big data analysis with signal processing on graphs, http://users.ece.cmu.edu/~asandryh/papers/spm14.pdf
16. Data Mining Tutorial: Process, Techniques, Tools & Applications. Meet Guru99 – Free Training Tutorials & Video for IT Courses, www.guru99.com/data-mining-tutorial.html#1
17. D.J. Hand, Data Mining Based in Part on the Article "Data Mining" by David Hand, Which Appeared in the Encyclopedia of Environmetrics. – Hand – 2001 – Major Reference Works. Wiley Online Library, American Cancer Society, 15 Jan 2013, onlinelibrary.wiley.com/doi/full/10.1002/9780470057339.vad002.pub2
18. I.H. Witten, E. Frank, M.A. Hall, C.J. Pal, *Data Mining: Practical Machine Learning Tools and Techniques* (Morgan Kaufmann, Amsterdam, 2019)
19. A. Greenberg, J. Hamilton, D.A. Maltz, P. Patel, The cost of a cloud: Research problems in data center networks. ACM SIGCOMM Comput. Commun. Rev. **39**(1), 68–73 (2008)
20. W. Meng, Y. Wang, W. Li, Z. Liu, J. Li, C.W. Probst, Enhancing intelligent alarm reduction for distributed intrusion detection systems via edge computing, in *Australasian Conference on Information Security and Privacy*, (Springer, Cham, 2018, July), pp. 759–767
21. E. Cuervo et al., MAUI: Making smartphones last longer with code offload, in *Proceedings of the 8th International Conference on Mobile Systems, Applications, and Services, San Francisco, CA, USA*, (ACM, New York, 2010), pp. 49–62

22. N. Venkataramanan, A. Shriram, *Data Privacy-Principles and Practice* (Taylor and Francis, CRC Press, Boca Raton, 2017)
23. M. Satyanarayanan, P. Bahl, R. Caceres, N. Davies, The case for VM-based cloudlets in mobile computing. IEEE Pervasive Comput. **8**(4), 14–23 (2009)
24. F. Bonomi, R. Milito, J. Zhu, S. Addepalli, Fog computing and its role in the Internet of things, in *Proceedings of the First Edition of the MCC Workshop Mobile Cloud Computing, Helsinki, Finland*, (ACM, New York, 2012), pp. 13–16
25. W. Shi, J. Cao, Q. Zhang, Y. Li, L. Xu, Edge computing: Vision and challenges. IEEE Internet Things J. **3**(5), 637–646 (2016)
26. S.G. Reenu, S. Sabitha, Data anonymization and integrity checking in cloud computing, in *2013 Fourth International Conference on Computing, Communications and Networking Technologies (ICCCNT)*, (IEEE, Piscataway)
27. C.C. Aggarwal, P.S. Yu, A general survey of privacy-preserving data mining models and algorithms, in *Privacy-Preserving Data Mining – Models and Algorithms*, (Springer, London/New York, 2010)
28. S. Kavitha, E. Sivaraman, P. Raja Vadhana, A suvery on k-anonymity generalization algorithms. Int J Adv Res Comput Commun Eng **2**(11), 8471–8474 (2012)
29. T. Ernst, R. Guillemaud, P. Mailley, J.P. Polizzi, A. Koenig, S. Boisseau, E.P. Vaujour, C. Plantier, G. Delapierre, E. Saoutieff, R. Gerbelot-Barillon, E.C. Strinati, Sensors and related devices for IoT, medicine and smart-living, Université Grenoble Alpes, Grenoble, France, 2018
30. R.M. Aileni, G. Suciu, C.A. Valderrama, S. Pasca, IoT performability for medical wearable device by data privacy and fault-tolerance, in *Smart Systems for E-Health*, ed. by H. Idouti, T. Val, (Springer, Cham, 2019)
31. R.M. Aileni, S. Pasca, C. Valderrama, Cloud computing for big data from biomedical sensors monitoring storage and analyze, ROLCG 2015, Cluj-Napoca, Romania
32. C. Neuhaus, A. Polze, Cloud security mechanisms, Potsdam University, 2014
33. B. Kenig, T. Tassa, A practical approximation algorithm for optimal k-anonymity. Data Min. Knowl. Discov. **25**(1), 134–168 (2012)
34. https://drivesncontrols.com/news/fullstory.php/aid/5553/1_in_4_wireless_IIoT_connections_will_use_LPWA_by_2025.html
35. H. Malik, M.M. Alam, Y. Le Moullec, Q. Ni, Interference-aware radio resource allocation for 5G ultra-reliable low-latency communication, in *Globecom Workshops (GC Wkshps) 2018*, (IEEE, Piscataway, 2018), pp. 1–6
36. T. Shinzaki, I. Morikawa, Y. Yamaoka, Y. Sakemi, IoT security for utilization of big data: Mutual authentication technology and anonymization technology for positional data. Fujitsu Sci. Tech. J. **52**(4), 52–60 (2016)
37. M. Templ, B. Meindl, A. Kowarik, S. Chen, Introduction to statistical disclosure control (SDC), IHSN Working Paper No 007, 2014
38. S. Vijayarani, A. Tamilarasi, Data transformation technique for protecting private information in privacy preserving data mining. Adv. Comput. Int. J. **1**(1), 1–9 (2010)
39. H. Patel, K. Suthar, A novel approach for securely processing information on dew sites (Dew computing) in collaboration with cloud computing: An approach toward latest research trends on Dew computing, in *2017 Nirma University International Conference on Engineering (NUiCONE)*, (IEEE, Piscataway, 2017), pp. 1–6
40. N.K. Singh, D.S. Tomar, Privacy preservation of social media services, in *Exploring Enterprise Service Bus in the Service-Oriented Architecture Paradigm*, (IGI Global, Hershey, 2017)
41. A. Machanavajjhala, D. Kifer, J. Gehrke, M. Venkitasubramaniam, *l*-Diversity: Privacy beyond k-anonymity. ACM Trans. Knowl. Discov. Data **1**(1), 3-es (2007)
42. V. Puri, S. Sachdeva, P. Kaur, Privacy preserving publication of relational and transaction data: Survey on the anonymization of patient data. Comput. Sci. Rev. **32**, 45–61 (2019)
43. D.M. Rodríguez, J. Nin, M. Nuñez-del-Prado, Towards the adaptation of SDC methods to stream mining. Comput. Secur. **70**, 702–722 (2017)

44. A. Hundepool, J.D. Ferrer, L. Franconi, S. Giessing, R. Lenz, J. Naylor, E.S. Nordholt, G. Seri, P.P. De Wolf, Handbook of Statistical Disclosure Control (ESSnet on Statistical Disclosure Control, 2010), Statistics Netherlands, https://ec.europa.eu/eurostat/cros/system/files/SDC_Handbook.pdf
45. E. Fayyoumi, B.J. Oommen, A survey on statistical disclosure control and micro-aggregation techniques for secure statistical databases. Softw. Pract. Exp. **40**(12), 1161–1188 (2010)
46. J. Domingo-Ferrer, F. Sebé, A. Solanas, A polynomial-time approximation to optimal multivariate microaggregation. Comput. Math. Appl. **55**(4), 714–732 (2008)
47. H. Wang, M.S. Mahmud, H. Fang, C. Wang, Architecture, in *Wireless Health, Springer Briefs in Computer Science*, (Springer, Cham, 2016), pp. 27–37
48. A.B. Jemmaa, H. Ltifi, M.B. Ayed, Multi-agent architecture for visual intelligent remote healthcare monitoring system, in *HIS 2016: Hybrid Intelligent Systems*, (Springer, Cham, 2016), pp. 211–221
49. X. Yue, H. Wang, D. Jin, M. Li, W. Jiang, Healthcare data gateways: Found healthcare intelligence on blockchain with novel privacy risk control. J. Med. Syst. **40**(10), 1–9 (2016)
50. U. Raza, P. Kulkarni, M. Sooriyabandara, Low power wide area networks: An overview. IEEE Commun. Surv. Tutor. **19**(2), 855–873 (2017)
51. G. Suciu, I. Petre, A. Pasat, A. Vulpe, S. Obreja, Retrieving health and environmental data using secure IoT cloud platform, in *2018 International Symposium on Electronics and Telecommunications (ISETC)*, (IEEE, Piscataway, 2018), pp. 1–4
52. V. Poulkov (ed.), *Future Access Enablers for Ubiquitous and Intelligent Infrastructures: 4th EAI International Conference, FABULOUS 2019, Sofia, Bulgaria, March 28–29, 2019, (Vol. 283) Proceedings* (Springer, Cham, 2019)
53. K.H. Park, J. Park, J.W. Lee, An IoT system for remote monitoring of patients at home. Appl. Sci. **7**(3), 260–282 (2017)
54. Y.K. Kim, H. Wang, M.S. Mahmud, Wearable body sensor network for health care applications, in *Smart Textiles and their Applications*, (Woodhead Publishing, Amsterdam, 2016), pp. 161–184
55. T.N. Gia, I.B. Dhaou, M. Ali, A.M. Rahmani, T. Westerlund, P. Liljeberg, H. Tenhunen, Energy efficient fog-assisted IoT system for monitoring diabetic patients with cardiovascular disease. Futur. Gener. Comput. Syst. **93**, 198–211 (2019)
56. D. N. Le, C. M. Bhatt, M. Madhukar (eds.), *Security Designs for the Cloud, IoT, and Social Networking* (John Wiley & Sons, Hoboken, 2019)
57. G. Manogaran, D. Lopez, Health data analytics using scalable logistic regression with stochastic gradient descent. Int. J. Adv. Intell. Paradig. **10**(1–2), 118–132 (2018)
58. A. Abraham, S. Han, S. Al-Sharhan, H. Liu, Hybrid intelligent systems, in *HIS 2016. Advances in Intelligent Systems and Computing (Vol. 420)*, (Springer, Cham, 2016)
59. A.K.M. Jahangir Alam Majumder, Y. Amr ElSaadany, R. Young Jr., D.R. Ucci, An energy efficient wearable smart IoT system to predict cardiac arrest. Adv. Hum. Comput. Interact. **2019**, 1–21 (2019)
60. A.B. Jemmaa, H. Ltifi, M.B. Ayed, Multi-agent architecture for visual intelligent remote healthcare monitoring system, in *International Conference on Hybrid Intelligent Systems*, (Springer, Cham, 2016), pp. 211–221
61. Project Establish Itea 3, D2.2 Technical requirements and high-level specifications v2, Available: https://itea3.org/project/workpackage/document/download/4988/15008-ESTABLISH-WP-2-D22Technicalrequirementsandhighlevelspecificationsv2
62. EVERSENSE CGM System, https://www.eversensediabetes.com/
63. Guardian Connect Continous glucose Monitoring, https://www.medtronicdiabetes.com/products/guardian-connect-continuous-glucose-monitoring-system
64. MoMe Kardia, https://infobionic.com
65. PillCam™ SB 3 System, https://www.medtronic.com/covidien/en-us/products/capsule-endoscopy/pillcam-sb-3-system.html
66. Nexadia System, https://www.bbraun.com/en/products-and-therapies/extracorporeal-blood-treatment/hemodialysis/renal-it-solutions.html

67. A. Davenport, Portable and wearable dialysis devices for the treatment of patients with end-stage kidney failure: Wishful thinking or just over the horizon? Pediatr. Nephrol. **30**(12), 2053–2060 (2015)
68. M. Wester, K. G. Gerritsen, W. H. Boer, J. A. Joles, and J. P. Kooman, The wearable artificial kidney: a promise for the future?. Nederlands tijdschrift voor geneeskunde, **157**(52), A6965–A6965 (2013)
69. V. Gura, A.S. Macy, M. Beizai, C. Ezon, T.A. Golper, Technical breakthroughs in the wearable artificial kidney (WAK). Clin. J. Am. Soc. Nephrol. **4**(9), 1441–1448 (2009)
70. C. Ronco, L. Fecondini, The Vicenza wearable artificial kidney for peritoneal dialysis (ViWAK PD). Blood Purif. **25**(4), 383–388 (2007)
71. G. Yang, J. Deng, G. Pang, H. Zhang, J. Li, B. Deng, Z. Pang, J. Xu, M. Jiang, P. Liljeberg, and H. Xie, An IoT-enabled stroke rehabilitation system based on smart wearable armband and machine learning, IEEE journal of translational engineering in health and medicine, **6**, 1–10
72. R. Ani, S. Krishna, N. Anju, M. S. Aslam, and O. S. Deepa, Iot based patient monitoring and diagnostic prediction tool using ensemble classifier, in 2017 International Conference on Advances in Computing, Communications and Informatics (ICACCI), pp. 1588–1593 (September 2017, IEEE)
73. M. Subramaniyam, S. H. Hong, J. Yu, and S. J. Park, Wake-Up Stroke Prediction through IoT and Its Possibilities, in 2017 International Conference on Platform Technology and Service (PlatCon), pp. 1–5 (February 2017, IEEE)
74. A. J. A. Majumder, Y. ElSaadany, M. ElSaadany, D. R. Ucci, and F. Rahman, A wireless iot system towards gait detection in stroke patients, in 2017 IEEE International Conference on Pervasive Computing and Communications Workshops (PerCom Workshops), pp. 449–454 (March 2017)
75. J. Chen, Y. Dai, and S. Gao, A Piezoelectric Flexible Insole System for Gait Monitoring for the Internet of Health Things, 2020
76. V. Sharma, S. Som, and S. K. Khatri, (2019) Future of wearable devices using IoT synergy in AI, in 2019 3rd International conference on Electronics, Communication and Aerospace Technology (ICECA), (IEEE)
77. L. Plant, B. Noriega, A. Sonti, N. Constant, K. Mankodiya, Smart E-textile gloves for quantified measurements in movement disorders, in *Proceedings of the 2016 IEEE MIT Undergraduate Research Technology Conference (URTC)*, (IEEE, Piscataway, 2016)
78. A. Zieger, Kaiser reports second data loss incident, HEALTHCAREDIVE, 2013
79. B. Farahani, F. Firouzi, V. Chang, M. Badaroglu, N. Constant, K. Mankodiya, Towards fog-driven IoT eHealth: Promises and challenges of IoT in medicine and healthcare. Futur. Gener. Comput. Syst. **78**, 659–676 (2018)
80. Ponemon Institute LLC, Third annual benchmark study on patient privacy & data security, Research Report, 2012
81. I.-C. Surdu, Cybersecurity. Risks, threats, and trends of manifestation in Romania, in International Conference RCIC (Redefining Community in Intercultural Context), Bucharest, 2018
82. H. Kupwade Patil, R. Seshadri, Big data security and privacy issues in healthcare, in *IEEE International Congress on Big Data, USA*, (IEEE, Piscataway, 2014)
83. H.K. Patil, R. Seshadri, Big data security and privacy issues in healthcare, in *2014 IEEE International Congress on Big Data*, (IEEE, Piscataway, 2014), pp. 762–765
84. W. D. de Mattos and P. R. L. Gondim, M-Health Solutions Using 5G Networks and M2M Communications, in IT Professional, vol. 18, no. 3, pp. 24–29, (May–June 2016, IEEE)
85. W.D. de Mattos, P.R.L. Gondim, M-health solutions using 5G networks and M2M communications. IT Prof. **18**(3), 24–29 (2016)
86. IMT-2020 (5G) Promotion Group, White paper on 5G concept (2013), http://www.imt-2020.org.cn/en
87. Y. Yang, J. Xu, G. Shi, C.X. Wang, *5G Wireless Systems* (Springer, Berlin, 2018)
88. J. Cao, M. Ma, H. Li, Y. Fu, X. Liu, EGHR: Efficient group-based handover authentication protocols for mMTC in 5G wireless networks. J. Netw. Comput. Appl. **102**, 1–16 (2018)

89. Q. Wang, M. Zhu, C. Hou, 5G key technologies on mooc, in *2015 10th International Conference on Computer Science & Education (ICCSE)*, (IEEE, Piscataway, 2015), pp. 431–433
90. S. Shi, W. Yang, J. Zhang, Z. Chang, Review of key technologies of 5G wireless communication system, in MATEC Web of Conferences 2015,Vol. 22, p. 01005, (EDP Sciences, 2015) https://doi.org/10.1051/matecconf/20152201005
91. D. Huo, Z. Zhang, Q. Zhou, Q. Tang, Z. Huo, Ergodic secrecy sum rate maximization in downlink sparse code multiple access network, in *2017 20th International Symposium on Wireless Personal Multimedia Communications (WPMC)*, (IEEE, Piscataway, 2017), pp. 11–15
92. I. Chih-Lin, Seven fundamental rethinking for next-generation wireless communications. APSIPA Trans. Signal Inf. Process. **6**, e10 (2017)
93. B. Wang, K. Wang, Z. Lu, T. Xie, J. Quan, Comparison study of non-orthogonal multiple access schemes for 5G, in *2015 IEEE International Symposium on Broadband Multimedia Systems and Broadcasting*, (IEEE, Piscataway, 2015), pp. 1–5
94. Q. Han, S. Liang, H. Zhang, Mobile cloud sensing, big data, and 5G networks make an intelligent and smart world. IEEE Netw. **29**, 40–45 (2015)
95. J. Gouweleeuw, P. Kooiman, L. Willenborg, P.P. de Wolf, Post randomization for statistical disclosure control: Theory and implementation. J. Off. Stat. **14**(4), 463–478 (1998)
96. http://www.hamilton.ie/privacy2014/josep.pdf
97. D. Lu, T. Liu, The application of IoT in medical system, December 2011, https://doi.org/10.1109/ITiME.2011.6130831
98. S.L. Peng, S. Pal, L. Huang, *Principles of Internet of Things (IoT) Ecosystem: Insight Paradigm* (Springer, Cham, 2020)
99. https://internetofthingsagenda.techtarget.com/feature/Can-we-expect-the-Internet-of-Things-in-healthcare

Part II
Data Analysis

A Decision Support System for Smart Health Care

Khouloud Dorgham, Hajer Ben-Romdhane, Issam Nouaouri, and Saoussen Krichen

1 Introduction

In the area of urban development, Internet of Things (IoT) and Information and Communication Technology (ICT) are important building blocks for creating an intelligent infrastructure to administer, serve, and support the ever-growing urban population. Therefore, the Smart City is an economically sustainable urban development plan that provides a high standard of living for its residents. Technology has a key role to play in building Smart Cities that need technological efficiency in different domains of human life such as transport and mobility, communication, citizen relationship, and especially health care.

In recent years, the pressure on health-care system increases more and more due to an aging population and a rise in chronic illness. Consequently, most issues in hospitals are mainly due to resource shortage and resource assignment efficiency. As a result, health care requires efficient and reliable monitoring on daily operations, service, and resources. Hence, to build Smart Cities, we need smart hospitals that can remotely flow patients, provide emergency services quickly, facilitate patient admission, provide preventive measures, analyze patient data, and use them in

---

K. Dorgham (✉)
Institut Supérieur de Gestion de Tunis, Université de Tunis, LARODEC, Tunisie

Laboratoire de Génie Informatique et d'Automatique de l'Artois (LGI2A), Université Artois, Béthune, France

H. Ben-Romdhane · S. Krichen
Institut Supérieur de Gestion de Tunis, Université de Tunis, LARODEC, Tunisie

I. Nouaouri
Laboratoire de Génie Informatique et d'Automatique de l'Artois (LGI2A), Université Artois, Béthune, France

N. Gupta, S. Paiva (eds.), *IoT and ICT for Healthcare Applications*, EAI/Springer Innovations in Communication and Computing,
https://doi.org/10.1007/978-3-030-42934-8_6

better research practices. Today, combining operational research with health-care management is required to provide smart and efficient health-care systems owing to the fact that the future of Smart Cities will be incomplete without the proper health and wellness centers for its residents.

ICT has a huge impact on many industries. But there are few domains where its potential benefits are as important as in health care: they can improve patient outcomes and quality of life for patients and enable new approaches to precision medicine. Wherefore, operational research committee recall different well-known combinatorial optimization problems over years to deal with real-world problems in a health-care system, i.e., scheduling operating rooms [2], admission patient scheduling [13], hospital capacity planning [5], etc.

In this study, we are particularly concerned with the hospital bed management, which become one of the most studied decision health-care problem with notable interest at the different levels of decision-making pyramid and especially when we speak about smart hospitals. Patient bed assignment problem (PBAP) is an important subproblem of the operational level that seeks not only for a place to sleep for patients but the allocation of appropriate crucial resources (beds, medical facilities, treatments, etc.) and then impacts the overall patients flow. The objective is to assign patients to suitable beds in the appropriate departments in compliance with bed availability, medical constraints, and patient's needs and preferences. The PBAP is known to be an $\mathcal{NP}$-hard problem [24].

In this paper, we deal with the dynamic variant of the PBAP (D-PBAP), in which a real-time management of bed assignment is performed to take into consideration a sequence of urgent patients arriving over time, without any prior information. The main research contributions of the present paper are (1) to propose an appropriate model for the dynamic PBAP, (2) to propose a hybrid evolutionary approach to deal with the D-PBAP that combines the genetic algorithm (GA) with the simulated annealing (SA), and (3) to develop a decision support system embedding the proposed approach for assisting bed assignment managers.

The remainder of this paper is organized as follows: In Sect. 2, we will present the role of Iot and ICT in the health-care system. The next section presents an overview of both static and dynamic version of the PBAP addressed in the literature. The problem formulation is described in Sect. 4. Section 5 addresses the proposed approach and the DSS developed in this study. The computational results and the benchmark instances are reported in Sect. 6, and finally, a conclusion with some future research studies is given in Sect. 7.

2 Role of IoT and ICT in Health-Care System

Health care is one of the main priorities for all governments [6]. Despite the advancement of medicine, technologies, and their role in improving the lives of human beings, it has also created new types of diseases and health problems in the urban population. IoT and ICT will have a huge impact on our health-care life system which will revolutionize our habits in hospitals and care centers. Indeed, much of our work will be automated which will allow us to gain in productivity and efficiency in the health-care sector.

In a smart health-care system, different actors in the hospital can be concerned: patients, families, physicians, and even hospital administrative staff and hospital managers [6]. Combining health care, IoT, and ICT can reduce health-care costs, improve treatment results, increase patient engagement and satisfaction, reduce the length of stay of patients in hospitals, well organize patient admission, prevent readmissions, and perform a better resources allocation.

Some hospitals and health centers use IoT and ICT to enhance their daily operations and help the managers to improve the patient safety and satisfaction. For example, in New Zealand health-care provider is taking advantage of IoT technology to improve the efficiency of its ambulance service [22]. The condition of the patient can be monitored and transmitted to the hospital, so that the medical staff is prepared with the right resources for the arrival of the patient and can offer better treatment. It also helps the hospital in terms of timing and prioritization, so staff can optimize results for all patients more effectively.

Several researchers are interested by the field of IoT and ICT in health-care system. For example, in [10], authors presented a decision support and home-based medical health monitoring system for neurological disabled patients to facilitate the assistance of physicians in medical treatment, prescriptions, diagnosis, rehabilitation, and patient progress. Gaur et al. [12] propose a multilevel Smart City architecture that combines smart health, smart environment, smart energy, smart security, smart office and residential buildings, smart administration, smart transport, and smart industries. Therefrom, we can say that with Smart City and ICT we can create a more secure and convenient infrastructure for better human living [12]. In [3], authors consider that the mainly IoT issues in health care today are turning around the problem of critical treatments which have high risk of life and the routine medicine/check-up/admission of the patient in hospitals. And this constitutes the main motivation of this study.

3 Patient Bed Assignment Problem: Related Work

In recent decades, several studies in the literature have focused on scheduling of patients and hospital resources. Above all, the patient bed assignment problem (PBAP) is of particular significance, since its solution impacts essentially all other health care-related problems. In [11], Demeester et al. presented a PBAP modeled as an offline challenging problem, in which hospitals with a central planning unit have to be managed, and they proposed a tabu search method to solve it. This combinatorial optimization problem aims to find optimal bed assignments for elective patients. In their paper, [4] proposed a hyper-heuristic approach to improve the results in [11]. Authors in [7] improved the best known results on all of the available PBAP benchmarks by presenting a new formulation of the problem and proposing a simulated annealing algorithm to solve it.

A two-mixed integer programming (MIP)-based heuristics is proposed in [23], where the authors propose a decomposition of the problem instances into sub-problems. Recently, in [14], a meta-heuristic solution framework called FiNeMath is proposed. It combined fix and optimize heuristic with the large neighborhood searches and ILP solvers to generate new best known solutions in the PBAP. In [1], an exact method is proposed to solve a MIP formulation of the PBAP. The method does not require pre-processing, and it generates new best solutions for 9 out of 13 tested benchmark instances.

Besides to the static variants of the PBAP, several authors have shown a special interest in studying its dynamic variant. In static problems, a stream of patients are present simultaneously for the assignment, and the length of their stays is known a prior. However, dynamic PBAPs deal with a sequence of urgent patients arriving over time, without any prior information, so that patients are assigned in an online manner. Subsequently, the LoS is unknown because patient admission and discharge dates are not fixed a prior.

Several studies on PBAP have extended the original formulation of the problem to an online version to manage some dynamic aspects. In [25] authors proposed models that aim respectively to assign patients to rooms as they arrive and account for future planned arrivals. The model was tested on four benchmark instances of the PBAP by adding randomly a registration and a departure date. Also, in [7] an application of the technique proposed before was applied to the dynamic case for daily scheduling.

One year later, a reconsideration of the PBAP with a new formulation of the problem is proposed in [8]. It focuses on real-world situations in hospitals that considers (1) emergency patients, (2) uncertain length of stay, and (3) delays of admissions. In 2016, with the aim to make the PBAP more real and better suited to be applicable in real-world situations, authors in [9] proposed to add the operating theater to the previous model developed in [8], which is one of the scarcest resources in a hospital and considered a flexible planning horizon with a more complex notion for the delays of admissions.

In [21], a decision support system was proposed to deal with the PBAP on its online version and in a flexible way at the same time: the user can specify the hard and soft constraints depending on his preferences. The authors propose a mixed-integer goal-programming modeling approach to solve the PBAP; the model was tested on a real case study in Mount Sinai Medical Center in New York. Finally, in [18], the authors provided an improved method for solving the dynamic version of the PBAP. They consider the uncertainty in the length of stays of patients and deal with emergency patients. Their improved method is based on a simulated annealing framework utilized by an adaptive large neighborhood search procedure.

Table 1 summarizes the different solution approaches used in the literature to deal with the PBAP in both its static and dynamic versions.

Table 1 Overview of the solutions approaches dealing with PBAP in the reviewed papers

Approaches	PBAP	
	Static	Dynamic
Tabu search	[11]	–
Hyper-heuristic	[4]	–
F&O + F&R	[23]	–
LNS + Fix-and-Relax	[14]	–
Exact method	[1]	–
Simulated annealing	[7]	[7–9, 18]
Exact method	–	[21, 25]
Hybrid simulated annealing	–	This work

4 Problem Formulation

Nowadays, in view of the increase of patient demands in hospitalization process with regard to bed availability, bed assignment process needs very long waiting times to be well scheduled, especially when dealing with emergency patients needing immediate admission besides to the currently assigned patients.

In this context, we introduce the dynamic version of the PBAP that involves assigning a set of patients that arrive at random – without any prior appointment with the physician – to hospital beds while maximizing several quality measures reflecting patient comfort and treatment efficiency.

Patient requiring medical treatments should be assigned a bed. Each patient needs one or more specialism (cardiology, orthopedics, etc.) depending on his/her pathology. In the hospital, beds are located in rooms, and each room belongs to an appropriate hospital department. Both rooms and departments have a different degree of expertise in multiple specialism. Furthermore, rooms, with different room capacity, can be equipped by various available properties such as telemetry, nitrogen, and oxygen. Depending on the needed cure, patient can require or desire a certain room equipment's and a certain room capacity preferences.

The dynamic PBAP has many similarities with problems addressed in the dynamic manufacturing process environments. Urgent patients are similar to jobs that arrive continually, and they have to be assigned to an equipped bed (machine) in a well-defined planning horizon. The dynamic aspect of the problem is shown by a variation of the number of arrived patients. Indeed, when the assignment of planned patients that are given statically is performed then new patients are received for urgent hospitalization, the assignment has to be rearranged. The aim is to allocate a suitable bed for each patient throughout their stay, in the way to ensure their comfort. Furthermore, satisfying room policies and department policies as well as patients treatment requirements and patient preferences are also major goals for the problem.

In what follows, belonging to the terminology presented in next section, different constraints are defined and discussed. We enumerate and summarize the different notations that we use for the mathematical formulation.

4.1 Sets and Parameters

- $t \in T$: Time period.
- $p \in P$: Set of patients $(P_M \cup P_F)$
- P_M: Set of male patients.
- P_F: Set of female patients.
- $r \in R$: Set of rooms.
- $d \in D$: Set of planning days.
- D_p: Set of days in which the patient p is present in the hospital.
- $C_{p,r}$: Penalty of assigning patient p to room r.
- RC_r: Capacity of room r in number of available beds.
- W_{RG}: Weight of room gender policy constraint.
- W_{Tr}: Weight of transfer constraint.

4.2 Decision Variables

-
$$x_{p,r,d}(t) = \begin{cases} 1 \text{ if patient } p \text{ is assigned to a bed in room } r \text{ in day } d \text{ at time } t. \\ 0 \text{ otherwise.} \end{cases}$$

-
$$\text{Trans}_{p,r,d}(t) = \begin{cases} 1 \text{ if patient } p \text{ is transferred from room } r \text{ in day } d \text{ at time } t. \\ 0 \text{ otherwise.} \end{cases}$$

-
$$m_{r,d}(t) = \begin{cases} 1 \text{ if there is at least one male patient in room } r \text{ in day } d \text{ at time } t. \\ 0 \text{ otherwise.} \end{cases}$$

-
$$f_{r,d}(t) = \begin{cases} 1 \text{ if there is at least one female patient in room } r \text{ in day } d \text{ at time } t. \\ 0 \text{ otherwise.} \end{cases}$$

-
$$b_{r,d}(t) = \begin{cases} 1 \text{ if there are both male and female patients in room } r \text{ in day } d \text{ at time } t. \\ 0 \text{ otherwise.} \end{cases}$$

The mathematical model of the dynamic PBAP, based on the reformulation of the PBAP proposed in [8], is presented as follows.

$$\min \sum_{p\in P, r\in R, d\in D_p} C_{p,r}.x_{p,r,d}(t) + \sum_{r\in R, d\in D} W_{RG}.b_{r,d}(t) + \sum_{p\in P, r\in R, d\in D} W_{Tr}.\text{Trans}_{p,r,d}(t) \tag{1}$$

$$\sum_{r\in R} x_{p,r,d}(t) = 1, \forall p \in P, d \in D_p \tag{2}$$

$$\sum_{p\in P} x_{p,r,d}(t) \leq RC_r, \forall r \in R, d \in D \tag{3}$$

$$x_{p,r,d}(t) \leq f_{r,d}(t) \forall p \in P_F, r \in R, d \in D \tag{4}$$

$$x_{p,r,d}(t) \leq m_{r,d}(t), \forall p \in P_M, r \in R, d \in D \tag{5}$$

$$m_{r,d}(t) + f_{r,d}(t) - 1 \leq b_{r,d}(t), \forall r \in R, d \in D \tag{6}$$

$$x_{p,r,d}(t) - x_{p,r,d+1}(t) \leq \text{Trans}_{p,r,d}(t), \forall p \in P, r \in R, d \in D_p - 1 \tag{7}$$

The objective function (1) aims to minimize the total penalties associated with assigning patients to rooms during their period of stay. Composed by three different cost components: (1) the cost of assigning patients to rooms with a combined penalty of constraints, then (2) the cost of violating the room gender policy, and (3) the cost associated with a transfer.

Constraints (2) ensure that among his/her admission and discharge dates, every patient need to be assigned to one room. Equation 3 preserves the room capacity constraints, i.e., the number of available beds in a room had to be greater than the number of patients assigned to it. Equations 4 and 5 seizure the gender of the patient assigned to the dependent-gender room, and constraints (6) ensure that in each day, only male or female patients can be assigned together. Constraints (7) guarantee that the transfer and the displacement of patients between rooms during their period of stay in the hospital should not be allowed.

5 Proposed Approach

In this section, as first part, we propose a hybrid approach which combines simulated annealing with the genetic algorithm for solving the PBAP. Then we present the implemented decision support system (DSS) which embeds the proposed approach.

5.1 Hybrid Simulated Annealing Algorithm

So far, various meta-heuristic approaches were proposed to solve the PBAP in its dynamic version. In our research study, we propose to develop a hybrid meta-heuristic based on the simulated annealing algorithm that is a Monte Carlo model firstly used in [19] and developed to solve combinatorial optimization problems in [15].

In the following, we present the outline of the proposed approach HSA that combines both GA and SA advantageous to escape the drawbacks of each of the two techniques and improve the quality of the obtained solutions. Hybrid algorithms combining SA and GA have been previously proposed to solve several optimization problems, such as the multi-constraint zero-one knapsack problem and the traveling salesman problem [16]. In view of the high performance of this approach, we propose such a hybrid meta-heuristic algorithm for the D-PBAP.

Since searching from a population of solutions increases the probability of finding a better solution than searching from a unique one, the first step of our approach to solve the PBAP is to start proceeding from a population that is generated randomly. Afterward, GA and SA algorithms are processed in an iterative manner. The genetic operators (selection, crossover, and mutation) are applied to the initial population, and a new population of solutions is generated. Moving to the annealing part to improve the obtained results, the set of solutions produced by the GA presents the input of the annealing process that will be executed for each unique chromosome in the population. In a given generation, when the SA is applied for all chromosome solutions of the GA, the best obtained solutions obtained from SA will be considered as input for the next generation of GA. The algorithm keeps proceed until the termination criteria are met.

When the environment changes, new patients arriving at the hospital in real time are to be taken into consideration. Therefore, the current assignment needs to be adjusted immediately to adapt to the environment change (add the new patients in the schedule) and respect the system constraints. The proposed method is given by Algorithm 1.

5.2 DSS Architecture Integrating a HSA Algorithm

We develop a decision support system (DSS) in order to find the best patients assignment. The developed DSS integrates the proposed HSA algorithm. The purpose of using a DSS is to give hospital administrators a flexible application to manage the patient admission process in a smart health-care system and help them to make decisions about problems that may be rapidly changing and not easily specified. The DSS developed is detailed in Fig. 1. It is based on three different parts: (1) data inputs that can be immediately done when new patient arrived and should be scheduled, (2) solution process, and (3) best solution display. In order to generate

Algorithm 1 Pseudo-code of the Hybrid Simulated Annealing

Input: $PopulationSize, P_{Crossover}, P_{Mutation}, T_0 > 0, T_{min}, \alpha$
Output: Best Solution
Randomly generate an initial population X_0
Apply genetic operators to create X_1
Calculate the fitness value and the cost for each solution in X_1
$T \leftarrow T_0; k \leftarrow 1$
while $T > T_{min}$ **do**
  if Environmental change **then**
    Re-adapt solutions in X_k
  else
    $n \leftarrow 0$
    $CurrentSolution \leftarrow LowstCost$ in X_k
    while $n \leq populationSize$ **do**
      Pick a neighborhood solution $NewSolution$ from the $CurrentSolution$
      $\Delta C \leftarrow cost(NewSolution) - cost(CurrentSolution)$
      if $\Delta C \geq 0$ **then**
        $CurrentSolution \leftarrow NewSolution$
      else
        Generate a random number $r \in (0, 1)$
        if $min(1, \exp(-\Delta C/T)) > r$ **then**
          $X_{k'} \leftarrow NewSolution$; $CurrentSolution \leftarrow NewSolution$
        end if
      end if
      $n \leftarrow n + 1$
    end while
    Apply genetic operators to $X_{k'}$ to create X_{k+1}
    Calculate the fitness value and the cost for each solution in X_{k+1}
    if LowstCost in $X_{k+1} < CurrentSolution$ **then**
      $CurrentSolution \leftarrow LowCost$ inX_{k+1}
    end if
    $T_{k+1} = \alpha * T_k \; (0 \leq \alpha \leq 1); k \leftarrow k + 1$
  end if
end while

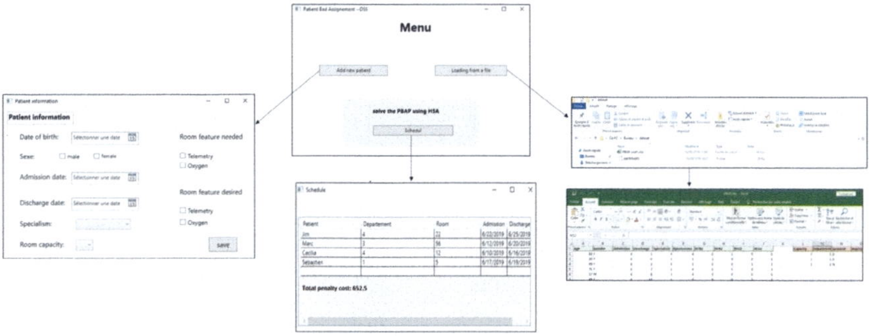

Fig. 1 A HSA based DSS for solving the PBAP

the appropriate assignment and matching as much as possible patient preferences with rooms, we run the DSS, integrating the proposed HSA approach. Finally the best solution that satisfies at maximum the patients' requirements is displayed.

6 Computational Experiments

In what follows, we first introduce the benchmark instances with the experimental settings used, and then we present a discussion for the computational results obtained.

6.1 *Problem Instances*

The experiments are carried out on 13 existing benchmark instances provided in [4] and [7]. Due to the problems related to the patients privacy, real-world data are difficult to be collected. Therefore, after interviewing several patients schedulers in the Belgian hospital, [4] generate automatically 6 different instances. In [7], 7 more complex instances were generated. All data sets that exist in the literature are realistic and have been confirmed by specialists in health-care domain to present well the real-life patients' situation.

The different characteristics of each data set are presented in Table 2. According to their properties, the PBAP instances can be divided in two sets with different sizes. Instances from 1 to 7 have a small size compared to those from 8 to 13. The first set of instances has a planning horizon of 14 days and a small number of elective

Table 2 Characteristics of the benchmark instances

Instance	#Beds	#Patients	#Rooms	#Departments	#Plannin horizon
1	286	652	98	14	14
2	465	755	151	14	14
3	395	708	131	14	14
4	471	746	155	14	14
5	325	587	102	14	14
6	313	685	104	14	14
7	472	519	168	14	14
8	441	895	148	21	21
9	310	1400	105	28	28
10	308	1575	104	56	56
11	318	2514	107	91	91
12	310	2750	105	84	84
13	368	907	125	28	28

patients comparing to instances from 8 to 13, where the number of elective patients is quite higher and the planning horizon is between 21 and 91 days. Furthermore, looking at the number of departments and room features of all data sets, we can observe that instances 1–6 have a homogeneous situations comparing to instances 7–13. The existing benchmarks are frequently used in the literature for the static variant of the PBAP. To adapt this benchmark for the dynamic problem, we introduce the notion of *Effective Degree of Dynamism (edod)*. In our system only three parameters are relevant: (1) the number of static patients, (2) the number of dynamic arrived patients, and (3) the arrival times of the dynamic patients. We apply the method proposed in [17]. A number of patients was selected based on the *edod* as presented in Eq. 8 to perform the initial assignment.

$$edod = \frac{\sum_{i=1}^{n_A} \left(\frac{t_i}{T}\right)}{n_A + n_P} \quad (8)$$

The number of dynamic arrivals during the planning horizon is denoted n_A, the number of advance planned patients is denoted n_P, and ti represents the time in which patient i arrives. The PBAP is consider as a pure dynamic problem if the $edod = 0$ and as a pure static problem if the $edod = 1$ [20].

6.2 Experimental Results

We present in this section the obtained results of the computational experiments performed over different data sets as shown in Table 2. Furthermore, we introduce a comparison study between the dynamic proposed approach and the dynamic SA and then between the two different results obtained by our proposed method to solve both dynamic and static variant of the problem. Our purpose is to test the efficiency of the HSA to solve both versions of the PBAP and improve the quality of the found solutions.

As it can be seen from the summarized experimental results in Table 3, our hybrid proposed approach has a good performance on most benchmark instances comparing to the classic SA approach in solving the D-PBAP. So we can say that the hybridized SA produces a greatest satisfaction and a lowest cost values in many cases. We consider the improvement percentage as a metric of performance that measures the difference between the best found solution obtained by the proposed HSA and the best solution obtained by the classic SA.

We start by solving instances from 1 to 7 that have a lower bed occupancy rate comparing to the remaining seven instances, and they are more flexible to be scheduled. As demonstrated above, five instances give best results among seven small data sets with an average improvement of 10.1%.

Then we consider large instances (8–13), where the bed occupancy rate is very high comparing to the remaining first seven ones, which implies a difficulty in

Table 3 Comparison results of the algorithms for benchmark problem

	SA	HSA		
Instance	Dynamic	Static	Dynamic	Improvement (%)
1	795.2	652.5	**652.5**	18
2	1334.0	1185.0	**1310.2**	2
3	**773.0**	759.4	795.4	0
4	1415.0	1167.3	**1350.0**	5
5	802.6	631.0	**694.2**	14
6	2160.7	834.3	**900.0**	32
7	**1207.2**	1165.7	1240.6	0
8	5120.3	4200.4	**4285.3**	16
9	2320.1	21135.6	**2215.6**	5
10	**7985.7**	7930.5	8043.2	0
11	**15140.7**	11984.0	11985.4	0
12	27540.1	24740.6	**24950.4**	10
13	**13572.0**	9230.0	9294.1	0

finding a solution especially with a large planning horizon. Experimental results indicate the good performance of the proposed approach. With instances 8, 9, and 12, the proposed HSA algorithm performs best results with improvement of 5.3% and gives the highest satisfaction and lowest cost values.

This proves that our dynamic approach is better than the classic SA algorithm in finding the minimum penalty cost of the assignment and maximizes as much as possible the patient comfort while satisfying all new arrived demands, which indicates the effectiveness of our method.

Finally, a comparison between different results generated by the static and the dynamic versions of the problem shows that in our case, the improvement degree decreases with the size of the problem and the length of the planning horizon. But the GAP between both variants is not significantly high. These results confirm that the emergency situation is more crucial, i.e., the arrival of new patients that are not planned disturbs the scheduling.

7 Conclusion and Future Work

The hospital bed management is one of the most challenging organizational task in the health-care resource planning. It becomes a central topic of clinical operational research. Especially, with the increase of patient demand, the hospital bed management becomes a crucial issue. For this reason, hospital managers should more concentrate on how to ensure patients satisfaction and provide their comfort among their period of stay in a smart health-care system.

In this work, we tackled this problem on its operational level, i.e., the patient bed assignment that is the most classic and common variants of the problem; even in its basic version, the problem is considered as a complex combinatorial optimization one.

In our work, we dealt with the PBAP in its dynamic version; it consists on assigning emergency patients to beds within different hospital units. It aims to maximize treatment efficiency and patient comfort while satisfying all necessary medical constraints and taking into consideration patient preferences as much as possible. A DSS is developed to assist the hospital staff in the assignment process. To this purpose, we applied a hybrid simulated annealing algorithm that merge the features of genetic algorithm with those of simulated annealing that enhances the initial solution obtained by the GA in an iterative mode. After running our method on the different benchmark instances with different sizes, the computational experiments show a very acceptable result.

As future work, we aim at first to test our approach on more large-scale instances, using a real case study within a hospital group of territory. In this case, we move from a simple optimization problem to a big data optimization one. Then we opt to consider intensive care departments that incorporate patients with severe illnesses and injuries, to improve the patient assignment task, and finally adding to the model an extra real-world constraint such as workload balance of the health-care professionals.

References

1. L.S. Bastos, J.F. Marchesi, S. Hamacher, J.L. Fleck, A mixed integer programming approach to the patient admission scheduling problem. Eur. J. Oper. Res. **273**(3), 831–840 (2019)
2. J. Beliën, E. Demeulemeester, Building cyclic master surgery schedules with leveled resulting bed occupancy. Eur. J. Oper. Res. **176**(2), 1185–1204 (2007)
3. Y. Bhatt, C. Bhatt, Internet of things in healthcare, in *Internet of Things and Big Data Technologies for Next Generation Healthcare* (Springer, 2017), Cham, pp. 13–33
4. B. Bilgin, P. Demeester, M. Misir, W. Vancroonenburg, G.V. Berghe, One hyper-heuristic approach to two timetabling problems in health care. J. Heuristics **18**(3), 401–434 (2012)
5. M. Brandeau, F. Sainfort, W.P. Pierskalla, *Operations Research and Health Care: A Handbook of Methods and Applications*, Kluwer's International Series, Dordrecht, 70 (Springer Science & Business Media, 2004)
6. G. Carnaz, V.B. Nogueira, An overview of IoT and healthcare (2016)
7. S. Ceschia, A. Schaerf, Local search and lower bounds for the patient admission scheduling problem. Comput. Oper. Res. **38**(10), 1452–1463 (2011)
8. S. Ceschia, A. Schaerf, Modeling and solving the dynamic patient admission scheduling problem under uncertainty. Artif. Intell. Med. **56**(3), 199–205 (2012)
9. S. Ceschia, A. Schaerf, Dynamic patient admission scheduling with operating room constraints, flexible horizons, and patient delays. J. Sched. **19**(4), 377–389 (2016)
10. I. Chiuchisan, O. Geman, An approach of a decision support and home monitoring system for patients with neurological disorders using internet of things concepts. WSEAS Trans. Syst. **13**(1), 460–469 (2014)
11. P. Demeester, W. Souffriau, P. De Causmaecker, G.V. Berghe, A hybrid tabu search algorithm for automatically assigning patients to beds. Artif. Intell. Med. **48**(1), 61–70 (2010)

12. A. Gaur, B. Scotney, G. Parr, S. McClean, Smart city architecture and its applications based on IoT. Proc. Comput. Sci. **52**, 1089–1094 (2015)
13. P. Gemmel, R. Van Dierdonck, Admission scheduling in acute care hospitals: does the practice fit with the theory? Int. J. Oper. Prod. Manag. **19**(9), 863–878 (1999)
14. R. Guido, M.C. Groccia, D. Conforti, An efficient matheuristic for offline patient-to-bed assignment problems. Eur. J. Oper. Res. **268**(2), 486–503 (2018)
15. S. Kirkpatrick, C.D. Gelatt, M.P. Vecchi, Optimization by simulated annealing. Science **220**(4598), 671–680 (1983)
16. F.T. Lin, C.Y. Kao, C.C. Hsu, Applying the genetic approach to simulated annealing in solving some np-hard problems. IEEE Trans. Syst. Man Cybern. **23**(6), 1752–1767 (1993)
17. K. Lund, O.B. Madsen, J.M. Rygaard, *Vehicle Routing Problems with Varying Degrees of Dynamism* (IMM Institute of Mathematical Modelling, 1996)
18. R.M. Lusby, M. Schwierz, T.M. Range, J. Larsen, An adaptive large neighborhood search procedure applied to the dynamic patient admission scheduling problem. Artif. Intell. Med. **74**, 21–31 (2016)
19. N. Metropolis, A.W. Rosenbluth, M.N. Rosenbluth, A.H. Teller, E. Teller, Equation of state calculations by fast computing machines. J. Chem. Phys. **21**(6), 1087–1092 (1953)
20. V. Pillac, M. Gendreau, C. Guéret, A.L. Medaglia, A review of dynamic vehicle routing problems. Eur. J. Oper. Res. **225**(1), 1–11 (2013)
21. B.G. Thomas, S. Bollapragada, K. Akbay, D. Toledano, P. Katlic, O. Dulgeroglu, D. Yang, Automated bed assignments in a complex and dynamic hospital environment. Interfaces **43**(5), 435–448 (2013)
22. C.E. Turcu, C.O. Turcu, Internet of things as key enabler for sustainable healthcare delivery. Proc.-Soc. Behav. Sci. **73**, 251–256 (2013)
23. A.M. Turhan, B. Bilgen, Mixed integer programming based heuristics for the patient admission scheduling problem. Comput. Oper. Res. **80**, 38–49 (2017)
24. W. Vancroonenburg, D. Goossens, F. Spieksma, On the complexity of the patient assignment problem. Technical report (2011)
25. W. Vancroonenburg, P. De Causmaecker, G. Vanden Berghe, Patient-to-room assignment planning in a dynamic context, in *Proceedings of the 9th International Conference on the Practice and Theory of Automated Timetabling (PATAT-2012)* (Citeseer, 2012), pp. 193–208

Real-Time Processing and Monitoring in Health Care

Sudakshina Chaudhury, Shreya Roy, Ishika Agarwal, and Niranjan Ray

1 Introduction

1.1 IoT in Health Care

Internet of Things (IoT) is a dynamic network infrastructure consisting of a billion end-user devices connected via the Internet. It has advanced from RFID-based networks to WBAN networks using IP-based protocols such as 6LoWPAN. It has the power to produce unprecedented data, which are stored and processed in cloud. It is also referred to as cloud of things (CoT). Thus, it integrates various technologies such as WBAN, RFID with fog computing, networking along with big data and cloud computing. IoT has numerous applications in different areas; one of them is health care.

The current hospital-centric healthcare system poses various challenges with the rise in the ageing population and patients with chronic and lifestyle diseases requiring constant monitoring and care. The population of people suffering from different ailments has increased, while healthcare resources have remained limited. Thus, various issues arise like medical negligence, ill-timed prognosis and rise in medical costs. The need of the hour is quality health care. Tasks such as patient monitoring, resource management and time-sensitive decision-making are still done manually. Such practices may lead to tragic errors. Automation of these tasks is proposed in a smart healthcare system.

S. Chaudhury (✉) · S. Roy · I. Agarwal · N. Ray
School of Computer Engineering, KIIT Deemed to be University, Bhubaneswar, India
mailto: 1605610@kiit.ac.in

N. Gupta, S. Paiva (eds.), *IoT and ICT for Healthcare Applications*, EAI/Springer Innovations in Communication and Computing,
https://doi.org/10.1007/978-3-030-42934-8_7

According to [1] efficiency of healthcare systems can be improved with the integration of IoT. The multi-layered architecture of IoT can be tailored to enhance the healthcare system. The lowest layer in IoT can be viewed as the sensing layer, which consists of e-health mobile apps, health monitoring wearable devices containing sensors, etc. The data are then transmitted through the networking layer where routing algorithms are implemented. The next layer can be seen as the services layer which will process the data accordingly. The output can be summary reports or predictions and preventive measures (actionable insights). The last layer is the application layer that can be used as an Application User Interface (APIs) where the data can be viewed or shared among patients, physicians and clinics.

IoT-enabled health care has many applications such as clinical care, telemedicine and remote patient monitoring [2]. It discusses the constant monitoring of patient data which are processed and stored in the cloud. It can be later accessed by physicians to make decisions remotely. Prognosis of diseases can be done readily, thus reducing the risk of critical ailments and the cost of treatment. Thus, integrating IoT with health care is beneficial to the society. Figure 1 illustrates an e-health system.

The popularity of IoT and bio-devices and E-health apps has exponentially increased the network of end-user devices. In this type of system, the large volumes of data from the network of sensors and healthcare wearables constitute big data. Thus, the storage and processing are shifted to the cloud layer, as it is cheaper and follows a 'pay-as-you-go model', which reduces the complexity of computations and cost of the system. Hence, Cloud of Things (CoT) is introduced to further enhance the healthcare system.

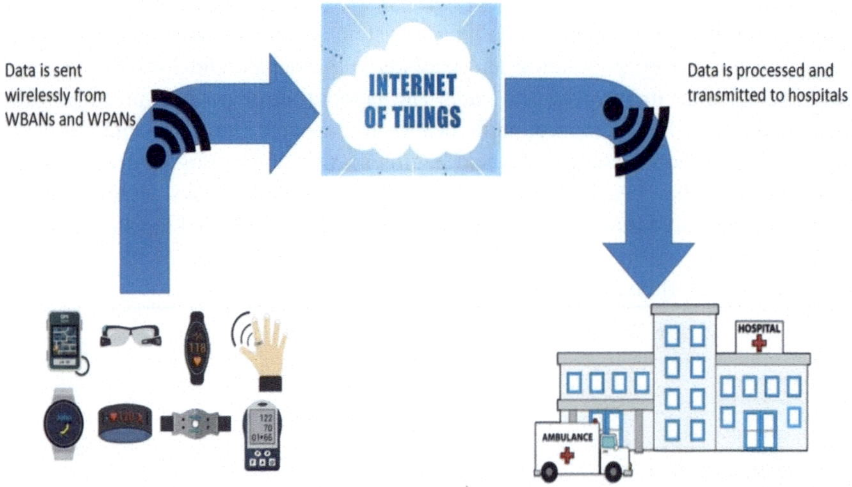

Fig. 1 E-health system

1.2 Cloud of Things

Cloud provides services like Infrastructure as a Service (IaaS), the lowest layer of the cloud computing platform. It provides virtual machines and storage capabilities for processing vast amount of data. The next is Platform as a Service (PaaS); it provides for database, which can be queried and accessed through application software defined in the Software as a Service (SaaS) layer. SaaS provides the APIs through which patients and doctors can view and send data to each other. Thus, the integration of cloud with IoT resolves the storage and processing issues along with global access of data. It makes the computational processes simpler to interpret. It provides payable services as and when requested from the network, making it cost-effective. Figure 2 shows the cloud stack.

As stated earlier, data from millions of connected e-health wearables in the IoT system will culminate to big data. The processing power and storage available in IoT devices are limited. Increasing the capacity by scaling would be very costly and can form a complicated mesh of networks. Hence, cloud is proposed where the data storage and processing are performed centrally and the cost is reduced considerably. All the complications of computing and storing data are taken away from the end devices.

However, CoT has some issues such as batch processing and latency. In the healthcare sector, decisions are required to be taken quickly and are time-sensitive and precision-oriented. The batch processing nature of cloud computing does not support real-time analysis of data. The large amount of data being transmitted may lead to faulty data. Due to the increase in heterogeneity in healthcare wearables and the centralized nature of the cloud, interoperability and ubiquitous monitoring of data are also difficult. Thus, to process data in real time and to reduce latency, the idea of bringing cloud services to the edge of the network is proposed in literature.

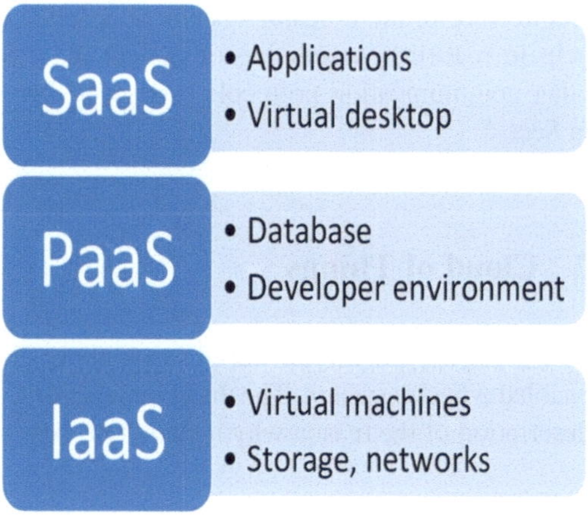

Fig. 2 Cloud stack. (Source: Internet)

1.3 Fog Computing

Fog computing decentralizes the cloud platform. It brings the services of the cloud to the edge of the network. The fog layer comprises nodes. The nodes can provide services such as storage, computing and network connectivity. Nodes act like gateways or routers. It supports a variety of network protocols and can connect to heterogeneous devices. This improves interoperability.

The fog layer comprises these smart gateways which lie within the local area of the sensor network and end-user devices. After data processing, the processed data are stored in the cloud. As the amount of data transmitted becomes less, the network congestion is also greatly reduced. This reduces the latency of storing and retrieving data considerably and also improves the network bandwidth. This, intelligent pre-processing of data enables the fog layer to act as a smart gateway between the IoT and the cloud layer.

1.4 6LoWPAN

IoT uses RFID-based communication protocols to identify devices and sensors. Non-IP-based protocols like ZigBee, Z-Wave and Bluetooth provide low power-consuming communication services. Few studies find that IP-based protocols like 6LoWPAN increase interoperability and can connect to any device or node as long as it is connected to the Internet. The use of IPv6 provides numerous IP addresses as compared to its over-utilized predecessor IPv4. Hence, 6LoWPAN can help in globally connecting doctors and patients bringing about remote patient monitoring and a ubiquitous healthcare system.

The remaining chapter discusses the following. Section 2 discusses cloud on things and Sect. 3 discusses fog computing. Basically, it focuses on different architecture in the e-health system. It describes how its properties and orientation help in real-time processing of patient data. Section 4 is about 6LoWPAN and other communication protocols used in IoT networks and the conclusion is made in Sect. 5.

2 Cloud of Things

This section gives a brief introduction about cloud computing and discusses CoT-enabled e-health system. The disadvantages of CoT are discussed, which give a brief description of the reason why it cannot process data in real time.

The cloud infrastructure provides data management and global access functionalities. It provides many facilities like on-demand storage and software applications, and manages shared resources and infrastructure in a transparent manner [4]. It is a model that enables convenient on-demand network access to resources shared within the network. These resources can be provided with minimal service effort from the platform providers. The access of these resources over the network supports the heterogeneous nature of client devices. Hence, it supports interoperability. The resources provided by the platform include virtual machines, network bandwidth, among others. The main advantage lies in the way users can reprovision technological resources in an agile manner.

There are various well-known cloud providers such as Dropbox and Amazon Web Services (AWS). These cloud providers support many services and applications. But they are not suitable for the healthcare sector.

2.1 CoT in E-Health

Wearable devices like smart watches, phones and sensors form a network of heterogeneous devices in the smart healthcare system. This network of devices constitutes a mesh of WSNs and WBANs. The amount of these devices connecting to the Internet is increasing exponentially. These networks comprise heterogeneous devices following varied communication protocols. The data generated from these devices are high in volume and are highly varied. The rate at which the data are generated can also be observed as high. Thus, the data generated can be termed as Big Data. Conversion of these data to actionable insights is what the e-health system is all about. The devices connected to the Internet also transmit their data to the Internet. The data from patient devices are then transmitted to a cloud computing platform connected to the network. The cloud platform stores and manages the data. It processes the data to convert them to insights and transmits them to the necessary destination like hospitals and physicians. The data can also be accessed by patients and doctors alike.

Many studies have also discussed the improvement of IoT systems with integration of cloud. In [5], cloud computing and IoT integration provide new storage, processing, scalability and networking capabilities, which were so far limited in the IoT due to its characteristics.

A cloud-centric system is proposed in [6]. It explains that Internet-centric system integrated with cloud at the centre can prove to be cost-effective and highly scalable. Figure 3 illustrates a CoT-enabled e-health system.

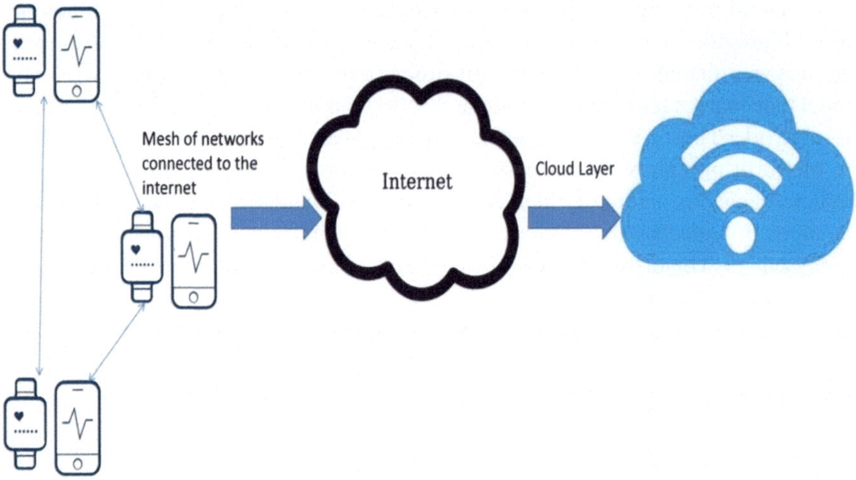

Fig. 3 E-health system with cloud layer. (Ref. [6])

2.2 Challenges in Cloud of Things

CoT does have some issues and challenges. Few are discussed here briefly.

Mobility As explained in [3], the cloud suffers with mobility. The mobility of the smart devices affects the network connection to the cloud. In the healthcare system, immobility of the device is not a viable option. By mobility, users cannot perform any movement at the time of wearing sensors. This hinders the process of detecting many ailments and is not consistent with usage in remote patient monitoring where details of all health and ailments are required to make medical decisions.

Scalability and Latency Scalability of end-user devices increases the latency. The more the number of end devices connected to the same network implies more data generation and transmission. This increases the congestion in the network which in turn delays the transmission of data.

Batch Processing In the healthcare sector, various scenarios may arise where decisions need to be made as and when data arrive. For instance, during emergencies like heart attack; the abnormal ECG details about the patient need to be sent to the doctor, who in turn needs to act upon it immediately and send the required help or facilities. It is observed that data need to be processed in real time to aid in time-sensitive decision-making. Batch processing is done during analysis and transmission of data from the cloud layer to end devices. This causes a certain time

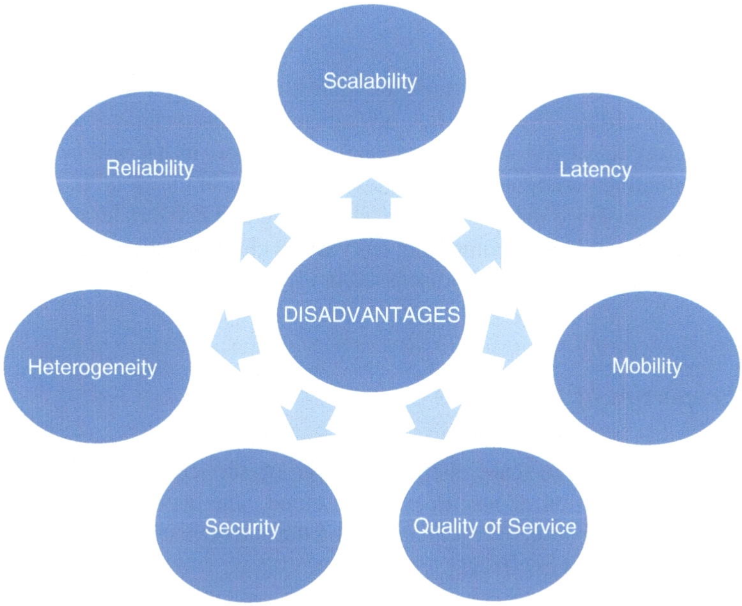

Fig. 4 Challenges of CoT: overview

delay in processing data received from the patient sensors and in transmitting data from the cloud to the physician. This latency is not feasible when it comes to making decisions which are time-sensitive.

Reliability The wireless nature of the network makes transmission of a large amount of data unreliable. Transmission of a large amount of data may generate a bit error. As error checking in wireless systems is not very robust, erroneous data will be transmitted. At the end of the analysis, decision-making will be performed on these data, resulting in fatal accidents. Thus, the non-reliability of data transmitted in the cloud defeats the purpose of improving the healthcare system.

Heterogeneity and Quality of Service (QoS) The heterogeneous nature of the sensor network also poses many problems. Each device has its own protocol stack and network architecture. This causes variation in the Quality of Service (QoS). According to [6], it is not easy to provide QoS guarantees in wireless networks, as segments often constitute gaps in resource guarantee due to resource allocation and management ability constraints in shared wireless media. As discussed, time-sensitive decisions and real-time processing along with reliable transmission of data are required in healthcare systems. Security is another concern in CoT. Figure 4 shows the challenges of CoT.

3 Fog Computing

This section discusses the characteristics of fog computing, which enables real-time processing of data. The properties, advantages and disadvantages are discussed along with examples of various healthcare scenarios where its use will be beneficial.

It is a computing paradigm that provides the services of the cloud to the edge of the network. Any device that has storage, computing and network connectivity can be treated as a fog node. Fog computing has various characteristics that improve the centralized cloud-based system as discussed in [7, 8].

3.1 Properties of Fog

The fog layer is a large-scale distributed layer of services and applications as opposed to centralization of resources in the cloud. The distributed nature of the fog helps in the distribution of resources among the end devices. The characteristics of the fog are discussed in [10–11].

Location Awareness Gateways can locate the devices using IP and MAC addresses of the devices within its network. This helps health institutions and doctors to locate patients in case of emergencies.

Latency and Real-Time Notification The fog nodes act as a dynamic access point between the device and the cloud layer. The proximity to the device layer helps in reducing the latency to process data. As opposed to batch processing in the cloud, data are computed in real time in fog. It is known that real-time signals are always transmitted from devices to the fog and then to the cloud. The real-time notification service provided by the system also helps in addressing abnormal situations which may arise in any healthcare situation. Abnormal situations can be of various types. A real-time signal is sent to a remote cloud which sends the notification to the respective destination devices.

Heterogeneity and Interoperability In a cloud infrastructure, homogeneous resources are centrally deployed. In fog computing, heterogeneous resources are deployed in a distributed manner. Heterogeneity is in terms of different manufacturers, models with follow and support of varied communication protocols. They are required to be deployed in different platforms and operating systems. Fog is platform-independent and supports many operating systems. It also provides protocol conversion methods. Thus, it has the ability to be used in different platforms and can thereby inter-operate between various devices and sub-networks. Figure 5 provides an overview of the properties.

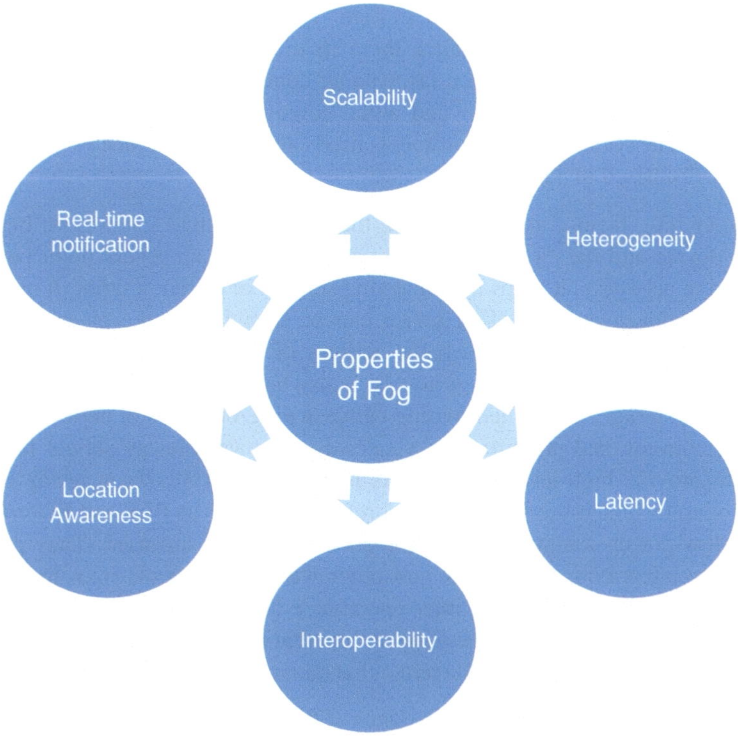

Fig. 5 Properties of fog: overview

3.2 *Fog Architecture in Health Care*

The basic unit of a healthcare system is the Wireless Body Area Networks (WBAN). This network generates a huge amount of data which are beyond the processing capacity of the devices. Thus, the data are connected to the cloud layer through the Internet. Due to its shortcoming as discussed in the previous section, a new layer is introduced between the existing two layers. This layer is known as the fog layer, which overcomes many existing shortcomings of the previous system. One of the main focuses in this chapter is the fact that it helps in real-time processing of data. The system comprises three layers, namely sensor network (WBAN), fog layer and cloud layer.

WBAN Its nice properties such as low cost, power efficiency, low bandwidth and processing power play an important role in a health monitoring system. The end devices comprising sensors, mobile and actuators form the device layer. It basically comprises wearable and implantable sensors. The device layer is responsible for collecting, sensing and extracting data from patients. It helps by monitoring patient status and health. The collected data are then sent to the next layer, that is, the fog layer for further processing and storage. The device layer then receives data and should also be able to visualize them in the appropriate format to doctors and health institutions. The device layer being connected to other devices and the Internet through various wireless and wired communication protocols constitutes the WBAN. The communication protocols used can be WiFi, Zigbee or 6LoWPAN, which are discussed in a later section.

Fog Layer Fog layer communicates by the help of a smart gateway in between the end device and the cloud layer. The gateway acts as an access point between two discrete networks. It helps in connecting sensors of different types. It may also provide location awareness. The device can be traced from the gateway, which acts as the local node from which the address of the device can be found. The gateway also translates different protocols. It also has other functionalities like data aggregation, storage, ltering and routing between different types of networks. Apart from the conventional properties of a gateway, a smart gateway should have the ability to provide high-level services. As mentioned in [10], the gateway should consist of embedded routers, which support protocols such as BLE and Ethernet. To support low-power communication protocols, sink nodes are used. They provide support for integrating 6LoWPAN and other wireless protocols into the system. Thus, it helps in achieving an inter-operable e-health system. Each embedded component also consists of an operating system that has functionalities to propagate the working of the fog and cloud layer so that data can be sent in real time.

It is seen that gateways bring the services of the cloud to the edge of the network. The fog nodes can be designed to be within the local area of the device and sensor network. Due to the proximity of the gateway, latency in transmission is greatly reduced. The time and bandwidth that were previously required to transmit the big data, generated from the end-user devices, were very considerable. Transmission of such a large amount of sensitive data over a wireless medium could jeopardize the integrity and security of data. But with the introduction of fog, the probability of such occurrences is greatly reduced. The pre-processing of data reduces the amount of data needed to be stored in the cloud, thus reducing the chances of sending erroneous data. The required data can also be temporarily stored in the fog layer and can be utilized in real time for analysis and transmission. This enables real-time processing of data to make time-sensitive decisions in health care. These error-free data are then transmitted to the cloud layer.

Cloud Layer It provides the backend support to the e-health system. It provides a cost-effective solution for storing and processing data. It provides services such as broadcasting, data warehousing and analytics along with a centralized database system. The data retrieval is done by the fog nodes from the cloud paradigm. The cloud also provides various application software tools and acts as a graphical user interface for a web client. The data collected in the cloud can be processed and used for statistical research and predictions.

The cloud layer only interacts with the fog nodes. Most of the processing and transmission are done between the fog and the sensor network. Thus, the complexity

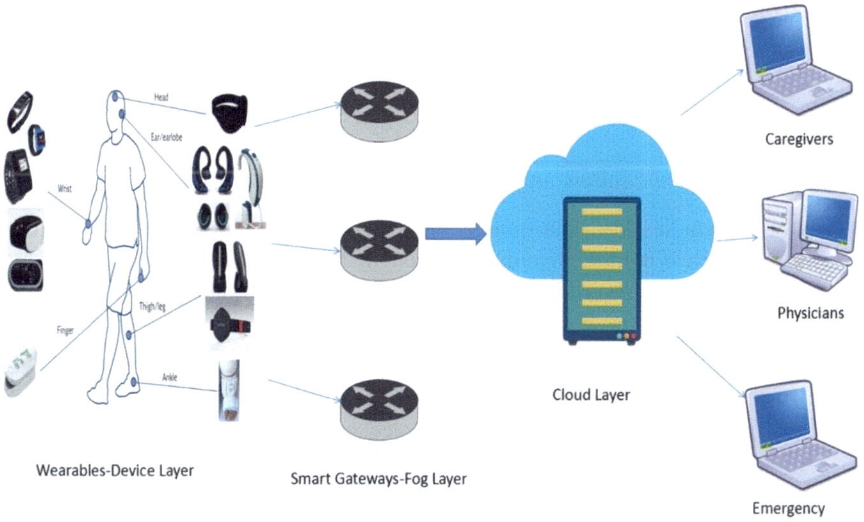

Fig. 6 E-health system with fog layer acting as smart gateways

of the services needed and computational capacity is greatly reduced for the cloud layer. This reduces the bandwidth and network congestion. Figure 6 shows an e-health system implemented using fog and cloud.

3.3 Limitations of Fog Computing

The major limitation of fog computing is in security [9]. Each fog node consists of sensitive information like that of patient data, which can be easily hacked. Providing security is complex for each and every node. The distributed set-up of fog computing increases the energy consumption as compared to the centralized cloud computing. In [12], the dynamic nature of IoT devices is highlighted. The configuration of handheld devices undergoes frequent upgradation, which necessitates the change in the internal properties of fog nodes. Hence, intelligent fog nodes need to be designed that can reconfigure according to the topological changes in the network.

4 6LoWPAN

This section focuses on ubiquitous monitoring of remote patients using 6LoWPAN. It discusses various communication protocols used in the IoT network and gives a brief analysis on them. In the later subsections, 6LoWPAN in the healthcare sector is extensively discussed.

4.1 Communication Protocols in IoT-Based Health Care

Many protocols have been introduced to enable communication among smart devices. They can be categorized into two parts: (1) LPWAN (2) Short Range Network.

LPWAN covers the following:

1. SigFox: It handles low data transfer speeds of 10–1000 bits per second and is used for wireless communication in sensors and M2M applications.
2. Cellular: Applications which need high-throughput data and require operation over large distance use this communication protocol.
3. 6LoWPAN: It is used for addresses of different lengths. Mesh and star topology are some of the major topologies that are supported by 6LoWPAN. Details of this will be discussed later.
4. Zigbee: The various topologies supported by this protocol are mesh, star and tree network topology.
5. BLE: Also known as Bluetooth Smart, this protocol is used for short-range, low-bandwidth and low-latency applications. Star network topology having unlimited nodes is supported by this protocol.
6. RFID: It supports P2P network technology. Smart shopping, health care, national security and agriculture are some of the areas where this protocol is used.
7. NFC: It is a very short-range communication technology. Similar to RFID, it also supports P2P network topology. IoT devices can be easily connected and controlled in different environments by using NFC.
8. Z-Wave: It is a low-power MAC protocol, which can be used to connect 30–50 nodes. It supports mesh network topology. It is best suited for small messages in IoT applications.

4.2 6LoWPAN

6LoWPAN stands for IPv6 over low-power wireless personal area networks. It is an IP-based standard internetworking protocol. It is the most commonly used standards. The number of addresses supported by this protocol is 2128. It is a network protocol that encapsulates and uses header compression mechanism.

Properties of 6LoWPAN.

1. It uses IPv6 over low-power wireless networks.
2. It is ideal to create mesh networks since it carries IPv6 over 802.15.4 standard.
3. It provides seamless connectivity to a variety of networks and also provides end-to-end IP.
4. It provides low data rates. It provides a basic transport mechanism in order to produce complex control systems. It can also communicate with the various complex system in a cost-effective way through low-power wireless network.

5. It also uses AES-128 link layer security which can provide link authentication and encryption. In addition to this, other security mechanisms such as TLS or digital signature can also be implemented.

Figure 6 shows an e-health system implemented using fog and cloud.

4.3 6LoWPAN in the E-Health Sector

As the population is increasing day by day, it is necessary that healthcare monitoring should be omnipresent. There are many people in the world who suffer from various illness because they do not have proper access or reach to hospitals and health services on time. Death statistics indicated that death rate around the world caused by non-communicable or age-related chronic conditions is predicted to rise from 59% in 2002 to 66% in 2030 as mentioned in [16]. As the medical expenditure is increasing with each passing day, it is important that cheaper methods should be developed so that better treatment can be provided to patients by monitoring patients continuously rather than only when their health condition is at a critical stage. These challenges are met by ubiquitous healthcare systems, as they can help in real-time monitoring of patients and provide the required feedbacks on time (Fig. 7).

With the advancement in technology, it has been made possible to monitor patients remotely without the actual need of visiting hospitals physically. This has been made possible due to various wireless solutions that are connected to IoT.

For developing ubiquitous healthcare systems, various short-range protocols such as zigbee and bluetooth were introduced. Short-range protocols are used for short distances only. In such protocols, the remote nodes are linked to very short distances that can be few metres. Interoperability is a feature which helps in the exchange of data and good communication among the various components of a system without

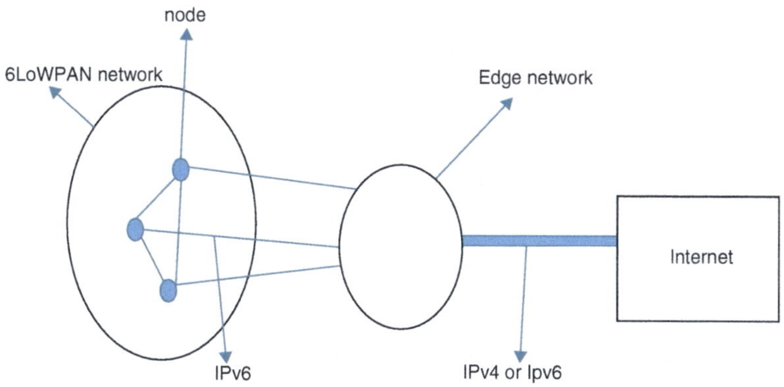

Fig. 7 Efficient working of 6LoWPAN

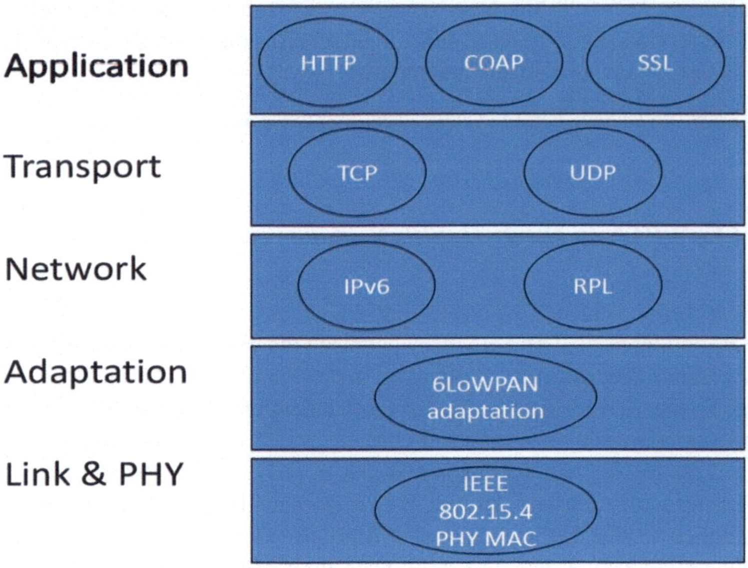

Fig. 8 Protocol stack of 6LoWPAN. (Source: Internet)

taking into consideration its manufacturer or specifications. So, these short-range protocols failed to exhibit this important feature of interoperability of devices. This issue of interoperability was resolved by 6LoWPAN, as it provides a common platform for all the devices. The features of 6LoWPAN are numerous. It provides proper header compression of the packets coming from different layers, and the network can be automatically configured. 6LoWPAN is connecting more devices and nodes to the cloud and so it is heavily used in IoT applications (Fig. 8).

6LoWPAN has been used to change the IoT landscape thoroughly. Bluetooth and other personal systems use a complex gateway to connect to the Internet. This issue was resolved by 6LoWPAN as an adaption layer was introduced for transmitting datagrams over 802.15.4 radio links. This layer was introduced between the data link and the network layer (Fig. 9).

In remote monitoring health care, wearable devices and sensors are used. IPv6 and 6LoWPAN are used in the transmission of data over the 802.15.4 protocol in these sensors and wearables. Sensor nodes then send back their data with the help of User Datagram Protocol (UDP).

For achieving a specific task in a given time period, biomedical sensors are associated with 6LoWPAN devices.

Medical data are acquired continuously by the sensors on 6LoWPAN devices. A 6LoWPAN network comprises an edge router that helps in connecting 6LoWPAN and ipv6 network by acting as a gateway. The packets are exchanged between these networks with the help of the adaption layer and header of the data packets is also compressed. IPv6 supports 128 bits of addresses: 64 bits for network address and

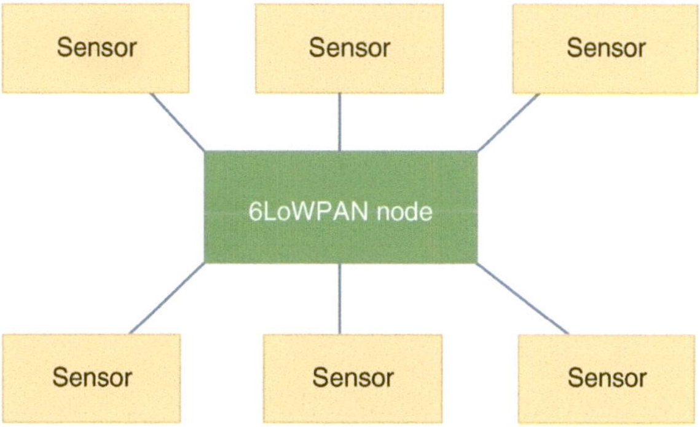

Fig. 9 Sensors and wearables connected through 6LoWPAN

rest 64 bits for host address. The compression in 6LoWPAN removes the 64 bits for network, as this will always remain same for any given LoWPAN network. The number of host bits is reduced to 16. A specific IPv6 gateway is integrated with the network of 6LoWPAN. This gateway is highly linked with the biomedical sensors in 6LoWPAN networks. The biomedical sensors extract an individual's health data like heart rate, blood pressure and other fitness-related data from the individual's body area network and send them to the doctor's mobile phone through this gateway in 6LoWPAN. However, the nodes in 6LoWPAN network send data to the gateway through an edge router. Sensor nodes then send back their data with the help of User Datagram Protocol (UDP).

Various routing mechanisms are then used by the gateway to initiate the resource solicitation. Subsequent frame techniques are then used by the data packets. However, there were many hurdles for 6LoWPAN networks in healthcare. Few of them are as follows:

1. It was difficult to maintain a proper balance between different biomedical sensors in Body Area Network.
2. Patient's movement should not affect the transmission of biomedical data to the gateway.
3. Quality of Service is the technology that is responsible for managing data traffic so that delay and packet loss can be reduced in the network.

In [13], a routing protocol-based on 6LoWPAN is proposed. This approach enables reliable and fast transmission of messages as required in healthcare systems. In [14] the various headers present in 6LoWPAN are discussed. It also gives a comparative analysis on the routing algorithms like DYMO-low, LOAD and HiLow (hierarchical routing). A method to enhance the IoT healthcare architecture based on customized 6LoWPAN is discussed in [15] (Fig. 10).

It describes an energy-efficient and scalable architecture. It focuses on extended node battery life, temperature notification and patient notification, completely based on real-time functioning.

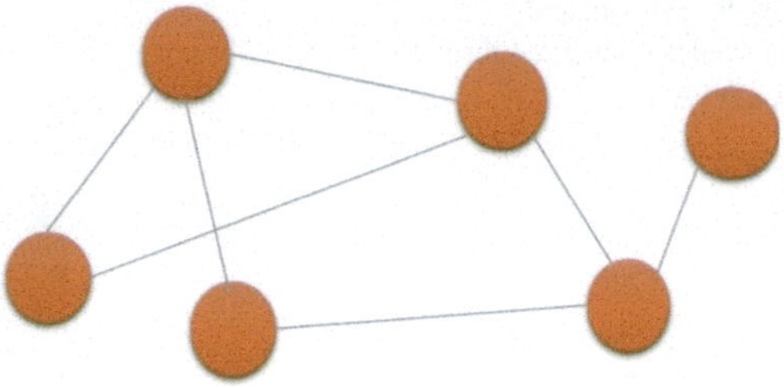

Fig. 10 Mesh topology used in 6LoWPAN network

4.4 Advantages of 6LoWPAN

Scalability In 6LoWPAN, the number of nodes can vary from a few nodes to millions of nodes. Also, the design should be such that there must be maximum utilization of the resources. For example, in any kind of home automation, the number of nodes in the network can be 250–300, whereas if sensor nodes are connected or clustered in a larger scale, then the number of nodes in each clustering region can range between 100 and 10,000; therefore, 6LoWPAN networks are highly scalable.

Mobility This is an important feature in Wireless Sensor Network applications. In legacy systems, static nodes were used. However, with the evolution of technology like IoT, the mobility of nodes has become a necessity. Mobility helps in connecting sensor network to other networks without the need of a gateway. In 6LoWPAN network, the need for mobility may arise due to a failure in a router or any kind of topology change.

Low Cost A 6LoWPAN network consists of low-cost and low-power devices. The topologies that are mainly supported by this network include start and mesh.

Low Power Consumption 6LoWPAN is best suited for mesh topology because it provides end to end connection. Apart from mesh it can be used for other networks using the same standards. It also includes direct connection with the Internet. The sensors that are used in body area networks are low-power devices and hence 6LoWPAN is ideal for these wireless sensors.

Security For security purposes, TLS or digital signature techniques (mainly used in electronically transmitted documents) can be used.

4.5 Limitations of 6LoWPAN

There can be several attacks on the security level of the network using 6LoWPAN protocol. The mobility of the nodes is highly affected if any kind of wrong information is received. Thus, authentication and confidentiality of information should be implemented.

A large number of resources are involved in 6LoWPAN networks, so connectivity and attacks in the networks should be taken care. The resource consumption of these networks is large. 6LoWPAN network requires more memory and more overhead.

5 Conclusion

In this chapter, we have discussed the different issues and challenges associated with IoT-based healthcare system. This chapter discusses different emerging technologies such as cloud computing, fog computing and few communication protocols that are associated with the healthcare system to provide better services to the end users more efficiently. Different challenges of existing technologies are also discussed briefly. We also briefly discuss analyses and some proposals in this area. It is observed that the universal nature of 6LoWPAN and the versatility of the smart gateways will enable real-time processing and remote patient monitoring more efficiently.

References

1. P. Castillejo, J.-F. Martinez, J. Rodriguez-Molina, A. Cuerva, Integration of wearable devices in a wireless sensor network for an E-health application. IEEE Wirel. Commun. **20**(4), 38–49 (2013)
2. A. Kulkarni, S. Sathe, Healthcare applications of the Internet of Things: a review. Int. J. Comput. Sci. Inform. Technol. **5**(5), 6229–6232 (2014)
3. P. Sethi, S.R. Sarangi, Internet of things: architectures, protocols, and applications. J. Electric. Comput. Eng. **2017**, 25 (2017). Article ID 9324035
4. C. Doukas, I. Maglogiannis. Bringing IoT and cloud computing towards pervasive healthcare. In 2012 Sixth International Conference on Innovative Mobile and Internet Services in Ubiquitous Computing, pp. 922–926. IEEE, 2012
5. M. Daz, C. Martn, B. Rubio, State-of-the-art, challenges, and open issues in the integration of Internet of Things and cloud computing. J. Netw. Comput. Appl. **67**, 99–117 (2016)
6. J. Gubbi, R. Buyya, S. Marusic, M. Palaniswami, Internet of Things (IoT): a vision, architectural elements, and future directions. Futur. Gener. Comput. Syst. **29**(7), 1645–1660 (2013)
7. S. Yi, Z. Hao, Z. Qin, Q. Li. Fog computing: platform and applications. In *2015 Third IEEE Workshop on Hot Topics in Web Systems and Technologies (HotWeb)*, pp. 73–78. IEEE, 2015
8. F. Bonomi, R. Milito, J. Zhu, S. Addepalli. Fog computing and its role in the internet of things. *In Proceedings of the first edition of the MCC Workshop on Mobile Cloud Computing*, pp. 13–16, ACM, 2012
9. A.M. Rahmani, T.N. Gia, B. Negash, A. Anzanpour, I. Azimi, M. Jiang, P. Liljeberg. Exploiting smart e-Health gateways at the edge of healthcare Internet-of-Things: a fog computing approach. *Future Generation Computer Systems*, 78(P2), pp. 641–658. Elsevier, 2018
10. T.N. Gia, M. Jiang, A.-M. Rahmani, T. Westerlund, P. Liljeberg, H. Tenhunen. Fog computing in healthcare internet of things: a case study on ECG feature extraction. *In 2015 IEEE International Conference on Computer and Information Technology; Ubiquitous Computing and Communications; Dependable, Autonomic and Secure Computing; Pervasive Intelligence and Computing*, pp. 356–363. IEEE, 2015

11. H. Atlam, R. Walters, G. Wills, Fog computing and the Internet of Things: a review. Big Data Cogn. Comput. **2**(2), 10 (2018)
12. T.H. Luan, L. Gao, Z. Li, Y. Xiang, G. Wei, L. Sun, Fog computing: focusing on mobile users at the edge. arXiv preprint arXiv:1502.01815 (2015)
13. D. Singh, H.-P. Kew, U.S. Tiwary, H.-J. Lee, W.-Y. Chung, Global patient monitoring system using IP-enabled ubiquitous sensor network. *In 2009 WRI World Congress on Computer Science and Information Engineering*, vol 1, pp. 524–528. IEEE, 2009
14. G.K. Ee, C.K. Ng, N.K. Noordin, B.M. Ali, A review of 6LoWPAN routing protocols. Proc. Asia-Pacic Adv. Netw. **30**, 71–81 (2010)
15. T.N. Gia, N.K. Thanigaivelan, A.-M. Rahmani, T. Westerlund, P. Liljeberg, H. Tenhunen. Customizing 6LoWPAN networks towards Internet-of-Things based ubiquitous health-care systems. In *2014 NORCHIP*, pp. 1–6. IEEE, 2014
16. C.D. Mathers, D. Loncar, *Updated Projections of Global Mortality and Burden of Disease, 2002–2030: Data Sources, Methods and Results* (World Health Organization, Geneva, 2005)

Health Data Analytics: Current Perspectives, Challenges, and Future Directions

Kavi Kumar Khedo, Shakuntala Baichoo, Soulakshmee Devi Nagowah, Leckraj Nagowah, Zahra Mungloo-Dilmohamud, Zarine Cadersaib, and Sudha Cheerkoot-Jalim

1 Introduction

The rise in data generation from electronic health systems is changing the landscape of healthcare as it is creating an immense demand for health data analytics, which aims at improving patients' care by deriving actionable insights from these datasets. The healthcare system worldwide is rapidly changing and adopting electronic health records. This is considerably increasing the quantity of clinical data that are available electronically. Moreover, an increasing number of mHealth tools are being adopted. Altogether, these are contributing to a rise in the generation of digital data. There is, therefore, a need for computational tools that can enable health practitioners to obtain new information from the massive data sets, known as medical big data. Immeasurable amounts of disparate medical data have become available in various healthcare institutions (consumers, providers, and pharmaceutical), ranging from patient records through medical imaging to clinical trials, among others.

Medical big data as compared to traditional big data are not very different. However, it is quite difficult to analyze medical big data; the size and intricacy of these datasets present great hurdles in analyses and consequent applications to a practical clinical environment. Additionally, medical data are difficult to access

K. K. Khedo · S. Baichoo · Z. Mungloo-Dilmohamud
Department of Digital Technologies, FoICDT, University of Mauritius, Réduit, Mauritius

S. D. Nagowah (✉) · L. Nagowah · Z. Cadersaib
Department of Software and Information Systems, FoICDT, University of Mauritius, Réduit, Mauritius
e-mail: s.ghurbhurrun@uom.ac.mu

S. Cheerkoot-Jalim
Department of Information and Communication Technologies, FoICDT, University of Mauritius, Réduit, Mauritius

N. Gupta, S. Paiva (eds.), *IoT and ICT for Healthcare Applications*, EAI/Springer Innovations in Communication and Computing,
https://doi.org/10.1007/978-3-030-42934-8_8

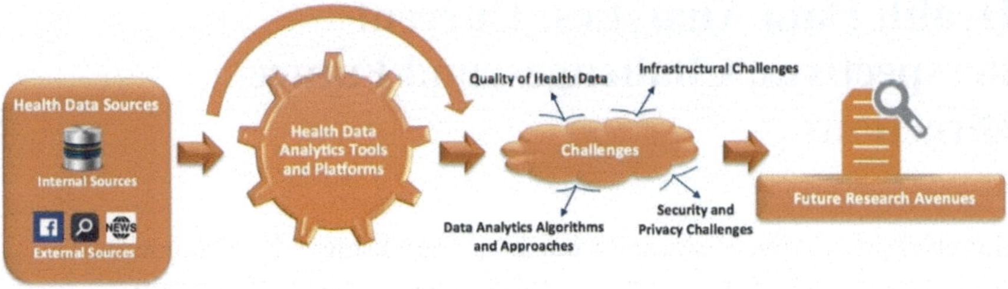

Fig. 1 Graphical summary

for the risks of misuse, as patient details are expected to remain confidential, unless patients' approval is sought. Despite these hurdles, big data technology is already being used in healthcare, for example, for predictive analytics [6] and to reduce healthcare costs [6]. Health data analytics utilizes numerous techniques, including data modeling, data mining, as well as machine learning, among others, to evaluate historical and real-time data in view of predicting future outcomes. It can be used to improve the overall management of healthcare systems from end to end.

This chapter provides an overview of the main application areas that can benefit from health data analytics, by providing practical examples and emphasizing on the type of data they use with their sources, as well as listing their benefits and challenges. A nonexhaustive list of tools and platforms used in the context of health data analytics is provided. Finally, a thorough discussion on the future and existing research opportunities in health data analytics is provided.

Methodology
In order to conduct the review of health data analytics in this chapter, the following methodology has been adopted.

1. The main application areas of data analytics in healthcare are discussed.
2. The different sources of health data are analyzed.
3. State-of-the-art health data analytics tools and platforms (based on popularity and adoption rate) are surveyed, reviewed, and compared.
4. Use cases for the application of health data analytics are described.
5. The main challenges for the use of health data analytics in healthcare institutions and possible solutions are discussed.
6. Future directions for health data analytics are elaborated.

Figure 1 provides a graphical summary of the chapter.

2 Application Areas for Health Data Analytics

Health data analytics are progressively being used by different healthcare institutions to improve effectiveness of the services being provided. This section describes the main application areas of data analytics in healthcare. A summary of these application areas is provided at the end of this section, which highlights the sources of data, benefits, and challenges.

2.1 *Data Analytics for Drug–Disease Association*

Currently, over 23 million published biomedical research articles, clinical case reports, and randomized controlled trials are available on MEDLINE, a US bibliographic database in life sciences, mostly geared toward biomedicine [60]. MEDLINE, therefore, contains biomedical information, including drug–disease associations, from various sources. However, most of the available knowledge is in textual format and has limited machine understandability. Several works, such as Chen et al. [15], Chiang and Butte [16], Frijters et al. [27], and Xu and Wang [88], have focused on different text mining techniques to extract co-occurring concepts, particularly drug–disease treatment associations, from the available biomedical literature. Many of the studies involved the use of knowledge sources such as UMLS (Unified Medical Language System) Metathesaurus and the comparison with established clinical knowledge sources, such as ClinicalTrials.gov [16, 88] and DRUGDEX [16].

2.2 *Data Analytics for Disease Outbreak Detection and Surveillance*

A number of disease outbreak detection and surveillance systems have been developed to help in the early detection of outbreaks and to assist in outbreak investigation and propagation. These systems heavily use data analytics. Lopez et al. [53] proposed a framework to predict H1N1 influenza epidemic based on H1N1 prevalence and climatic data accumulated from previous years in India. Sources of data include huge repositories of health and geospatial data. By applying analytics to these huge sources of data, risk factors and geospatial vulnerabilities have been determined. The result of these analytics can help in developing control and preventive strategies for influenza epidemics, allowing effective use of limited resources in the public health sector. Another model is proposed by Culotta [18], where the author analyzes messages posted on a social website (Twitter) to determine the correlation between search term frequency with influenza statistics with the aim of responding rapidly to health epidemic. Twitter has been chosen mainly because of its diversity of users. Five hundred thousand Twitter messages were collected from a 10-week period. Simple linear regression and multiple linear regression models were used to predict influenza-like illness (ILI) rates based on the frequency of messages related to specific keywords.

2.3 Data Analytics for Pharmacovigilance

Adverse drug reactions (ADRs) are those that are generally caused by the intake of medicines. Pharmacovigilance refers to all the activities that relate to the detection, assessment, understanding, and prevention of adverse effects associated with prescribed drugs [76]. Nikfarjam et al. [64] designed a machine-learning-based approach to extract indications of adverse drug reactions from the unstructured text in social media. User posts were collected from two different social media sites, mainly Twitter and DailyStrength. Two experts then independently annotated the posts as being an adverse drug reaction, an outcome of interest, an indication, and any other mention of symptoms. A supervised sequence labeling conditional random field (CRF) classifier was then used to extract the adverse drug reaction concepts from user posts.

Pharmacovigilance programs mostly rely on spontaneous reporting. The US Food and Drug Administration (FDA) adverse event reporting system (AERS) is such an example. There is, however, a need to exploit other nontraditional resources generated by patients on the Internet. These include social media and the logs of popular search engines, which might contain useful health information. White et al. [87] used such search logs for systematic signal reporting and also assessed the potential of using these search logs for the generation of early warnings about adverse drug reactions. On another note, Harpaz et al. [33] stated that due to certain limitations, the early stage of development, and the complexity of adverse drug reactions, it is still not possible to make clear-cut statements about the ultimate utility of the concept.

2.4 Data Analytics for Healthcare Management

Everyday healthcare institutions around the world capture information electronically ranging from hospital data to patient data. The amount of data generated is huge, and data analytics can play a pivotal part in enhancing human health and reducing costs in healthcare institutions [86]. One main example of application of data analytics in healthcare management is monitoring consequences like infections at surgical sites, rate of readmission of patients, patient waiting times, and room utilization in healthcare institutions. Another example is the identification of patients and staff [42, 90] across the hospital using radiofrequency identification (RFID). The data captured using RFID can be used to model patient flow in hospitals as well as care delivery processes and to monitor compliance with hospital policies [68]. For disease management, Moore [62] has proposed the use of data analytics in the monitoring of patients with chronic diseases like diabetes and hypertension, for reaching specific blood sugar and blood pressure targets using smart devices embedded with RFID which communicate with their providers. Other example applications include the monitoring of elderly patients to prevent falls [72].

2.5 Data Analytics for Clinical/Medical Research

Diseases evolve over years and the same drug cannot effectively combat a mutated form of the same disease. In the healthcare and pharmaceutical industries, there is a surge in structured and unstructured biomedical data generation from several sources, including the output from surveys and research experiments gathered by pharmaceutical companies, patients, healthcare providers, and social media [14]. In order to make good use of the data, it is very important to integrate all these heterogeneous data and devise smart algorithms to link them. Integration of disparate data can help in identifying and establishing the worthiness of new drug targets, support early identification of safety and efficacy issues, and improve patients' treatment.

Antibiotics are normally used to treat bacterial infections. However, there is increasingly a rise in resistance to antibiotics that once cured ailments across the spectrum, and this is turning into a potential source of prolonged illness known as "super-bugs," which can lead to disability and death. Antibiotic resistance (ABR) occurs either due to natural mutations of diseases or because of unregulated and frequent use of antibiotics, whereby the pathogens develop resistance against a specific antibiotic. Big data analytics can also be helpful in managing the problem of ABRs and in tracking the production to the distribution of antibiotics in the retail market. The data generated can also be used for developing statistical models to show the relationship between antibiotic consumption and associated resistance.

2.6 Data Analytics for Clinical Practice

Cancer is a very complex disease with only one tumor having billions of cells each capable of mutating. The genetic makeup of the cells needs to be captured to understand how they evolve and adapt. These data need to be acquired often to provide a clear picture of the disease, and hence, huge quantities of data are generated. The volume of published cancer research has increased significantly along with a rapid increase in patient-specific information such as genomic data. Clinical trials work by testing new treatments in small cohorts at initially, looking at how well the treatment works and to identify any adverse effects. If the results of a trial prove encouraging, it is expanded to include larger groups of people. However, in most of the cases, enrolled patients do not represent the general population of patients with cancer. Moreover, real-world patients with cancer tend to be older, sicker, and more ethnically diverse than the typical study patients who tend to be in good health – except for the cancer. CancerLinQ project (https://cancerlinq.org/) by American Society of Clinical Oncology (ASCO) aims at providing oncologists with data from a large number of patients, with lifelong learning opportunities and a common platform for all cancer stakeholders. CancerLinQ is making unparalleled use of massive amounts of medical records of patients with cancer to discover patterns and trends and to measure their care against that of similar products and recommended guidelines.

2.7 *Summary of Application Areas for Health Data Analytics*

In this section, a summary of the six application areas for health data analytics is given. The type of data analytics used, the sources of data, the benefits and challenges for each application area are highlighted in Table 1.

As it can be observed from Table 1, there is a wide range of data analytics techniques being used in the healthcare sector ranging from simple regression methods to sophisticated machine-learning algorithms. One of the main sources of data remains the electronic patient records as it contains valuable information about the diagnosis, disease, and treatment. However, it is noted that there is a growing number of online open databases for diseases, drugs, and genes that are becoming important sources of data for health data analytics. Even social media such as Facebook and Twitter are becoming important sources for health data. In all the application areas investigated, it is clear that health data analytics are increasingly being used and several benefits are being derived, such as the discovery of new knowledge, improving quality of services, early detection of disease outbreak, new drug development, and cost-effective operations. However, there are still a number of challenges to overcome in the area of health data analytics. The reliability and accuracy of the health data analytics need to be improved, and there are integration issues to be handled for these large volumes of data emanating from diverging sources.

3 Health Data Sources and Types

Health data used for analytics can be obtained from various sources as presented in Sect. 2. These sources can be categorized into internal sources (such as data from electronic patient records) and external sources (such as online open databases) [56]. Each of the sources contains different data types with varied level of details. Moreover, the reliability, accuracy, and accessibility of the data vary. In this section, the different sources of health data, categorized as internal and external, together with their characteristics, are detailed.

3.1 *Internal Data Sources*

Healthcare providers routinely produce a large amount of data as part of their normal operations. Since these data are produced from internal data management systems,

Table 1 Summary of application areas for health data analytics

Application area	Type of data analytics	Sources of data	Examples	Benefits	Challenges
Drug–disease association	Sequence, structure, pathway, text analysis, pattern learning, guilt-by-association, and statistical techniques	DrugBank database, CORUM database, FunDO database for diseases, PPI network data, MEDLINE, and DRUGDEX reference standard	Drug repurposing, novel drug use suggestions	Application of known drugs to treat new diseases (better than new costly drug discovery which is lengthy and risky)	Reliability and accuracy of the data analytics techniques
Disease outbreak detection and surveillance	Regression analysis, multiple linear regression models, keyword matching, and classifier	Geospatial databases, health records, climatic bulletins, social media	Prediction of H1N1 influenza epidemic, prediction of influenza-like disease	Effective prevention and control strategies, optimized resource allocation	Prediction model affected by other factors (e.g., biological and sociobehavioral attributes)
Pharmacovigilance	Supervised sequence labeling CRF classifier, self-controlled study designs	DailyStrength, Twitter, Internet search logs	Extract complex medical concepts with relatively high performance, ADR detection	Analysis of highly informal text in social media, ADR detection	Misspellings, abbreviations, and phrase construction irregularities
Healthcare management	Text analysis techniques, pattern matching, empirical/statistical analyses, and data mining techniques	Healthcare institutions records (patients, rooms allocation, surgical instruments RFID)	Optimal allocation of surgical instruments and rooms to patients	Better service provision, reduce waiting time for patients, and cost-effective operations	Large amount of data to analyze and multiple systems
Clinical/medical research	Predictive analytics, statistical modeling, survival analysis, genomic analytics and machine learning (including images)	Sequencing and gene expression data, drug data, protein and drug interaction data, clinical trial, medical images, and electronic patient record data	Diagnostic of diseases based on image analysis, prediction of new potential drugs, drug repositioning	Fast development of new drugs, determining most appropriate treatments using medical images	Integration of disparate sources of data, need of complex data mining algorithms specially for medical images
Clinical practice	Machine learning	Patient data, prescription data, data related to observational studies	Uncover patterns and trends in medical records of patients, connects and analyzes real-world data from different sources	Improving medication adherence and patient care	Mining heterogeneous information from different sources

they are considered as internal data sources. Although these data are collected routinely, they are seldom collected with data analysis in mind and, hence, need to be processed to be useful [82]. Data from internal sources often reside on separate information systems (ISs), and therefore, correlations between those data cannot be made and useful insights cannot be derived. Table 2 lists the various internal data sources identified, their data types, and characteristics.

Internal data are obtained from disparate information systems. The data are normally structured and reliable. Internal data contain mainly individual information about patients and information about internal processes of the healthcare institutions. Internal data are normally obtained routinely, and therefore, over time, there is a large volume of data which are regularly archived. Data analytics may be applied to internal sources in order to gain insights into diseases trends, effectiveness of treatments, cost-effectiveness of operations, proper management of resources, and performance of individual departments.

Table 2 Internal data sources

Internal data sources	Types of data	Characteristics of the data
Clinical data	Medical imaging, diagnosis, physician advice, and notes and medication nonadherence	Data routinely collected. No exact standard or format for storing physician notes or medical imaging. Medical imaging not always part of main IS (e.g., separate electronic files)
Electronic patient records (EPRs)/electronic medical records (EMRs)	Patient details, medical history, appointments, diagnosis, treatment details, and prescription requests	Patient data are routinely updated. Main source of patient information. Data stored using common medical record standards
Health monitoring data	Heartbeat, blood pressure, food intake, daily walking distance, and activities	Fine-grained individual data. Often not recorded on ISs. Volume of data is huge
Pharmacy information system	Data specific to drug prescription (e.g., patient name, ID, date, time and duration, drug dosage, administration route, frequency of intake), drug stock information, and expiry dates	Routinely updated. Data stored using common formats. Usually a separate IS from main healthcare provider system
Laboratory information management system (LIMS)	Data related to laboratory samples and results of different tests	Data usually stored in a separate IS from the healthcare provider's main system. Data are of technical nature and require expert knowledge for interpretation
Real-time location tracking systems	Location of assets and medical equipment, staff, and patients	Location information is collected routinely. Information collected from localization technologies (e.g., RFID, beacons, tags, and sensors)

3.2 External Data Sources

While internal sources of data reflect those data that are under the control of the healthcare institutions and are heavily regulated, external data are not generated in the healthcare environment. However, external data supplement internal data and can give useful insights into health-related conditions as well as help into making important decisions in this sector. Table 3 provides a nonexhaustive list of external data sources. The data types and characteristics are also described.

Increasingly, there is a huge amount of data that is being created at an unprecedented rate from external sources such as the Internet. There are a number of open online databases, from which very useful and reliable health data can be obtained. While data from internal sources are highly localized data from individual healthcare institutions, external sources mainly provide data from all over the world. However, data from external sources are mainly unstructured and are normally not stored using common standards and formats. Moreover, accuracy of the data cannot be guaranteed in most of the cases. Data analytics on external data sources can provide very important insights into disease outbreaks around the world, behavioral patterns of patients, and health events in different countries.

4 Health Data Analytics Tools and Platforms

As already elaborated in the previous sections, healthcare institutions are generating large volumes of data from various sources. Therefore, these healthcare organizations are seeking to harness the potential of their data. In recent years, numerous health data analytics tools and platforms have been developed. The capabilities of these tools and platforms are also evolving rapidly. They are increasingly being used for drug discovery, better decision support, predictive analysis, and evidence-based treatment. This section explores different tools and platforms (including a comparative analysis), types of data manipulated by these tools, data analytics techniques used, and the trend of these tools.

4.1 Tools and Platforms

In this section, state-of-the-art health data analytics tools and platforms are analyzed. These tools and platforms have been selected based on their popularity and adoption rate.

Table 3 External data sources

External data sources	Type of data	Characteristics of the data
Biomedical literature	Journals, online databases, books, proceedings, newsletters, technical reports, web pages, citation database for peer-reviewed literature from life sciences, health sciences, and so on	Embase by Elsevier is subscription based. Medline by US National Library of Medicine (NIH) is open access. Both are regularly updated. Large volume of data from multiple sources
Clinical Trials (ClinicalTrials.gov)	Experiments and observations done in clinical research	ClinicalTrials.gov (US) database: publicly and privately supported clinical studies conducted on participants from many countries, open access, and regularly updated
DRUGDEX	Reference standard for drug use review	Covers FDA-approved and investigational prescription and nonprescription drugs, as well as non-US preparations. DRUGDEX updated weekly and in each Metathesaurus release as part of RxNorm (standardized nomenclature for clinical drugs for humans, produced by US National Library of Medicine)
Spontaneous reporting systems (SRSs). E.g., US Food and Drug Administration (FDA) Adverse Event reporting System (AERS)	Reports of suspected adverse drug reactions (ADRs) that typically capture suspected and concomitant drugs, indications, suspected events, and limited demographic information	Raw data consisting of individual case safety reports extracted from the FAERS database are available to public through creation of relational databases. AERS has millions of drug–event combinations.
Health social networking sites (SERMO, PatientsLikeMe, and MedHelp)	Patients' profiles, health conditions, medications. PatientsLikeMe has an openness policy (users can agree to share all their health data). Collects and stores two types of data from users: shared data (e.g., biography, conditions, treatments, symptoms, outcomes, laboratory results) and restricted data	Data are mostly unstructured from dispersed geographical areas. Large volume of data posted daily. Accuracy of data cannot be guaranteed
Generic social networks such as Twitter	Individual posts about health, medications, doctors' visits	Data are unstructured. Requires data mining and natural language processing for capturing health-related information
Users' web search logs	Time stamp information, location of user, disease information, drug information, behavioral data, health-related symptoms, and frequency of search (to indicate seriousness of health condition)	Can be categorized into two main types, namely: search logs (user raises query to a search engine and selectively clicks on the search results returned) and browse logs (user visits a web page other than a search result page)
Large repositories of geospatial and health data, e.g., HealthMap	Latest real-time disease outbreak information, H1N1 prevalence, and climatic data accumulated from previous years	Real-time data from large geographical areas. Reliable disease outbreak information. Regularly updated.

4.1.1 Philips HealthSuite Digital Platform

Philips HealthSuite Digital Platform [37] is an open platform available on the cloud, which aims at collecting, integrating, and analyzing clinical and other data from different internal and external sources, including sensors and devices. It uses Amazon Web Services (AWS) to securely store about 15 PB of patient data that can be analyzed to give valuable insights to patients and healthcare specialists. The platform uses contextual predictive analytics, decision support algorithms, and machine learning to analyze data to provide useful insights on patients' health.

4.1.2 EMIF (European Medical Information Framework) Platform

The European Medical Information Framework [24] Platform is a common information platform that links up clinical information of about 50 million patients around Europe. It integrates and harmonizes data from various data sources, such as population-based registries, national registries, biobanks, and hospital-based registries, to allow reuse of data. Additionally, the platform makes data available for browsing and exploitation by the user, analyses data, and provides support for data visualization. A number of techniques such as Spearman correlation analysis, hierarchical clustering, K-means clustering, principal component analysis, linear regression, survival analysis, and statistical methods have been used by the platform to analyze data.

4.1.3 Hortonworks

Hortonworks [36] is bringing a revolution, which aims at transforming big data analytics in healthcare and medicine by making healthcare data available and accessible. It offers open-source connected data platforms, which are based on Apache Hadoop and Apache NiFi. The potential benefits of Hortonworks Data Platform (HDP) include the use of accumulated information over time for healthcare predictive analytics with feeding algorithms predicting the probability of an emergency earlier than it could be detected with traditional bedside visits.

4.1.4 IBM Analytics

IBM Analytics (https://www.ibm.com/analytics/us/en/) solutions enable healthcare organizations to connect information from disparate sources into a single-trusted view with the eventual delivery of insights into the data. Big data analytics are mainly possible due to the accuracy and performance of the IBM InfoSphere Master Data Management. The Watson Health Cloud collates large amounts of medical data into a centralized cloud-based repository. The IBM platform brings advanced analytics, by reading 200 million pages of text in 3 s, which help turn the raw and disparate data into vital insights that are beneficial to the patients.

4.1.5 Sagitec HealHub Digital DataOps Platform for Healthcare and Life Sciences

HealHub [74] is a DataOps platform for healthcare organizations. It enables real-time collaboration between data analysts and the teams involved in the day-to-day running of the hospital. It restructures quality information from disparate sources to provide a better understanding and for increased revenue. When data are ingested into the platform, the information is automatically cleaned and scrutinized according to protocols, standards, and master data. The system accepts data in all formats. It can capture data from health and medical devices, which is then integrated into the system. In addition to sharing operational data with internal analysts, HealHub also integrates with external systems, such as customer relationship management, clinical data management, and data warehouses; with online libraries and websites like PubMed and ClinicalTrials.gov; with ratings from doctors; with reviews from patients and Facebook/Twitter; and with other data channels. It provides visualization capabilities as well as what-if analysis.

4.1.6 Sisense

Sisense Business Intelligence [35] software provides hospitals and other healthcare organizations with a robust business analytics solution to structure, analyze, and visualize complex data. It supports healthcare organizations in providing enhanced services to patients or clients by closely monitoring various metrics and key performance indicators (KPIs), with timely and accurate data which can be medical, financial, and administrative. The dashboards in the software provide detailed information to track individual performance. Sisense has robust Extract, Transform and Load (ETL) and data modeling capabilities, advanced analytics and statistics, and an intuitive user interface for creating dashboards, generating reports, and providing visualizations.

4.1.7 EHDViz: Clinical Dashboard Development Using Open-Source Technologies

EHDViz [5] is a clinical dashboard development framework toolkit for generating web-based, real-time clinical dashboards for visualizing dissimilar biomedical, healthcare, and well-being data. It is developed as an extensible toolkit that uses R packages for data management, normalization, and producing high-quality visual representations over the web using R/Shiny web server architecture. EHDViz can integrate data from different sources, such as biomedical and healthcare data visualization for a combined health assessment. EHDViz can also be used as a toolkit to emulate EHR environment to increase simulation-based learning. Hospitals and healthcare systems are emerging as learning health systems (LHS), and thus, data capture, smart clinical dashboards, and adjustive visual analytics can play an important role in managing the patient population.

4.1.8 ICDA: A Platform for Intelligent Care Delivery Analytics

Intelligent Care Delivery Analytics platform (ICDA) [29] conducts risk assessment analytics by processing large collections of ever-changing electronic medical data to identify high-risk patients, in view of improving patient outcomes and managing costs. ICDA works by integrating large volumes of data into an integrated data model, then effecting a collection of analytics that identify at-risk patients. It also provides an interactional environment through which users can access and critically evaluate the analytics results. Additionally, ICDA provides APIs via which analytics results can be extracted to interface in external applications.

4.1.9 EMC Healthcare Analytics

Dell EMC Services provides healthcare institutions with the capabilities to assess, prove, and deploy data analytics use cases, as well as the technology, capabilities, and platform required to support them. The EMC Healthcare Analytics Solution [23] allows the aggregation, analysis, and visualization of data to make informed patient care and operational decisions. Dell [19] describes some sample use cases such as reduction of hospital readmission and improving patient care.

4.1.10 OpenML Platform

Open Machine Learning is a collaboration platform, which is designed to organize datasets, machine-learning workflows, models, and their evaluations [84]. Scientists and researchers can share machine-learning data sets, code, and experiments for more effective, large-scale, real-time collaboration. Its easy-to-use APIs help to automate many tedious research tasks. It is also easily integrated into several machine-learning tools. Researchers can share their latest results, thus speeding up research through combined and linked results. OpenML can be accessed through some interfaces, like R and WEKA. Some researchers have proposed extensions of the OpenML platform in applications like healthcare predictive analytics, which allow secure sharing of data, workflows, and evaluations.

4.1.11 Generic Tools and Platforms: Google Health, Apple HealthBook, and Microsoft HealthVault

Information captured through new types of sensors and systems is integrated in databases facilitating data mining and revealing new insights. Several companies have created health toolkits, such as Google Health [81], Google Fit, Samsung S.A.M.I, Microsoft HealthVault, and Apple HealthBook [43]. Google Health can be used to automatically import health records, test results, and prescription history. Other services include appointment scheduling, prescription refills, as well as wellness tools. Microsoft HealthVault, also known as the "PayPal for health information," consists of two main products, namely an electronic repository for health data and a specialized search engine for health information. Comparing Microsoft HealthVault and Google Health showed that they were both extendible platforms where functionalities vary on configurations used as well as affordability toward services. No deductions could be made as to which of these two systems are better in terms of functionality. Besides, Microsoft HealthVault can only be used in the United States [81].

According to Schulz and Neumann [77], applications such as Apple HealthBook can only be considered as a data diary (e.g., for blood pressure) since there are no certified sensors or secure data transmissions are available [77, 81]. Moreover, the plausibility of data provided through these media still has to be validated. "Full-profile search" was also missing in Microsoft HealthVault and Google Health [81]. The study by Jovanov [43] reveals that health data analytics tools have even penetrated the smartwatch industry. In fact, these smartwatches, being the most popular wearable sensors, provide for continuous measurement of physiological parameters including heart rate and temperature. Being in close contact with the skin, these devices can be used to collect biological, environmental, and behavioral information about user activities and potentially provide for user identification.

4.2 Comparative Analysis of Health Data Analytics Tools and Platforms

In this section, a comparative analysis of the reviewed tools and platforms is carried out. The objective is to identify the commonalities and differences between those tools and platforms. This comparative analysis will also allow to uncover the trends in data analytics techniques used in latest tools and platforms. The comparison is based on four main criteria: features and capabilities, data analytics techniques used, types of data used, and the areas of application.

4.2.1 Features and Capabilities

Table 4 shows a comparison between the different tools and platforms described in this section, in terms of features and capabilities.

Based on the feature comparison in Table 4, all the health data analytics tools compared integrate data from various sources, derive new insights from the data, and provide data visualization capabilities. Additionally, most of the tools provide some predictive analysis capabilities that allow healthcare institutions to predict important healthcare parameters and, thus, allow timely decision-making. Some tools are not flexible and do not provide APIs for further development of the data analytics.

4.2.2 Types of Data

In this section, the data types and data sources used by the different tools and platforms are surveyed, and the results are displayed in Table 5. The majority of data analytics tools process semistructured and unstructured data. Moreover, most of the tools integrate data from both internal and external sources.

4.2.3 Areas of Application

Six main areas of application for data analytics in healthcare were identified in Sect. 2. In this section, the application areas of each of the tools and platforms are surveyed and summarized in Table 6. It can be observed that all health data analytics tools are used for healthcare management, that is, analyzing daily operations of the healthcare institution. Moreover, data analytics tools and platforms are increasingly being used for clinical practice in order to improve patient care. It can be seen that IBM Analytics cover the whole spectrum of the application areas and can be used for pharmacovigilance and medical research among others.

4.2.4 Data Analytics Techniques

The data analytics techniques used by each of the tools and platforms are summarized in Table 7. Most of the tools use predictive analytics and machine-learning algorithms in order to derive new insights from the health data. Most tools perform common data analytics tasks such as classification, clustering, and correlation. Moreover, artificial neural networks and decision trees are increasingly being used for data analytics.

Table 4 Feature comparison of health data analytics tools and platforms

Tools and platforms	Data integration (various sources)	Data discovery and analysis	Data visualization/dashboard	Scalable and flexible	Provision of APIs	Predictive analysis/machine learning
Philips HealthSuite Digital Platform	✓	✓	✓	✓	✓	✓
EMIF Platform	✓	✓	✓			✓
Hortonworks	✓	✓	✓			✓
IBM Analytics	✓	✓	✓	✓	✓	✓
HealHub	✓	✓	✓			✓
Sisense	✓	✓	✓	✓	✓	
EHDViz	✓	✓	✓	✓	✓	
Intelligent Care Delivery Analytics platform (ICDA)	✓	✓	✓	✓	✓	✓
EMC Healthcare Analytics	✓	✓	✓	✓		✓
OpenML platform	✓	✓	✓	✓	✓	✓

Table 5 Comparison of data sources and data types

Tools and platforms	Data type Structured	Semistructured	Unstructured	Data source Internal	External
Philips HealthSuite Digital Platform	✓	✓	✓	✓	✓
EMIF Platform	✓	✓	✓	✓	✓
Hortonworks	✓	✓	✓	✓	✓
IBM Analytics	✓	✓	✓	✓	✓
HealHub	✓			✓	✓
Sisense	✓		✓	✓	✓
EHDViz	✓	✓	✓	✓	✓
Intelligent Care Delivery Analytics platform (ICDA)	✓	✓		✓	
EMC Healthcare Analytics	✓	✓	✓	✓	✓
OpenML platform	✓	✓	✓	✓	✓

Table 6 Comparison of areas of application

Tools and platforms	Disease outbreak and surveillance	Drug–disease association	Pharmaco vigilance	Healthcare management	Clinical/ medical research	Clinical practice
Philips HealthSuite Digital Platform				✓	✓	✓
EMIF Platform				✓		
Hortonworks		✓		✓	✓	
IBM Analytics	✓	✓	✓	✓	✓	✓
HealHub		✓	✓	✓		✓
Sisense				✓		✓
EHDViz				✓		✓
Intelligent Care Delivery Analytics platform (ICDA)				✓		✓
EMC Healthcare Analytics		✓		✓	✓	✓
OpenML platform				✓		✓

4.3 Trends in Health Data Analytics Tools and Platforms

Based on the tools and platforms reviewed in this section, it can be deduced that most health data analytics tools and platforms provide a complete set of services to perform the analytics process, which includes data warehousing, batch processing,

Table 7 Comparison of data analytics techniques

Tools and platforms	Data analytics techniques
Philips HealthSuite Digital Platform	Contextual predictive analytics, decision support algorithms, and machine learning
EMIF Platform	Spearman correlation analysis, hierarchical clustering, K-means clustering, principal component analysis, linear regression, survival analysis, statistical methods
Hortonworks	Combination of Hadoop predictive analytics with a number of data science and iterative machine-learning techniques
IBM Analytics	Predictive analytic solutions comprising of techniques such as artificial neural networks and decision trees
HealHub	Machine-learning techniques
Sisense	Machine-learning techniques, Bots, Natural Language Processing
EMC Healthcare Analytics	Predictive analysis, decision trees, time series, neural networks
OpenML platform	Classification, regression, clustering, data stream classification, learning curve analysis, survival analysis
EHDViz	Predictive analysis, more specifically machine learning and risk algorithms
Intelligent Care Delivery Analytics platform (ICDA)	Predictive analysis (risk assessment analytics)

business intelligence, data workflow orchestration, and machine learning. These tools allow the integration of data from various sources and of different types efficiently without having to learn complex data processing platform such as Hadoop. Indeed, health data analytics tools are integrating data from a wide range of health information systems, including health management systems and electronic health records systems in order to create a unified data model for healthcare institutions. The unified data model is allowing healthcare administrators to have 360° view of the institution.

Moreover, data analytics platforms are increasingly being based on open architectures which allow them to integrate existing infrastructure easily. The open architectures significantly improve the scalability and flexibility of the platforms and support the broadest spectrum of data sources. Much emphasis is also being placed on ensuring high level of security and manageability of data analytics. Health data analytics tools are also focusing on the provision of advanced visualization and dashboards that allow analysts to effortlessly visualize and analyze data from multiple angles without having to depend on developers. The visualization components are providing important features such as interactive analysis, zooming, and drill through capabilities that are allowing greater insight into the health data.

Health data analytics tools and platforms are being used to derive completely new insights that were not possible before. These tools are used to determine adherence to medication, identify high-risk patients, and determine possibility of readmission and morbidity. The use of health data analytics tools is also helping healthcare institutions to predict disease outbreaks, reduce costs of operation, and avoid preventable diseases [71].

5 Use Cases for Health Data Analytics

Healthcare system in many developed countries like the United States is rapidly adopting electronic health records, which is generating a large quantity of electronic health data. Research in health data analytics is also emerging as a new trend in view of improving healthcare service provision. As a result of this advancement, there are unprecedented opportunities to use big data to reduce the costs of healthcare in the United States as well as other developed countries. The following subsections elaborate on some use cases in the area of health data analytics.

5.1 *Using Data Analytics to Predict Number of Patients in Hospitals*

Hospitals do not have the same number of admissions of patients at all times of the day. One of the major problems consists of deploying the right number of staff at specific times. In order to ensure a very good customer service, ideally hospitals should deploy maximum personnel at peak times and less at other times so as to avoid unnecessary staffing costs. According to a Forbes article (2016), a few hospitals in Paris are using big data analytics to solve the problem of staffing, in view of optimizing costs as well as improving their customer service. These hospitals, which are part of the *Assistance Publique-Hôpitaux de Paris*, have been mining data from a number of internal and external sources – including 10-years' worth of hospital admission records, in order to predict the number of patients at each hospital at given times of the day. They use machine learning to build a predictive model using the past data and predict admission rates at different times of the day. The resulting tool is a web-based application (according to Forbes) that helps healthcare professionals to forecast visits and admission rates for a period of 15 days ahead and plan deployment of staff accordingly. When higher numbers of patients are expected to visit the hospitals, more staff are scheduled in order to reduce waiting times.

5.2 *Using Data Analytics to Predict Sepsis Risk and Mortality*

Researchers at the University of California Davis have found that routine information such as blood pressure, respiratory rate, temperature, and white blood cell count found in the electronic health records (EHRs) of admitted patients can be used to predict the early stages of sepsis (an immune system response to infection that can damage organs and cause permanent physical and mental disabilities), a leading cause of death and hospitalization in the United States. Moreover, it was also determined that only three parameters, namely lactate level, blood pressure, and respiratory rate, can also predict whether a patient is likely to die from the disease. They used electronic health records pertaining to 741 adult patients (at the University of California Davis Health System) who met at least two systemic inflammatory response syndrome criteria, to associate patients' vital signs and white blood cell count (WBC) to sepsis occurrence and mortality [31]. Machine-learning algorithms such as generative and discriminative classification (naïve Bayes, support vector machines, Gaussian mixture models, and hidden Markov models) were used to integrate disparate patient data and create a predictive tool for the inference of lactate level and mortality risk.

5.3 Using Data Analytics to Predict Readmissions in Case of Heart Failures

In this study, Bayati et al. [8] have used data from 1172 hospital visits for heart failure to construct a classifier to predict the chance that a patient has of being rehospitalized within 30 days of discharge. The classifier was modeled using data from 793 hospital visits and was tested using data from 379 additional hospital visits. All data were de-identified patient data from the EHR of a hospital. The authors were able to use their model to predict the readmission rate. For this case study, the mean cost of readmissions was 13,679 with a standard error of 1214. The authors claim that if their proposed method would be used, there will be an 18.2% reduction in rehospitalizations and a 3.8% reduction in costs. They also found that the construction and use of predictive models that are custom tailored to populations have a higher accuracy than applications based on the same general rules across different hospitals. The use of local data for the construction and testing of the model also ensured highest predictive performance. Since the readmission patterns and prevalence vary for different hospitals and regions, a model needs to take these into account.

5.4 Using Data Analytics to Predict Decompensation

Decompensation refers to the failure of an organ (especially the liver or heart) to compensate for the functional overload resulting from disease. Often before decompensation, there is a period in which the physiological data of a patient can be used to determine whether she/he is at risk for decompensating [7]. Patients who are critically ill are placed in intensive care units (ICUs) so that they can be closely monitored. With the advancement in technologies, there are also facilities to monitor patients for risk of decompensation in general care units, nursing homes, as well as homes. Some of these technologies were available for many years, such as electrocardiographic monitoring and oxygen monitoring, while others are newer, such as end-tidal monitoring and monitors that allow detection of whether or not a patient is moving. These technologies, however, face the problem of noise which can raise false alarms. Fortunately, monitors that can compare multiple data streams are also becoming available, and coupled with analytics in the background, it can be determined whether a signal is valid. An example of such a monitor is one that is placed under the mattress of a patient and collects data about the patient's respiratory rate and pulse as well whether the patient is moving or not [12].

6 Challenges in Health Data Analytics

In this section, the challenges for health data analytics are explored. In spite of the several tools and platforms available for health data analytics, there remain a number of challenges that need to be addressed so that the full potential of the available data in healthcare institutions can be harnessed.

6.1 Quality of Health Data

In the context of healthcare, challenges with respect to data quality include inconsistency, irregularity (noise, incompleteness), low standard, unreliability (both primary and secondary data), inaccuracy (or erroneous inclusions), insufficient detail, large volumes, and variety of data and high-dimensionality [50, 69]. As identified by Powell et al. [69], the success depends on "high-quality" data provided by different sources. Over the past two decades, data quality has attracted much research including Total Data Quality Management (TDQM). Several challenges related to quality of data and information include both technical challenges, such as data integration from multiple sources, and nontechnical challenges, such as ensuring the right recipient receives the right data/information in the right format at the right place and at the right time [57]. The quality of data has a direct impact on the effectiveness of the operations being performed on data, such as capture, storage, searching, sharing, analysis, and visualization [51]. Low-quality data, such as missing data, inaccurate, or incomplete data, disturb the proper functioning of the different processes. Additionally, irregularity of data incorporates both noisiness and incompleteness. Noise is often incorporated in data itself where data cannot be interpreted and are meaningless. Healthcare data are often incomplete as not all data are being recorded properly [50].

The high volume of data generated from internal and external sources gives rise to a large number of data elements stored in them, leading to high dimensionality when dividing the healthcare data [50]. Processes like data mining, therefore, become difficult in healthcare applications due to corrupted, inconsistent, missing, or unstandardized data. For example, information might be recorded in different formats in varying data sources [10, 22, 47]. Powell et al. [69] provided two scenarios showing data quality issues in healthcare. In the first scenario, there was clinical evidence that diagnosis for only a few diseases was detected due to incomplete data. In the second scenario, accurate data were not provided in all cases; for example, there were accurate data for tumor registries but inaccurate data for outpatient treatment [69].

In the work by Berman et al. [10] related to the Protein Data Bank, new data are integrated in the existing data archive. It is, therefore, important that standard formats are used. This work relates to issues related to standardization and integration. Two approaches have been used in this work to address this issue: file-by-file analysis and batch processing [10]. Edwards et al. [22] further extend on the issue related to standardization in terms of representation of data. To allow efficient data storage, it is imperative that healthcare data be in a standardized format and make use of standard vocabularies for purposes of concept representation and communication [22].

6.2 Infrastructural Challenges

A fault-tolerant and scalable physical infrastructure is vital for the operation of a big data analytics project. This infrastructure is usually based on a distributed model, where the data can be physically stored in various locations connected through high-speed networks [54]. Demchenko et al. [20] identified the following infrastructure requirements for emerging big data analytics projects:

- Support the running of long experiments with large data volumes produced at high speed
- Confidentiality, integrity, and accountability of data
- Multitier interlinked data distribution and replication
- On-demand infrastructure provisioning
- Trusted environment for data storage and processing

AbuKhousa et al. [1] identified several challenges faced by an e-health cloud infrastructure which inherits the challenges of processing sensitive medical data of health information systems and cloud computing. As most health data analytics platforms use cloud computing technologies, the underlying infrastructures for these platforms will, therefore, face challenges such as availability, data management, scalability, flexibility, interoperability, security, and privacy.

6.3 Data Analytic Algorithms and Approaches

As discussed in Sect. 2, data analytics can be applied to a variety of different areas in healthcare, namely drug–drug association, disease outbreak detection and surveillance, pharmacovigilance, healthcare management, clinical/medical research, and clinical practice. However, these areas are very diverse with some directly and others indirectly related to the medical field. Hence, different approaches are used by each.

Newer methods, such as data mining, Bayesian statistics, optimization modeling, social network analysis, and agent-based simulation, have been combined to the previously well-established techniques, such as biostatistics and epidemiologic analysis, causal modeling, and Monte Carlo and discrete-event simulation [86].

In their review paper on data mining of big data in health analytics, Herland et al. [38] have categorized health informatics into different levels, and for each of these, they have provided a set of tools that can be used for the analysis [4, 13, 25, 32, 65, 75, 89]. Belle et al. [9] have also classified big data in medicine under different areas and have provided a set of tools for each. They have proposed the MapReduce framework, support vector machines (SVM), and wavelet analysis. As it can be observed, several methods have been proposed since the data analytics methods are very problem specific, and there is a need to combine data from disparate sources, for example, combining patient data, physician profile, and environmental variable features, to improve the management of risk-stratified patients to receive the most appropriate care [55].

The criteria for the evaluation of a big data analytics platform for healthcare include availability, continuity, ease of use, scalability, ability to manipulate at different levels of granularity, privacy and security enablement, and quality assurance [71]. Since healthcare data are rarely standardized, one major challenge of data analytics algorithms in healthcare is the need to deal with incompatible formats of highly fragmented data. Moreover, unstructured textual data like clinical notes can be difficult to understand in the right context [70]. Real-time big data analytics are key requirements for healthcare. Much research efforts are geared toward continuous data acquisition and cleansing and addressing the lag between data collection and processing [71].

Some data mining methods may yield good performance for one type of problems, while other methods are more appropriate for other types of problems [58]. To decide on an appropriate algorithm, it is important to understand the appropriateness of the method and its performance for a specific problem. The chosen algorithm needs to be extensively evaluated and experimented before applying it in a real medical system. There is no such algorithm which is best suited for all problems in the medical domain, since each algorithm is applicable to specific problems [58].

6.4 Security and Privacy Challenges

Health data analytics have enormous potential to reduce cost of operations, improve clinical practice, and improve the quality of the overall healthcare. However, this opportunity raises a series of security, privacy, ethical, and legal challenges [17]. Hence, we have the concept of Protected Health Information (PHI) which refers to individually identifiable health information transmitted by electronic media, maintained in electronic media, or transmitted or maintained in any other form or medium [63].

Security and privacy concerns related to health data that are accessed through remote locations such as cloud present a significant barrier to the adoption of big data analytics in the health sector. Some institutions have been experimenting with private clouds, which are, however, limited in terms of scalability. The concept of Cloud Service Providers has emerged as potential hosting spaces, and they provide additional security measures on top of the cloud service. In the long run, they can be better than private clouds as they can invest in measures against hacking and cyberattacks.

Often, identifiable health information cannot be disclosed without patient consent. In many cases, it is not practical to obtain individual patient consent because of the very large data sets involved. One approach to facilitate the use of health information for the purposes of predictive analytics is to de-identify data. The Privacy Rule of Health Insurance Portability and Accountability Act (HIPAA) of 1996 establishes minimum Federal standards for protecting the privacy of individually identifiable health information. The Privacy Rule establishes conditions under which covered entities can provide access to and use of PHI. The Privacy Rule permits covered entities to use and disclose data that have been de-identified without obtaining an authorization and without further restrictions on use or disclosure because de-identified data are no longer PHI and, therefore, are not subject to the Privacy Rule (https://www.hhs.gov/hipaa/for-professionals/privacy/index.html).

Predictive analytics in medicine can effectively help in identifying patients at high or low risk for serious complications, thus optimally allocating scarce clinical resources. Predictive analytics models make treatment recommendations that are designed to improve overall health outcomes among all patients, and these recommendations may conflict with physicians' ethical obligations to act in the best interests of individual patients. Health practitioners should be able to override computerized recommendations when they have sound reasons to believe that the predictive model did not capture some considerations. Such exceptions would allow treating physicians to play their traditional role as patient advocates.

The potential of health data analytics cannot be realized without collecting and analyzing vast amounts of heterogeneous data [49]. Moreover, the data involved may not all be owned or controlled by a single organization or institution. Sharing data might enable various organizations to identify and fill the gaps across their common coverage area, such as patterns of inappropriate use of antibiotic medications. However, competing healthcare institutions that treat common patients may be reluctant to share relevant data, fearing that others could use its data for their advantage.

7 Future Directions

Despite the enormous progress made in the field of health data analytics, there are still several challenges that need to be addressed so that healthcare institutions can leverage the full benefits of their accumulated data. Based on the studies reviewed in this chapter, it can be clearly observed that health data analytics has not yet matured and there is still a need for focused research. Several research avenues for health data analytics are, therefore, defined in this section in an attempt to provide researchers and practitioners with future directions in this domain.

7.1 Health Internet of Things

The Internet of Things is having a tremendous impact in the healthcare sector with 40% of IoT-related technology forecasted to be health related by 2020 [21]. Wearable IoT integrates wearable sensors that monitor human factors such as health, wellness, behaviors, and other useful information that have the potential to allow individuals to better manage their health [39]. However, a number of technical challenges need to be addressed in order for the Wearable Internet of Things (WIoT) to achieve multidimensional success. These include the generation of a flexible and robust framework for networking, storage, computation, and visualization while designing healthcare solutions which are clinically accepted and operational.

Hassanalieragh et al. [34] identified some challenges with respect to analytics with wearable sensor data. First, they argued that analytics on data from the sensors is challenging since there is a need to cope with streaming data with a number of missing values and varying dimensions and semantics of data. Although the machine-learning algorithms have matured, they are, however, not designed to deal with time-varying feature dimensionality, and incomplete data vectors, which if not catered for properly, can affect classification performance. Second, while there is an enormous amount of sensor data, these are completely untagged and need to be matched with the physician diagnoses. However, with the high load of physicians, this activity is nearly unfeasible. Third, sensor data and historical information available in clinical records are very different in nature. This heterogeneity is a major challenge for conventional machine-learning approaches that work primarily with homogeneous data.

According to Riazul Islam et al. [73], there is a need for a customized computing platform for the IoT-based healthcare systems. Moreover, they argue that libraries, for example, a specific class of disease-oriented libraries, and appropriate frameworks should be customized so that software developers for healthcare systems can make effective use of given classes, documents, templates, codes, and other useful data. A number of other limitations exist in the available platforms supporting data analytics for healthcare [71], and these should be addressed in order to ensure a large-scale adoption. Data analytics in healthcare platforms should be menu-driven, transparent, and user-friendly. Algorithms, models, and methods should be dynamically and easily accessible through drop-down menus. Since real-time analytics are vital in healthcare, the time lag between data collection and processing should be addressed and continuous data acquisition and cleansing should be favored. Other issues like ownership of data, incompatible and fragmented data, and standardization of data should also be addressed.

7.2 Precision Medicine

Precision medicine is an emerging approach for individualizing the practice of medicine [61] based on information about health and disease at the molecular, cellular, and organ levels [26]. It is also known as personalized, predictive, preventive, and participatory (P4) medicine [40]. Two major computational challenges [28] for precision medicine include the developing algorithms for, first, individual phenotyping, which refers to the annotation of patient records with disease conditions, and, second, the integration of electronic health records data with omics data in order to better understand the disease and its treatments. These challenges are mainly due to the noisiness, incompleteness, heterogeneity, and different formats of the electronic health records and genomic data.

There are a number of significant computational advances in precision medicine such as cloud-based toolkits and workflow platforms that provide for high-throughput processing and analysis of omics data [2]. Additionally, graphics processing units (GPUs), which provide faster computations compared to central processing units (CPUs), are being exploited to handle the exponentially growing data. However, given the heterogeneous nature of omics data and challenges in communications and synchronizations, future works are required to develop parallelization algorithms. Moreover, there is a need to create lightweight programming environments with a number of cloud-based utilities and to validate the reliability of the platforms before a large-scale adoption can be envisaged.

7.3 Data Analytics for Evidence-Based Medicine and Drug Repurposing

Evidence-based medicine (EBM) has, for a long time, been very successful in clinical decision-making in many countries [30]. It is the process of integrating individual clinical expertise, patients' preferences, and evidence from randomized clinical trials and research findings. The ethical care of the patient is a top priority for the success of EBM, and patients now demand better individualized evidence, presented and explained in a personalized way. However, in recent years, the implementation of evidence-based practice has become a major challenge, one of the reasons being that the volume of evidence is growing at an unexpected rate. Therefore, much research effort should be geared toward health data analytics techniques to extract the best relevant evidence from the ever-increasing volumes of clinical guidelines and research studies, for patients on a case-to-case basis, for better clinical decision-making. A major challenge for these techniques is to deal with data sets which are not only large but also usually complex, heterogeneous, high-dimensional, time-varying, noisy, and weakly structured or unstructured.

Drug repurposing, also known as drug repositioning, is the use of existing drugs to treat new diseases. In the past decades, this has become an increasingly important strategy for new drug discovery, given the very high cost and failure rate of new drug development [88]. With the increasing growth of drug-related data, new computational strategies and techniques for drug repositioning are emerging [52]. These methods integrate data from various sources. However, Li et al. [52] identify a few issues which need to be addressed. Factors like missing data, data bias, and technical limitations of computational methods need to be considered when applying the computational models into practical use. Moreover, the performance evaluation of the proposed models is quite hard, because of the lack of structured standard for drug repositioning. Gligorijević et al. [28] suggest the utilization of Topological Data Analysis methods to deal with the high dimensionality (volume) of biomedical data, the so-called anytime-algorithms to deal with the velocity of biomedical data, and Matrix Factorization-based methods to deal with the heterogeneity (variety) of biomedical data.

7.4 Improved Predictive Analytics

Predictive analytics have been used in different areas of healthcare as discussed in Sect. 2. According to Raghupathi and Raghupathi [71], big data analytics and applications in the field of healthcare are still at a nascent stage of development. However, they claim that this area is developing rapidly with the advances of new platforms and tools. There are still some healthcare areas where the full potential of predictive and big data analytics have not been exploited. Some of the areas are described further.

7.4.1 Identification and Management of High-Cost and High-Risk Patients

According to Bates et al. [7], the outcomes of predictive modeling and analytics currently come mostly from low-risk patients. It is, therefore, of utmost importance to apply concepts of predictive analytics for high-risk and high-cost patients' identification and management. Parikh et al. [67] also support this point by stating that existing systems have not progressed much in the field of risk stratification. As a future direction, the authors suggest that predictions models be integrated with clinical systems to assist health professionals to make decisions.

7.4.2 Readmissions of Patients

Readmissions of patients have been costly to healthcare institutions where patients get admitted again after receiving care at the particular institutions. Some work has already been done in the field [3, 8, 11, 44, 78, 92]. Bates et al. [7] identify readmissions of patients as a use case where additional research is required using analytics and big data to perform readmission risk prediction.

7.4.3 Triage of Patients

Another use case identified by Bates et al. [7] where there is potential for further research is triage, where risk complications are identified when a patient first visits a hospital. Few models exist in this domain. Sun et al. [79] present a predictive model where the need for a patient to be admitted in the emergency department (ED) is determined at the time of triage. However, the use of big data analytics has not been exploited by the model.

7.4.4 Treatment Optimization for Diseases Affecting Multiple Organ Systems

Bates et al. [7] argue that current approaches of Big Data Analytics focus more on use of analytics for patients with one condition rather than multiple conditions. Shams et al. [78] have proposed a hybrid prediction model that integrates classification and timing-based analytics models for patients suffering from four different conditions, though it has certain limitations. They plan to improve on their model in future.

7.5 Real-Time Health Data Analytics

While a number of real-time and stream analytics applications exist in healthcare [45, 48, 80, 85, 91], not all of them have been tested in real clinical conditions or with real data sets [45, 85]. Zhang et al. [91] implemented a Health Data Stream Analytics System which was still under evaluation at the time of publication. Future directions included proposed "formal clinical trials to evaluate the performance." Further research in this area is still required. In the context of real-time monitoring of patients and resources, it is important to integrate predictions with clinical systems "to help physicians and other healthcare professionals make decisions and track real-time quality" [67]. Future research directions in real-time healthcare analytics should not be disjoint from the challenges related to large volume streams, continuous data collection, cleansing, processing, near-real-time responsiveness, accuracy, and the impact on performance and scalability [1, 71, 85]. Moreover, although the technical aspects regarding the provision of real-time healthcare analytics might be satisfied, poor quality of data can still be a barrier. For instance, in some cases, this can be hindered by the unreliability, incompleteness, or unavailability of the relevant data. Therefore, data quality can be added to the research agenda related to real-time healthcare analytics.

7.6 New Data Analytics for Clinical and Medical Research

In the field of clinical and medical research, new data analytics techniques are being used extensively for the better understanding of diseases like cancer, heart diseases, and others. canSAR (http://cansar.icr.ac.uk) is a cancer-focused knowledge base, which has been developed to support cancer research and translate findings from fundamental research into medical practice and meaningful health outcomes as well as in drug discovery. canSAR integrates data from diverse sources like data on the genome, data on proteins, pharmacological data, drug and chemical data, as well as structural biology, and protein networks data. As on December 11, 2018, canSAR contains 556,825 proteins from 2148 organisms, data for 12,172 cancer cell and nontransformed cell line models, 6,367,677 experimental data points from patient-derived tissue samples, and 145,978 3D structures (http://cansar.icr.ac.uk/cansar/data-sources/) and is still growing. Although the system is already providing a number of functionalities, in the next phase it aims at improving the search and browsing power and developing expert tools [83]. The knowledge base aims at becoming, in the future, a system where scientists can simply ask a question pertaining to cancer and receive an answer very quickly.

Researchers are also using big data analytics on cancer data to investigate why some cancer patients relapse and others do not, and some initial results have already been published by Pan et al. [66], but there is still a lot that can be done. Yet, another work in the field of cancer and big data is the setting up of a database of cancer genome mutations to quickly identify which mutations in a tumor are most important. The project involves scanning cancer literature and performing constant database updates. The next step will involve crowdsourcing the data from more scientists who are willing to participate. Khurana et al. [46] have created a computational tool, FunSeq, that can mine terabytes of disregarded genomic data to find possible unknown drivers of cancer.

8 Conclusion

With the explosion of data being provided at unprecedented speed from multiple sources, the need for predictive analytics is being explored in various key sectors including healthcare. Predictive analytics can be used in conjunction with big data to allow valuable insights to be derived to enhance clinical practice and patients' outcomes as well as lower healthcare costs in conjunction with personalized medicine. This chapter visits some application areas in the healthcare sector which have been classified into drug–disease association, disease outbreak detection and surveillance, pharmacovigilance, healthcare management, clinical research, and clinical practice. Data analytics can be applied to both internal and external sources of data to gain insights on diseases trends, effectiveness of treatments, cost-effectiveness of operations, proper management of resources, performance of individual departments, disease outbreaks around the world, behavioral patterns of patients, and health events in different countries among others. The main sources of healthcare data, both internal and external, have been identified and analyzed.

Furthermore, different tools and platforms available in the healthcare sector have been analyzed. A comparative analysis is carried out for the different tools and platforms in terms of features and capabilities, types of data incorporated, data sources being integrated, areas of application, and data analytics techniques used. The main findings of the comparative study are that most health data analytics tools and platforms provide a complete set of services to perform the analytics process. Additionally, these tools integrate data from a wide range of health information systems, including health management systems and electronic health record systems, in order to create a unified data model for healthcare institutions. Data analytics platforms are increasingly being based on open architectures, which allow integration into existing infrastructure. These platforms emphasize on the provision of advanced visualization and dashboards that allow analysts to effortlessly visualize and analyze data from multiple angles without relying on technical experts. Health data analytics tools and platforms are being used to derive completely new insights that were not possible before.

Finally, the chapter presents a number of challenges that need to be addressed in terms of quality of health data, infrastructural challenges, data analytics algorithms and approaches, and security and privacy challenges so that the full potential of predictive analytics can be applied to healthcare. Future directions for health data analytics are discussed with respect to IoT healthcare and precision medicine, data analytics for evidence-based medicine and drug repurposing, improved predictive analytics, real-time health data analytics, and new data analytics techniques for clinical and medical research. More research needs to be carried out in these areas so that healthcare institutions and patients can leverage maximum benefits. The main contribution of this chapter is an attempt to review existing work in the area of health data analytics. It is expected that this review will help researchers and practitioners to shape new research directions to address the challenges in this domain.

Funding This manuscript is part of the work of a funded project by the Mauritius Research Council, Reference HPCRIG-A03.

References

1. E. AbuKhousa, N. Mohamed, J. Al-Jaroodi, e-Health cloud: Opportunities and challenges. Future Internet **4**(4), 621–645 (2012)
2. A. Alyass, M. Turcotte, D. Meyre, From big data analysis to personalized medicine for all: Challenges and opportunities. BMC Med. Genomics **8**(1), 33 (2015)
3. R. Amarasingham, R.E. Patzer, M. Huesch, N.Q. Nguyen, B. Xie, Implementing electronic health care predictive analytics: Considerations and challenges. Health Aff. (Project Hope) **33**(7), 1148–1154 (2014)
4. J. Annese, The importance of combining MRI and large-scale digital histology in neuroimaging studies of brain connectivity and disease. Front. Neuroinform. **6**, 13 (2012)

5. M.A. Badgeley, K. Shameer, B.S. Glicksberg, M.S. Tomlinson, M.A. Levin, P.J. McCormick, A. Kasarskis, D.L. Reich, J.T. Dudley, EHDViz: Clinical dashboard development using open-source technologies. BMJ Open **6**(3), e010579 (2016)
6. I. Bardhan, J. Oh, Z. Zheng, K. Kirksey, Predictive analytics for readmission of patients with congestive heart failure. Inf. Syst. Res. **26**(1), 19–39 (2015). https://doi.org/10.1287/isre.2014.0553
7. D.W. Bates, S. Saria, L. Ohno-Machado, A. Shah, G. Escobar, Big data in health care: Using analytics to identify and manage high-risk and high-cost patients. Health Aff. (Project Hope) **33**(7), 1123–1131 (2014)
8. M. Bayati, M. Braverman, M. Gillam, K.M. Mack, G. Ruiz, M.S. Smith, E. Horvitz, Data-driven decisions for reducing readmissions for heart failure: General methodology and case study. PLoS One **9**(10), e109264 (2014)
9. A. Belle, R. Thiagarajan, S.M.R. Soroushmehr, F. Navidi, D.A. Beard, K. Najarian, Big data analytics in healthcare. Biomed. Res. Int. **2015**, 370194 (2015)
10. H.M. Berman, T.N. Bhat, P.E. Bourne, Z. Feng, G. Gilliland, H. Weissig, J. Westbrook, The Protein Data Bank and the challenge of structural genomics. Nat. Struct. Biol. **7**(Suppl), 957–959 (2000)
11. J. Billings, J. Dixon, T. Mijanovich, D. Wennberg, Case finding for patients at risk of readmission to hospital: Development of algorithm to identify high risk patients. BMJ **333**(7563), 327 (2006)
12. H. Brown, J. Terrence, P. Vasquez, D.W. Bates, E. Zimlichman, Continuous monitoring in an inpatient medical-surgical unit: A controlled clinical trial. Am. J. Med. **127**(3), 226–232 (2014)
13. A.J. Campbell, J.A. Cook, G. Adey, B.H. Cuthbertson, Predicting death and readmission after intensive care discharge. Br. J. Anaesth. **100**(5), 656–662 (2008)
14. K.C.C. Chan, Big data analytics for drug discovery, in *2013 IEEE International Conference on Bioinformatics and Biomedicine*, (IEEE, Piscataway, 2013), pp. 1–1
15. E.S. Chen, G. Hripcsak, H. Xu, M. Markatou, C. Friedman, Automated acquisition of disease drug knowledge from biomedical and clinical documents: An initial study. J. Am. Med. Inform. Assoc. **15**(1), 87–98 (2008)
16. A.P. Chiang, A.J. Butte, Systematic evaluation of drug-disease relationships to identify leads for novel drug uses. Clin. Pharmacol. Ther. **86**(5), 507–510 (2009)
17. I.G. Cohen, R. Amarasingham, A. Shah, B. Xie, B. Lo, The legal and ethical concerns that arise from using complex predictive analytics in health care. Health Aff. (Project Hope) **33**(7), 1139–1147 (2014)
18. A. Culotta, Towards detecting influenza epidemics by analyzing Twitter messages, in *Proceedings of the First Workshop on Social Media Analytics – SOMA '10*, (ACM Press, New York, 2010), pp. 115–122
19. Dell EMC Data Analytic Services for Healthcare [Online]. Available from: https://www.ciossummits.com/Online_Assets_Dell_EMC_Service_Overview_-_Chicago.pdf. Accessed 5 Sept 2019
20. Y. Demchenko, P. Grosso, C. de Laat, P. Membrey, Addressing big data issues in Scientific Data Infrastructure, in *2013 International Conference on Collaboration Technologies and Systems (CTS)*, (IEEE, San Diego, California, 2013), pp. 48–55
21. D.V. Dimitrov, Medical internet of things and big data in healthcare. Healthc. Inform. Res. **22**(3), 156–163 (2016)
22. J.R. Edwards, D.A. Pollock, B.A. Kupronis, W. Li, J.S. Tolson, K.D. Peterson, R.B. Mincey, T.C. Horan, Making use of electronic data: The National Healthcare Safety Network eSurveillance Initiative. Am. J. Infect. Control **36**(3 Suppl), S21–S26 (2008)
23. EMC Healthcare Analytics Solution [Online]. Available at: https://www.emc.com/collateral/solution-overview/h11357-healthcare-analytics-so.pdf. Accessed 5 Sept 2019
24. EMIF, European Medical Information Framework [Online]. Available at: http://www.emif.eu/. Accessed 5 Sept 2019
25. F. Estella, B.L. Delgado-Marquez, P. Rojas, O. Valenzuela, B.S. Roman, I. Rojas, Advanced system for automously classify brain MRI in neurodegenerative disease, in *2012 International Conference on Multimedia Computing and Systems*, (IEEE, Piscataway, 2012), pp. 250–255

26. M. Flores, G. Glusman, K. Brogaard, N.D. Price, L. Hood, P4 medicine: How systems medicine will transform the healthcare sector and society. Pers. Med. **10**(6), 565–576 (2013)
27. R. Frijters, M. van Vugt, R. Smeets, R. van Schaik, J. de Vlieg, W. Alkema, Literature mining for the discovery of hidden connections between drugs, genes and diseases. PLoS Comput. Biol. **6**(9), e1000943 (2010)
28. V. Gligorijević, N. Malod-Dognin, N. Pržulj, Integrative methods for analyzing big data in precision medicine. Proteomics **16**(5), 741–758 (2016)
29. D. Gotz, H. Stavropoulos, J. Sun, F. Wang, ICDA: A platform for intelligent care delivery analytics. AMIA Annu. Symp. Proc. **2012**, 264–273 (2012)
30. T. Greenhalgh, J. Howick, N. Maskrey, Evidence Based Medicine Renaissance Group, Evidence based medicine: A movement in crisis? BMJ **348**, g3725 (2014)
31. E. Gultepe et al., From vital signs to clinical outcomes for patients with sepsis: A machine learning basis for a clinical decision support system. J. Am. Med. Inform. Assoc. **21**(2), 315–325 (2013)
32. T. Haferlach, A. Kohlmann, L. Wieczorek, G. Basso, G.T. Kronnie, M.-C. Béné, J. De Vos, J.M. Hernández, W.-K. Hofmann, K.I. Mills, A. Gilkes, S. Chiaretti, S.A. Shurtleff, T.J. Kipps, L.Z. Rassenti, A.E. Yeoh, P.R. Papenhausen, W.-M. Liu, P.M. Williams, R. Foà, Clinical utility of microarray-based gene expression profiling in the diagnosis and subclassification of leukemia: Report from the International Microarray Innovations in Leukemia Study Group. J. Clin. Oncol. **28**(15), 2529–2537 (2010)
33. R. Harpaz, W. DuMouchel, M. Schuemie, O. Bodenreider, C. Friedman, E. Horvitz, A. Ripple, A. Sorbello, R.W. White, R. Winnenburg, N.H. Shah, Toward multimodal signal detection of adverse drug reactions. J. Biomed. Inform. **76**, 41–49 (2017)
34. M. Hassanalieragh, A. Page, T. Soyata, G. Sharma, M. Aktas, G. Mateos, B. Kantarci, S. Andreescu, Health monitoring and management using Internet-of-Things (IoT) sensing with cloud-based processing: Opportunities and challenges, in *2015 IEEE International Conference on Services Computing*, (IEEE, New York, 2015), pp. 285–292
35. Healthcare Analytics & Business Intelligence | Sisense [Online]. Available at: https://www.sisense.com/solutions/healthcare/. Accessed 10 Sept 2019
36. Healthcare Predictive Analytics – Big Data Analytics in Medicine and Genomics | Hortonworks [Online]. Available at: https://hortonworks.com/solutions/healthcare/. Accessed 10 Sept 2019
37. HealthSuite Ecosystem | Philips Healthcare [Online]. Available at: https://www.usa.philips.com/healthcare/innovation/about-health-suite. Accessed 10 Sept 2019
38. M. Herland, T.M. Khoshgoftaar, R. Wald, A review of data mining using big data in health informatics. J. Big Data **1**(1), 2 (2014)
39. S. Hiremath, G. Yang, K. Mankodiya, Wearable Internet of Things: Concept, architectural components and promises for person-centered healthcare, in *2014 4th International Conference on Wireless Mobile Communication and Healthcare – Transforming Healthcare Through Innovations in Mobile and Wireless Technologies (MOBIHEALTH)*, (IEEE, Piscataway, 2014), pp. 304–307
40. L. Hood, S.H. Friend, Predictive, personalized, preventive, participatory (P4) cancer medicine. Nat. Rev. Clin. Oncol. **8**(3), 184–187 (2011)
42. P.L. Ingrassia, L. Carenzo, F.L. Barra, D. Colombo, L. Ragazzoni, M. Tengattini, F. Prato, A. Geddo, F. Della Corte, Data collection in a live mass casualty incident simulation: Automated RFID technology versus manually recorded system. Eur J Emerg Med **19**(1), 35–39 (2012)
43. E. Jovanov, Preliminary analysis of the use of smartwatches for longitudinal health monitoring, in *Conference Proceedings: Annual International Conference of the IEEE Engineering in Medicine and Biology Society 2015*, (IEEE, Piscataway, 2015), pp. 865–868
44. D. Kansagara, H. Englander, A. Salanitro, D. Kagen, C. Theobald, M. Freeman, S. Kripalani, Risk prediction models for hospital readmission: A systematic review. J. Am. Med. Assoc. **306**(15), 1688–1698 (2011)

45. H. Khazaei, N. Mench-Bressan, C. McGregor, J.E. Pugh, Health informatics for neonatal intensive care units: An analytical modeling perspective. IEEE J. Transl. Eng. Health Med. **3**, 3000109 (2015)
46. E. Khurana, Y. Fu, V. Colonna, X.J. Mu, H.M. Kang, T. Lappalainen, A. Sboner, L. Lochovsky, J. Chen, A. Harmanci, J. Das, A. Abyzov, S. Balasubramanian, K. Beal, D. Chakravarty, D. Challis, Y. Chen, D. Clarke, L. Clarke, F. Cunningham, M. Gerstein, Integrative annotation of variants from 1092 humans: Application to cancer genomics. Science **342**(6154), 1235587 (2013)
47. H.C. Koh, G. Tan, Data mining applications in healthcare. J. Healthc. Inf. Manage. **19**(2), 64–72 (2005)
48. E.I. Konstantinidis, A.S. Billis, L. Plotegher, G. Conti, P.D. Bamidis, Indoor location IoT analytics "in the wild": Active and healthy ageing cases, in *XIV Mediterranean Conference on Medical and Biological Engineering and Computing 2016. IFMBE Proceedings*, ed. by E. Kyriacou, S. Christofides, C. S. Pattichis, (Springer International Publishing, Cham, 2016), pp. 1231–1236
49. C.S. Kruse, R. Goswamy, Y. Raval, S. Marawi, Challenges and opportunities of big data in health care: A systematic review. JMIR Med. Inform. **4**(4), e38 (2016)
50. V. Kumar, H. Park, R. Basole, M. Braunstein, M. Kahng, D. Chau, A. Tamersoy, D. Hirsh, N. Serban, J. Bost, B. Lesnick, B. Schissel, M. Thompson, Exploring clinical care processes using visual and data analytics: Challenges and opportunities, in 20th ACM SIGKDD Conference on Knowledge Discovery and Data Mining Workshop on Data Science for Social Good, 2014
51. LexisNexis, Why data quality is the greatest challenge and opportunity for health care (2016). Available at: https://www.lexisnexis.com/risk/downloads/whitepaper/Data-Quality-POV.pdf
52. J. Li, S. Zheng, B. Chen, A.J. Butte, S.J. Swamidass, Z. Lu, A survey of current trends in computational drug repositioning. Brief. Bioinform. **17**(1), 2–12 (2016)
53. D. Lopez, M. Gunasekaran, B.S. Murugan, H. Kaur, K.M. Abbas, Spatial big data analytics of influenza epidemic in Vellore, India, in *IEEE International Conference on Big Data 2014*, (IEEE, Washington, DC, USA, 2014), pp. 19–24
54. D. Luna, J.C. Mayan, M.J. García, A.A. Almerares, M. Househ, Challenges and potential solutions for big data implementations in developing countries. Yearb. Med. Inform. **9**, 36–41 (2014)
55. G. Luo, B.L. Stone, F. Sakaguchi, X. Sheng, M.A. Murtaugh, Using computational approaches to improve risk-stratified patient management: Rationale and methods. JMIR Res. Protoc. **4**(4), e128 (2015)
56. Z. Lv, J. Chirivella, P. Gagliardo, Bigdata oriented multimedia mobile health applications. J. Med. Syst. **40**(5), 120 (2016)
57. S.E. Madnick, R.Y. Wang, Y.W. Lee, H. Zhu, Overview and framework for data and information quality research. J. Data Inf. Qual. **1**(1), 1–22 (2009)
58. P. Mahindrakar, D. Hanumanthappa, Data mining in healthcare: A survey of techniques and algorithms with its limitations and challenges. Int. J. Eng. Res. Appl. **3**(6), 937–941 (2013)
60. MEDLINE Fact Sheet [Online]. Available at: https://www.nlm.nih.gov/pubs/factsheets/medline.html. Accessed 10 Sept 2019
61. R. Mirnezami, J. Nicholson, A. Darzi, Preparing for precision medicine. N. Engl. J. Med. **366**(6), 489–491 (2012)
62. B. Moore, The potential use of radio frequency identification devices for active monitoring of blood glucose levels. J. Diabetes Sci. Technol. **3**(1), 180–183 (2009)
63. NIH HIPAA Privacy Rule and Its Impacts on Research [Online]. Available at: https://privacyruleandresearch.nih.gov/pr_07.asp. Accessed 10 Sept 2019

64. A. Nikfarjam, A. Sarker, K. O'Connor, R. Ginn, G. Gonzalez, Pharmacovigilance from social media: Mining adverse drug reaction mentions using sequence labeling with word embedding cluster features. J. Am. Med. Inform. Assoc. **22**(3), 671–681 (2015)
65. I. Ouanes, C. Schwebel, A. Français, C. Bruel, F. Philippart, A. Vesin, L. Soufir, C. Adrie, M. Garrouste-Orgeas, J.-F. Timsit, B. Misset, Outcomerea Study Group, A model to predict short-term death or readmission after intensive care unit discharge. J. Crit. Care **27**(4), 422.e1–422.e9 (2012)
66. H. Pan, Y. Jiang, M. Boi, F. Tabbò, D. Redmond, K. Nie, M. Ladetto, A. Chiappella, L. Cerchietti, R. Shaknovich, A.M. Melnick, G.G. Inghirami, W. Tam, O. Elemento, Epigenomic evolution in diffuse large B-cell lymphomas. Nat. Commun. **6**, 6921 (2015)
67. R.B. Parikh, M. Kakad, D.W. Bates, Integrating predictive analytics into high-value care: The dawn of precision delivery. J. Am. Med. Assoc. **315**(7), 651–652 (2016)
68. G. Perna, Shannon Medical Center. Leaders at one Texas-based hospital are not taking hand hygiene for granted. Healthc. Inform. **30**(1), 28–30 (2013)
69. A.E. Powell, H.T.O. Davies, R.G. Thomson, Using routine comparative data to assess the quality of health care: Understanding and avoiding common pitfalls. Qual. Saf. Health Care **12**(2), 122–128 (2003)
70. K. Priyanka, N. Kulennavar, A survey on big data analytics in health care. Int. J. Comput. Sci. Inf. Technol. **5**(4), 5865–5868 (2014)
71. W. Raghupathi, V. Raghupathi, Big data analytics in healthcare: Promise and potential. Health Inf. Sci. Syst. **2**, 3 (2014)
72. D.C. Ranasinghe, R.L. Shinmoto Torres, A.P. Sample, J.R. Smith, K. Hill, R. Visvanathan, Towards falls prevention: A wearable wireless and battery-less sensing and automatic identification tag for real time monitoring of human movements, in *Conference Proceedings: Annual International Conference of the IEEE Engineering in Medicine and Biology Society 2012*, (IEEE, Piscataway, 2012), pp. 6402–6405
73. S.M. Riazul Islam, D. Kwak, M. Humaun Kabir, M. Hossain, K.-S. Kwak, The internet of things for health care: A comprehensive survey. IEEE Access **3**, 678–708 (2015)
74. Sagitec HealHub [Online]. Available at: http://www.sagitec.com/healhub. Accessed 10 Sept 2019
75. R. Salazar, P. Roepman, G. Capella, V. Moreno, I. Simon, C. Dreezen, A. Lopez-Doriga, C. Santos, C. Marijnen, J. Westerga, S. Bruin, D. Kerr, P. Kuppen, C. van de Velde, H. Morreau, L. Van Velthuysen, A.M. Glas, L.J. Van't Veer, R. Tollenaar, Gene expression signature to improve prognosis prediction of stage II and III colorectal cancer. J. Clin. Oncol. **29**(1), 17–24 (2011)
76. A. Sarker, R. Ginn, A. Nikfarjam, K. O'Connor, K. Smith, S. Jayaraman, T. Upadhaya, G. Gonzalez, Utilizing social media data for pharmacovigilance: A review. J. Biomed. Inform. **54**, 202–212 (2015)
77. E.G. Schulz, C.L. Neumann, Interventional decentralized telemonitoring: Bridging the gap between patient's device and physician's needs in well selected indications. Kidney Blood Press. Res. **40**(2), 130–140 (2015)
78. I. Shams, S. Ajorlou, K. Yang, A predictive analytics approach to reducing 30-day avoidable readmissions among patients with heart failure, acute myocardial infarction, pneumonia, or COPD. Health Care Manag. Sci. **18**(1), 19–34 (2015)
79. Y. Sun, B.H. Heng, S.Y. Tay, E. Seow, Predicting hospital admissions at emergency department triage using routine administrative data. Acad. Emerg. Med. **18**(8), 844–850 (2011)
80. K.B. Sundharakumar, S. Dhivya, S. Mohanavalli, R.V. Chander, Cloud based fuzzy healthcare system. Procedia Comput. Sci. **50**, 143–148 (2015)
81. A. Sunyaev, D. Chornyi, C. Mauro, H. Krcmar, Evaluation framework for personal health records: Microsoft HealthVault vs. Google Health, in *2010 43rd Hawaii International Conference on System Sciences*, (IEEE, Piscataway, 2010), pp. 1–10
82. R.K. Thomas, *Health Services Marketing: A Practitioner's Guide* (Springer Science & Business Media, New York, 2007)

83. J.E. Tym, C. Mitsopoulos, E.A. Coker, P. Razaz, A.C. Schierz, A.A. Antolin, B. Al-Lazikani, canSAR: An updated cancer research and drug discovery knowledgebase. Nucleic Acids Res. **44**(D1), D938–D943 (2016)
84. J. Vanschoren, J.N. van Rijn, B. Bischl, L. Torgo, OpenML: networked science in machine learning. ACM SIGKDD Explor. Newsl. **15**(2), 49–60 (2014)
85. D. Wang, E. Rundensteiner, R. Ellison, H. Wang, Probabilistic inference of object identifications for event stream analytics, in *Proceedings of the 16th International Conference on Extending Database Technology – EDBT '13*, (ACM Press, New York, 2013), p. 513
86. M.J. Ward, K.A. Marsolo, C.M. Froehle, Applications of business analytics in healthcare. Bus. Horiz. **57**(5), 571–582 (2014)
87. R.W. White, R. Harpaz, N.H. Shah, W. DuMouchel, E. Horvitz, Toward enhanced pharmacovigilance using patient-generated data on the internet. Clin. Pharmacol. Ther. **96**(2), 239–246 (2014)
88. R. Xu, Q. Wang, Large-scale extraction of accurate drug-disease treatment pairs from biomedical literature for drug repurposing. BMC Bioinformatics **14**, 181 (2013)
89. H. Yoshida, A. Kawaguchi, K. Tsuruya, Radial basis function-sparse partial least squares for application to brain imaging data. Comput. Math. Methods Med. **2013**, 591032 (2013)
90. X. Yu, A. Ganz, Scalable patients tracking framework for mass casualty incidents, in *Conference Proceedings: Annual International Conference of the IEEE Engineering in Medicine and Biology Society 2011*, (IEEE, Piscataway, 2011), pp. 860–863
91. Q. Zhang, C. Pang, S. Mcbride, D. Hansen, C. Cheung, M. Steyn, Towards health data stream analytics, in *IEEE/ICME International Conference on Complex Medical Engineering*, (IEEE, Piscataway, 2010), pp. 282–287
92. K. Zolfaghar, N. Verbiest, J. Agarwal, N. Meadem, S.-C. Chin, S.B. Roy, A. Teredesai, D. Hazel, P. Amoroso, L. Reed, Predicting risk-of-readmission for congestive heart failure patients: A multi-layer approach (2013). Available at: https://arxiv.org/abs/1306.2094

Implementation and Computational Level Challenges for IoT in Healthcare Applications

Surjeet and Ravisankar Malladi

1 Introduction

The Internet of Things (IoT) technology and Information and Communication Technology (ICT) are ending up progressively in the healthcare industry. Their essential application in the field of savvy healthcare services incorporates the perception of material administration, digitization of medical data, and digitization of the medicinal procedures. However, the healthcare industry is in a condition of extraordinary gloom. What we are drawing closer is where essential services would end up distant to the common people. A huge area of society would go inefficient, attributable to aging and individuals would be progressively inclined to interminable sicknesses. While innovation cannot prevent the populace from aging or eliminate interminable infections without a moment's delay, it can at any rate make healthcare services simpler on a pocket and in terms of reachability. Diagnosis eats up huge amount of clinic bills. Innovation can move the schedules of medicinal checks from a clinic (emergency clinic-driven) to the patient's (home-driven). The correct analysis will likewise diminish the need for hospitalization.

IoT and ICT, the new ideal technologies, have broad appropriateness in various territories, including healthcare services. The full use of these innovative technologies in medicinal services zone is a common expectation since it enables restorative focus to work with all the more capability and for patients to get better

Surjeet (✉)
Department of Electronics and Communication Engineering, Bharati Vidyapeeth's College of Engineering, New Delhi, India
e-mail: surjeet.balhara@bharatividyapeeth.edu

R. Malladi
Department of CSE, Gudlavalleru Engineering College(A), Gudlavalleru, India
e-mail: mrsankar@giet.ac.in

N. Gupta, S. Paiva (eds.), *IoT and ICT for Healthcare Applications*, EAI/Springer Innovations in Communication and Computing,
https://doi.org/10.1007/978-3-030-42934-8_9

treatment. With the utilization of this innovation-based human services strategy, there are unrivalled advantages that could improve the quality and effectiveness of medications and likewise improve the strength of the patients.

IoT and ICT for healthcare applications can likewise be utilized for research purposes. This is because IoT empowers us to gather a monstrous measure of information about the patient's ailment, which would have taken numerous years in the event we gathered it physically. This information along these lines gathered can be utilized for factual examination that would bolster the restorative research. In this way, IoT and ICT do not just spare time, but also our cash, which would go in the examination. Consequently, it tends to be perpetually guaranteed that these innovations have an incredible effect in the field of medical research. It empowers the presentation of greater and better medication facilities. IoT is utilized in an assortment of gadgets that upgrade the nature of the services received by the patients. Indeed, even the current gadgets are presently being refreshed by IoT by basically installing smart chips. This chip extends the help and care that a patient requires. Until a couple of years back, medical clinics were a scary spot for the patients. Understanding the medical terms, counselling the specialists, and remaining in the unpleasant condition were very testing. Nevertheless, with the progressing time, the situation has totally changed. The healthcare applications industry began concentrating on computerized arrangements, which are not just changing the emergency clinic patient experience, but also getting the well-being data in a simpler manner. At present, the development of mobility arrangements is changing the essence of the healthcare applications industry and has turned into the common augmentation of healthcare services industry.

2 Related Research

The coming of the IoT has made tremendous changes and significant difficulties for any business hoping to take on the advanced change. In any case, no industry faces to a greater degree a test to roll out this improvement than human services. Associations that need to grasp IoT can battle for a long time in the quest for "going advanced" and still come up short. Emergency clinics and other medicinal services suppliers have all the operational complexities of different organizations with the additional obligations of guarding their patients, guaranteeing that well-being records are secure and keeping their offices operational every minute of every day. Furthermore, the medicinal services industry is an essential objective of progressively modern cybercriminals hoping to introduce ransomware to take tolerant well-being records or mischief patients with associated restorative gadgets, for example, insulin siphons or pacemakers. A healthcare office could adhere to the National Institute of Standards Technology's cybersecurity system precisely; however, its system may be as secure as the weakest passage point. Progressively, unbound therapeutic gadgets are utilized as footholds to access a clinic.

Each sort of associated medicinal gadget has its own arrangement of complexities that should be verified at the hour of item structure. Every gadget has an application programming interface (API), a UI, a URL, and regularly interfaces for HDMI, Bluetooth, or Wi-Fi, which may all be abused if not appropriately verified by the gadget maker. Lamentably, the real weight of obligation regarding verifying these gadgets at last falls on the medicinal services supplier. Cybercriminals more often than not concentrate on taking electronic well-being records (EHRs) because of their underground market estimation of $300–500 per record [1]. They may likewise introduce malware or ransomware on the emergency clinic, scrambling and handicapping the associated servers and frameworks, and making all out interruption the arrangement of consideration. The frameworks stay broken until the emergency clinic makes the payment, figures out how to subvert the encryption calculation – once in a while an inconsequential errand – or re-establishes frameworks from reinforcements, which could take a few days or more. The clinic's genuine expense of recouping from a ransomware assault for the most part keeps running in huge amount in USD. The expense to a littler medicinal practice is normally less; yet, this does exclude the disturbance of consideration and access to tolerant well-being data. Healthcare frameworks may likewise bring about overwhelming fines under General Data Protection guidelines for the burglary of EHRs and the rupture of individual well-being data (PHI). Therapeutic gadget makers that fabricate items to deal with, store, or transmit PHI may likewise be dependent upon considerable fiscal punishments whenever carelessness is demonstrated in a medical clinic rupture in which PHI was undermined.

Medicinal services associations are enduring the worst part of these devastating assaults. In the main quarter of 2018, a few clinics were assaulted and contaminated with ransomware and, as a rule, paid the payoff. As cybercriminals become all the more advanced, therapeutic gadget producers should be in any event similarly as responsive and complex in their endeavors to support their gadget security. Medicinal services associations, emergency clinics, facilities, and different suppliers are the significant clients and essential wellspring of income for therapeutic gadget makers. Next, we present the discoveries of the examination reviews being led by reaching around three dozen doctors, administrators, and different associations that help, or work in the interest of, healthcare suppliers. Here are their contemplations about cybersecurity:

- Unbound medicinal gadgets on healthcare systems can be utilized as footholds to penetrate the association's system and servers, permitting a cybercriminal to introduce malware or take EHR [2]. A strong 95% of the interviewees knew about this issue. Everybody concurred that medicinal gadget well-being and security must be top of the need list for every restorative gadget associated with healthcare systems.
- Practically all IT administrators and doctors met were putting considerably in securing their in-house systems. In any case, just bigger healthcare suppliers had the assets and the staff to organize official arrangements and techniques

to guarantee that secured therapeutic gadgets are made utilizing industry best practices for safe coding and cybersecurity.

Medicinal gadget makers should facilitate the weight on healthcare, improve the restorative gadget biological system, and give a lot more elevated level of patient security. From the underlying item structure, they should utilize thoroughly tried programming inside their gadgets to dodge item breakdown and vulnerabilities. System-associated therapeutic gadgets must expand upon FDA direction and utilize cybersecurity strategies equipped for protecting against complex, well-financed cybercriminals. Recently proposed SBOM direction should obviously advise the medicinal services industry regarding the dangers and abilities of associated therapeutic gadgets. In conclusion, there ought to be an increasingly even-handed circulation of obligation regarding verifying medicinal IoT. The therapeutic gadget producers that hold onto well-being and security as their top need will prevail at restorative IoT and above all, protect patients from digital mischief.

2.1 Healthcare IoT Statistics

The human services industry has been a fast adopter of IoT innovation. This accompanies numerous favorable circumstances and numerous dangers. Nevertheless, the general size and multifaceted nature of the market are an ideal case of the transformative intensity of IoT in a solitary, significant industry. The insights underneath uncover the utilization of IoT in social insurance and the general effect on the business [3].

- Nearly 60% of human services associations have brought IoT gadgets into their offices.
- 73% of human services associations use IoT for support and checking.
- 87% of human services associations intend to actualize IoT innovation by 2019, which is marginally higher than the 85% of organizations crosswise over different ventures.
- Nearly 64% utilization of IoT in the human services industry is in patient screening.
- 89% of human services associations have experienced IOT-related security rupture.

It is generally realized that interconnected gadgets are being utilized in such manners as social affair information from screens, blood glucose levels, electrocardiograms, and temperature screens. Nevertheless, a portion of these devices requires line up correspondence with a human services pro.

The human services IoT market is relied upon to be worth 534.3 billion by 2025 (Source: Grand View Research, Inc.).

- North America was the biggest market for IoT as of the close of 2017 (Source: Grand View Research, Inc.).

- The EU had around 870,000 dynamic medicinal services IoT gadgets in 2016, and will have anticipated 2.79 million gadgets before the finish of 2019. ETNO expects there to be more than 800 million human services IoT associations in Europe by 2025 (Source: ETNO).
- By the finish of 2016, 60% of human services associations were utilizing IoT gadgets (Source: IEEE).
- A greater part of medicinal services associations (87%) plans to actualize IoT tech before the finish of 2019. By examination, 85% of non-medicinal services organizations intend to do as such in that equivalent timeframe (Source: IEEE).
- 80% of social insurance business officials consider expanded to be as the greatest favorable position of IoT execution; 73% indicated cost investment funds, while 76% indicated "perceivability over the association" as key favorable circumstances [5, 6].
- The most normal social insurance IoT gadgets incorporate patient screens (64%), vitality meters (56%), and imaging gadgets (33%). Over 70% of medicinal services associations that utilize IoT gadgets use them for observing and upkeep (Source: IEEE).
- 50% of social insurance IoT is utilized for remote activity and control, and 47% associate their gadgets with area-based administrations (Source: Aruba).

Security ruptures are a huge disadvantage to IoT. As indicated by IEEE, over 80% of medicinal services associations that utilize IoT gadgets have endured a security break of their IoT gadgets or framework (Source: IEEE). The most widely recognized security dangers to IoT were malware (49%), human mistake (39%), and DDoS assaults (22%) (Source: Aruba). The well-being business is in charge of 6% of worldwide IoT ventures, with 55% of those happening in the Americas (Source: IoT Analytics).

2.2 *Cybersecurity in Healthcare*

One may contend that cybersecurity in healthcare, similar to the general medicinal services framework itself, is in basic condition. With the developing number of web-empowered restorative indicative gadgets, the extension of electronic medicinal records, and the commonness of organized planning and charging frameworks, and lacking worker preparing and affectability to cybersecurity issues has overpowered the human services condition. Healthcare representatives are frequently compelled to accomplish more with less so as to stay aware of mechanical upgrades. As they scramble to carry out their responsibilities, workers give less consideration to cybersecurity issues. An ongoing information break experienced by the University of California Davis Health framework is a normal case of the healthcare cybersecurity issue. As of late, it has been found that a representative reacted to a phishing email, giving programmers access to his login accreditations and that the programmers at that point used to bargain the records of in excess of 15,000 patients. This occasion cannot be expelled either as an exception or as an issue brought about by a reckless representative in light of the fact that comparable assaults have just tormented human services frameworks for quite a while. In light of the administrative commitments they face, and even with aggressive weights from other healthcare suppliers, emergency clinics and restorative focuses are starting to give more noteworthy consideration to cybersecurity matters. Their first request of business has been to improve endeavors to pull in and hold cybersecurity ability in the medicinal services field.

General industry patterns foresee critical deficiencies of cybersecurity representatives throughout the following 5 years. Perceiving this, the healthcare industry should improve pay rates for therapeutic cybersecurity pros so as to stay focused and to hold ability that will be baited to other industry divisions that guarantee more prominent pay rates and vocation openings. More significant compensations may make momentary budgetary issues for medicinal services frameworks yet apportioning increasingly budgetary assets to cybersecurity faculty will be supported in perspective on the greater expenses related to recouping from an information break. Building up a progressively strong healthcare cybersecurity workforce is one of six goals prescribed by the US Department of Health and Human Services. The other five goals incorporate streamlining cybersecurity authority, characterizing better desires for secure frameworks and gadgets, improving by and large gadget security, expanding worker instruction, and recognizing basic hazard exposures through better sharing of information and information break encounters all through the business. Producers in the medicinal IoT industry are now increasing their determination to conform to the basics to improve therapeutic gadget security. The center of those endeavors incorporates advancement of cybersecurity principles that all restorative IoT gadget makers will incorporate with their items for issues, for example, encryption, information stockpiling, client confirmation, programming and firmware updates, and the board of patches and bug fixes for recognized security dangers. These endeavors are accomplishing something other than keeping the patient alive and sound; at the end of the day, their impediment is a similar that is looked by all cutting-edge prescription. That is, similarly as medication can treat and fix the patient, but it cannot kill all occurrences of sickness, therapeutic cybersecurity activities can protect restorative innovation frameworks from hacking, yet it can never dispose of all digital dangers.

Since the digital danger hazard will consistently be available, healthcare cybersecurity protection is the last barrier to all medicinal cybersecurity endeavors. Cybersecurity protection will offer assurance to a medicinal focus, for instance, where regardless of all instruction endeavors, an indiscreet or tired representative snap on an email phishing join that opens the inside's system to a programmer's nosy eyes. That security incorporates money-related assets to enable an interface to recoup lost information and to reestablish harmed programming and gear. It can likewise cover liabilities to patients whose information was lost or taken during a cyberattack, and fines imposed by administrative experts for HIPAA and different infringements. In perspective on this, medicinal services cybersecurity protection is one of the best techniques that work in the healthcare industry.

2.3 Challenges of IoT in Healthcare

Human services is one of the most energizing verticals for IoT change in 2019. Sadly, 74% of IoT activities are fruitless [4]. Running from information security to inheritance framework may block medicinal services IoT activities, yet with a more profound comprehension of these deterrents, human services pioneers can push forward unhesitatingly into the new year. The IoT has considerably changed human services in a moderately brief time. For instance, associated gadgets enable more seasoned individuals to age set up securely for whatever length of time that is conceivable. They help specialists consult with experts over the world about complex cases, and they screen patients' interminable infections between office visits. In any case, any development in innovation carries with it the difficulties to be survived. Here are five of those difficulties confronting healthcare in 2019 alongside recommendations for what professionals should remember when they use human services IoT gadgets in their working environments. Figure 1 shows four most significant quadrants that ascribe the difficulties of IoT in medicinal services.

2.4 Security Threats

The basic stress for authoritative bodies is the security of personal health information, that are set and go on through related contraptions [7]. While various human administration affiliations guarantee that the sensitive data are taken care of in an ensured and mixed way, they do not have direction over the prosperity and security of the data and the ways being used to transmit the data. This creates a basic hazard that extends bit by bit reliant on the number of new devices related with the framework. IoT devices get and transmit data logically. In any case, by far, most of the IoT contraptions need data shows and standards. Despite that, there is basic unclearness regarding data ownership rule. All of these components

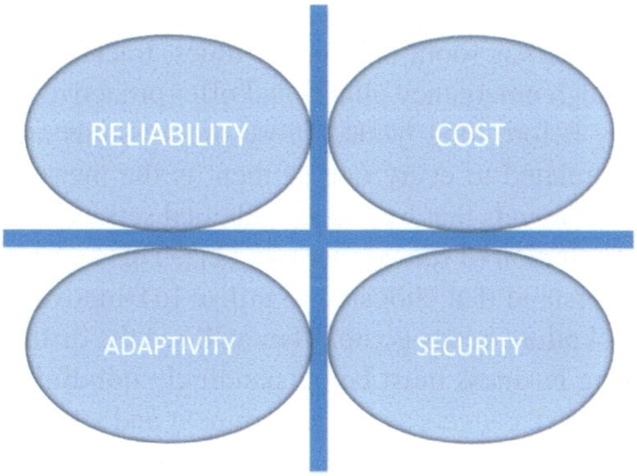

Fig. 1 Challenges of IoT in healthcare

make the data significantly frail to cybercriminals who can hack into the system and deal with personal health information (PHI) of the two patients similarly as pros. Cybercriminals can manhandle patient's data to make fake IDs to buy drugs and remedial equipment which they can sell later. Software engineers can, in like manner, report a tricky insurance in patient's name.

2.5 *Most IoT Initiatives Are Incomplete or Unsuccessful*

Accommodation and expedient information moves are two contemplations that may propel medicinal services associations to investigate IoT innovations. There is surely motivation to be amped up for IoT's potential. Nevertheless, a 2017 research from Cisco painted a not exactly exciting picture of IoT change endeavors. The exploration included getting input from in excess of 1800 individuals over the USA, UK, and India, who were partners in past or continuous IoT activities. Cisco's review uncovered that completed undertakings were just viewed as effective 26% of the time [8, 10]. Moreover, around 33% of the respondents considered their completed activities ineffective. Most ventures – 60% – experience inconvenience at the evidence of idea organization. Notwithstanding, it is significant that using outer associations (e.g., stage accomplices) was a significant factor for those associations that accomplished effective executions. In spite of the fact that this Cisco study did not concentrate its investigation on human services associations, that industry vertical was spoken to in the general study. The findings stress that associations ought to be mindful when arranging their IoT rollouts in 2019. For instance, they should begin little and organize ventures that line up with their most conspicuous business destinations or patient needs.

2.6 *Healthcare Will Generate a Tremendous Amount of Data*

Probably the most energizing healthcare IoT activities incorporate approaches to lessen crisis room holding up times, track resources and individuals that move all through emergency clinics and offer proactive alarms about restorative gadgets that may before long fizzle. Those progressions are in fact noteworthy; however, one test related to every one of them is the measure of information created. A gauge recommends that by 2025, medicinal services will be in charge of creating the most information of some other part [11]. This is the ideal opportunity for associations to understand that choosing to utilize IoT innovation will probably make information stockpiling needs go up. They will see the distinctions in 2019. Moreover, the well-being business must be outstandingly mindful so as to treat persistent information from IoT gadgets as per government and state guidelines. The surge of information made by the IoT contraptions and gadgets utilized in the medicinal services industry could likewise cause unanticipated issues if associations are not prepared to deal with it appropriately and check its quality.

2.7 IoT Devices Increase Available Attack Surfaces

The huge conceivable outcomes for utilizing IoT gadgets in medicinal services additionally present concerning vulnerabilities. As gadget use rises, so does the quantity of ways programmers could penetrate the framework and dig for the most significant information. One new hazard, a Zingbox concentrate, recognized is that programmers could find out about how an associated restorative gadget works by getting into the framework and perusing its mistake logs. The learning the programmer's addition could encourage breaking into an emergency clinic system or causing gadgets to distribute inaccurate readings that impact patient consideration. On an increasingly positive note, in any case, Zinbox's exploration indicated progress in sellers', suppliers', and producers' eagerness to team up. Those common endeavors could diminish patient dangers by shutting the holes that can shape between the layers of an IoT framework by fortifying models and normalizing secure conventions. It is impractical to realize how all the cybersecurity dangers well-being associations may look in 2019. In any case, offices wanting to actualize IoT innovation must take care to build familiarity with existing dangers and see how to shield systems and devices from programmers' endeavors.

2.8 Outdated Infrastructure Hinders the Medical Industry

Despite the fact that retrofitting can inhale new life into maturing framework, really exploiting IoT is dubious if an office's foundation is obsolete. Old framework is a known issue in human services. At the point when emergency clinics are in critical need of patched up foundation, they likewise experience issues procuring the staff to make redesigns. Technical ability is sought after. Forthcoming competitors might not have any desire to handle old framework.

2.9 IoT Poses Many Overlooked Obstacles

As indicated by research from Aruba Networks, the most widely recognized utilization of IoT innovation in healthcare is to apply it to patient checking frameworks. It is without a doubt helpful to adopt that strategy, yet something well-being associations frequently overlook is that which is not normal for sites, for instance, gadgets commonly cannot experience arranged times of vacation. Rather, refreshes need to happen persistently as individuals utilize the observing gadgets. Furthermore, emergency clinics regularly rely upon IoT-empowered stockpile cupboards to follow assets. When those frameworks are set up, the offices can frequently diminish past stock administration issues, yet even the most brilliant associated gadgets cannot dispose of human blunder. All things considered, we make those IoT frameworks. In the event that people are blunder-inclined, it pursues that our IoT frameworks can acquire mistake-inclined conduct from us. Seller appraisals are likewise urgent to overcome regularly unforeseen difficulties. A few makers are basically worried about beating contenders to the market with their items. In the surge, many do not incorporate security with their procedures from the beginning, so they should not be astounded when client databases are broken. Regardless of whether a medical clinic has better than expected cybersecurity resistances, patients may, in any case, be in danger from items that need satisfactory security. A rotten one can demolish the bundle. Cybersecurity must be solid and complete. Sadly, couple of current IoT frameworks are really secure by customary system security measurements.

2.10 *Regulatory Concerns*

IoT executions are gathering a lot of information that could conceivably be touchy or destructive whenever uncovered. This incorporates individual information about workers or clients, just as restrictive business information about tasks and interior procedures. On the administrative side, protection concerns, for example, explaining who can get to IoT information and how that information is utilized, must be attended to. Governments and industry bodies need to set measures and guidelines for the different businesses to guarantee that information is not abused.

2.11 *Scalability*

Organizations regularly effectively build up an IoT arrangement with numerous gadgets in a solitary area, but later find that versatility is an issue. It is along these lines that the structure of a framework cooks for extra assets.

2.12 *Technical Problems Facing Medical IoT*

In the therapeutic field, despite everything, we have to address numerous specialized issues confronting IoT. These issues incorporate Dynamic Networking and Node Mobility Management in Large-Scale Networks. When there is an extension of the observing framework to cover private networks, urban areas, or even whole nations, the size of the system will overpower, checking hubs and base stations all must be portable somewhat. In this manner, we need to plan a fitting system topology in the executive's structure and system versatility in the executives techniques.

2.13 *Data Completeness and Data Compression*

Hubs will now and again need to directly observe for 24 hours every day, gathering a monstrous measure of data that should be put away utilizing a pressure calculation to decrease stockpiling and transmission volume. In any case, conventional information pressure calculations are unreasonably expensive for sensor hubs. Besides, pressure calculations cannot lose the first information. In addition, the framework could misdiagnose the patient's condition. Remote sensor arranges hub's structure, a self-sorted out system, which is powerless against assaults and is, clearly, hazardous when managing persistent data that must be kept classified. The registering intensity of a sensor hub is incredibly lacking. Consequently, customary security and encryption innovation are not relevant to these situations. Along these lines, we should structure an encryption calculation appropriate to the abilities of a sensor hub.

2.14 *Cost*

Are you astounded to see cost contemplations in the test segments? I know a large portion of you would be; however, the main concern is that IoT has not made the medicinal services moderate to the normal man yet. The blast in the healthcare expenses is a stressing sign for everyone, particularly the created nations. The circumstance is to such an extent that it offered assent to "Therapeutic Tourism" in which patients with basic conditions get to medicinal services offices of the creating countries, which cost them as less as one-tenth. IoT in human services as an idea is interesting and promising.

3 Proactive Attitudes Leading to Better IoT Outcomes

In spite of the fact that this post talked about the various IoT-related dangers related to medicinal services, IoT activities – and keeping in mind that there are a lot more that we did not cover – healthcare experts should not feel disheartened about utilizing IoT in manners that bode well for the necessities, spending plans, and frameworks of their associations. Distinguishing the difficulties is the initial step to creating powerful arrangements. With a more profound comprehension of the difficulties confronting healthcare IoT activities, partners will be well while in transit to ending up well prepared enough to create and convey medicinal services IoT arrangements through 2019.

It flagged the mHealth [9] applications designed for sticking to HIPAA (Health Insurance Portability and Accountability Act) guidelines have turned into all-basic to improve patient consideration. How about we investigate how the joining of portable applications makes the patient consideration in the medical clinic better:

3.1 Making Patient's Records Go Paperless

Looking through the well-being records in the racks of the emergency clinic is a monotonous assignment and now and again, a little postponement in the data demonstrates to be deadly for the patient. That is the place where digitizing the patient's information is a panacea. The total well-being records such as X beam, vitals, solutions, medicinal history, past conferences and release notes of the patients are put away midway in the emergency clinic data framework and become available to the doctors on demand.

The helpful access to the patient's clinical information is progressive for crisis counsels. Additionally, the doctors can roll out the improvements in the well-being records, so the patient's well-being information consistently remains refreshed. As per the BMJ Quality and Safety study, "Supplanting paper wellbeing records with electronic well-being records have diminished the demise rates by 15%." The Medical Practice Management Software is an extra wherein catching the patient's well-being information, arrangement booking, computerized charging and report age is possible at one spot.

Progressively, the medical clinics are coordinating patient management programming with the goal that doctors can check the patient's past well-being records, look at the present well-being status, and propose the lab tests to perform, at the fingertips. Additionally, well-being conclusion applications empower the estimation of circulatory strain, pulse, glucose level, visual movement, and much more in only seconds. As indicated by Manhattan, "The dependence on electronic assets lessens the clinical basic leadership time considerably, limits the cost associated with treatment, and diminishes the pointless techniques that must be performed."

3.2 Patient's Management Digitally

Additionally, there are restorative mini-computer applications, which advise the doctors about hazard score subsequent to entering certain parameters. These are time-economical and demonstrated to be sans blunder in a quick-paced clinical condition to create dependable outcomes. Thus, more lives can be spared with helpful analysis, hazard score assurance, and rapid clinical basic leadership.

In the medical clinic, there are different areas, for example, inpatient wards, centers, outpatient administrations, labs, serious consideration units, and activity theater, which should be connected to one another. The versatile application advancement enables the individuals in the clinic to interface and team up through content, email or video call, which streamline and accelerate the treatment. For example, the data from labs can be quickly conveyed to the doctors in the activity theater without expecting to sit tight for an individual to physically bring the reports from the labs. The attendants in the inpatient wards can promptly counsel the doctor with the goal that no time shows signs of improvement and therapeutic choices can be made in the crisis cases. Moreover, there are some human services applications such as MediCall that empower fast meeting, determination, and solution from the doctor remotely through video gathering. The remote treatment works like a clinic in the pocket for the individuals.

3.3 Communication and Consulting Become Easier

Attendants in the clinic think that it is hard to recollect the movements, update with the assignments, go to patients, and give remedies which affect patient consideration. The strong portable application for attendants helps in the better administration of the patient's restorative timetable that empowers quicker recuperation and improved patient consideration. *NurseAlert* application is an enchantment shot that is splendid in staying up with the latest with move, never let the attendants neglect to go to the patient, and give drug on time through pill updates, and different errands can likewise be doled out and reminded to them. Moreover, medical caretakers can make and set the mechanized updates for the rehashed assignments.

3.4 Optimize the Nursing Activity

Medical caretakers in the emergency clinic think that it is hard to recall the movements, update with the errands, go to patients, and give solutions, which affect patient consideration. The durable versatile application for attendants helps in the better administration of the patient's therapeutic calendar that empowers quicker recuperation and improved patient consideration. *NurseAlert* application is an enchantment shot that is splendid in staying up with the latest with move of the executives, never let the attendants neglect to go to the patient, and give prescription on time through pill updates, and different undertakings can likewise be doled out and reminded to them. Furthermore, medical attendants can make and set the robotized updates for the rehashed assignments.

Figure 2 depicts different challenges in healthcare applications.

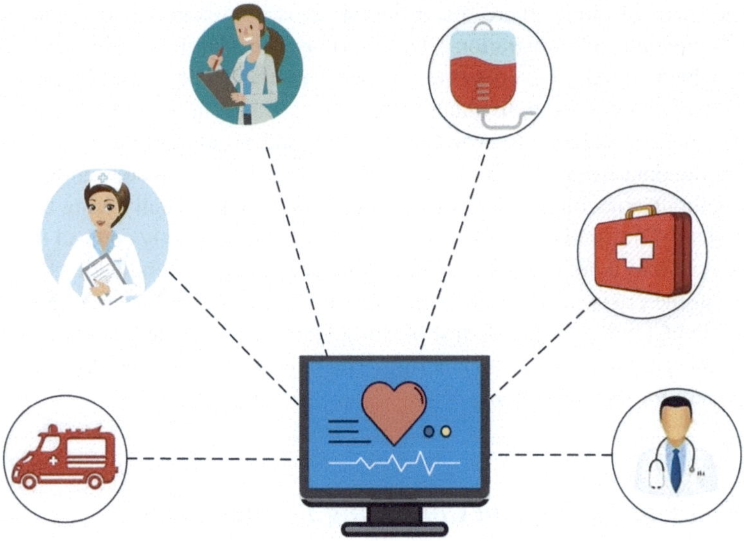

Fig. 2 Challenges in healthcare

4 Benefits

Innovation has disturbed each industry including human services, business, funding, and others. Healthcare remains the quickest to embrace innovative changes to upset the finding and treatment of the body. When we talk about the IoT, it offers a large number of advantages, for example, improving the adequacy and nature of administrations by conveying it in restorative gadgets.

4.1 Collecting and Understanding Medical Data

During a patient's remaining lifetime, they are tangled in medicinal gadgets including heart screens, blood siphons, respirators, and IVs. Nevertheless, the activity and recording of data from these gadgets take a great deal of time and are inclined to blunders for the benefit of parental figures. Today, with IoT a patient's information can be passed on through electronic health record frameworks. This strategy helps in expanding the exactness of information and enables medical attendants to invest more energy-giving consideration. Then again, specialists need to translate information to choose the remedy for patients. Because of the expansion of medicinal gadgets, it tends to be a challenge for specialists to think of an appropriate conclusion. For this, an IoT arrangement can be utilized to help well-being experts while joining IoT information from a huge number of therapeutic gadgets and increase bits of knowledge about patient's well-being, without dispersing the data.

4.2 *Patient Monitoring*

The development of wearable well-being devices such as the Apple iWatch has started assuming an essential job in the observing of a person's well-being. Overall, these items are not as exactly contrasted with general medicinal gear. Then again, wearable IoT gadgets can dissect and distinguish diverse well-being focuses, for example, circulatory strain, heartbeat, brainwaves, temperature, physical position, strides, and breathing. With the assistance of information gathered through IoT gadgets, specialists can share their criticism and give general recommendations because of a crisis. Despite the fact that the IoT is progressive in the medicinal services segment, there are not many difficulties also, which should be remembered. Let us examine them.

Some more benefits are listed below for a glimpse to the readers.

- Simultaneous reporting and monitoring
- End-to-end connectivity and affordability
- Data assortment and analysis
- Tracking and alerts
- Remote medical assistance
- Cutting cost through remote health monitoring [9]

5 Applications

Specifically, IoT technology in the field of material management has applications in the following areas. We can divide specific applications into the following areas:

5.1 *Medical Equipment and Drug Monitoring and Management*

With the assistance of RFIDs, IoT has started to discover more extensive applications in the field of medicinal material administration. IoT with RFIDs can help evade general medical issues by supporting in the creation, dispersion, and following of restorative gadgets and prescription. This builds the nature of restorative treatment while lessening the board costs. As indicated by the World Health Organization (WHO), the measure of fake meds on the planet adds up to over 10% of offers of medications around the world. Information from the Chinese Pharmaceutical Association demonstrates that, in China alone, at any rate, 200,000 individuals kick the bucket every year because of wrong or improperly utilized drug. Between 11% and 26% of patients utilize their meds inaccurately. This incorporates around 10% of wrongly recommended prescriptions. In this way, RFID innovation will assume a fundamental job in the following and checking of medications and hardware and the guideline of the market for restorative items.

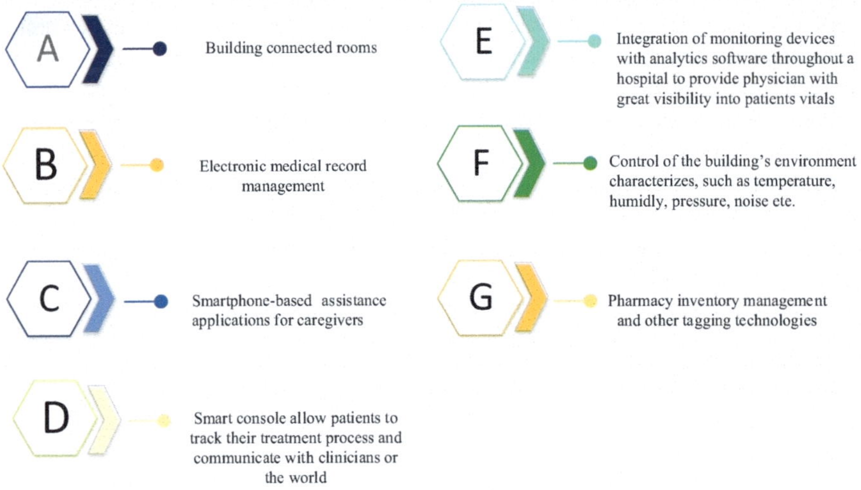

5.2 *Medical Device and Pharmaceuticals Anti-Counterfeiting*

The mark connected to an item will have a one of a kind personality that is very hard to fashion. This will assume a basic job in checking of data and hostile to falsifying, demonstrating a successful counter-measure against therapeutic extortion. For instance, it will be conceivable to transmit tranquilizer data to an open database from which a patient or medical clinic can check the substance of the name against the records in the database to effortlessly distinguish potential fakes.

5.3 *Complete Real-Time Monitoring*

From research to flow, the whole generation of procedures can use RFID labels to achieve a thorough item checking. This is particularly significant when the item gets dispatched. A peruser introduced on the generation line can naturally recognize each medication's data and transmit it to the database, as the item gets bundled. During the conveyance procedure, any halfway data get recorded whenever implying that it is conceivable to screen from start to finish.

5.4 *Medical Waste Information Management*

Collaboration by emergency clinics and transportation organizations will help build up a detectable therapeutic waste following framework utilizing RFID innovation. This will guarantee that the therapeutic waste gets appropriately moved to the treatment plant and will anticipate the unlawful dumping of biohazardous medicinal waste.

5.5 *Digital Hospital*

Web of Things has expansive application prospects in the field of medicinal data. At present, the interest for medicinal data for the executives in emergency clinics is chiefly in the accompanying angles: distinguishing proof, example acknowledgment, and restorative record recognizable proof. Distinguishing proof incorporates understanding recognizable proof, doctor ID, test ID (counting drug ID), restorative gear ID, lab ID, and medicinal record ID (counting indications and ID of malady).

5.6 *Patient Information Management*

The patient's family medicinal history, the patient's restorative history, different assessments, therapeutic records, tranquilizer sensitivities, and other electronic well-being documents can help specialists to create treatment programs. Specialists and medical caretakers can gauge the patient's indispensable signs, and, during medicines, for example, chemotherapy, they can utilize ongoing observing data to take out the utilization of wrong medications or wrong needles that can naturally remind attendants to complete medication checks and other work.

5.7 *Medical Emergency Management*

There are some strange conditions, for example, when there are enormous quantities of losses, a failure to arrive at relatives, or the fundamentally sick. In such situations, RFID advancements' solid and effective stockpiling and testing techniques will help with the quick recognizable proof of important subtleties, for example, the patient's name, age, blood classification, crisis contact, and past therapeutic history. This will accelerate confirmation strategies for crisis patients and leave all the more valuable time for treatment. Of specific significance is the establishment of 3G video hardware in ambulances. As patients are headed to the medical clinic, the crisis room is as of now getting acquainted with the patient's condition and can adequately get ready for crisis salvage. In the event that the area is extremely distant from the clinic, there is a likelihood of utilizing remote restorative imaging frameworks as a feature of the crisis salvage process.

5.8 *Drug Storage*

RFID innovation can robotize the entire chain of capacity, use, and assessment to decrease the necessary working hours, and streamline forms that recently got directed on paper. It can help forestall stock deficiencies and encourage the review of medications. It can likewise help maintain a strategic distance from perplexity emerging from comparative medication names or measurements sum and dose type. By and large, it will reinforce the board and guarantee that medicines are given and arranged immediately.

5.9 *Fighting Pharmaceutical Error*

Data the executives in the drug store will guarantee that medication is effectively appropriated and gotten. Until this point of time, drug store data the executives as of now gets actualized in such angles as giving medicines, altering measurements, nursing organization, understanding utilization of prescription, viability following, stock administration, acquisition of provisions, safeguarding of natural conditions, and assurance of timeframe of realistic usability. It is likewise used to affirm which sort of medication gets endorsed to the patient, and to not just record which medication the patient is taking and whether they have taken it but also the parcel number the medication originated from. This evades the plausibility of patients missing planned drugs, and, in case of a quality control issue, health sensitive patients get affected rapidly.

5.10 *Medical Equipment and Drug Tracking*

By precisely recording health updates and patient characters and giving solid help to mishap taking care of, accessible therapeutic gadgets and medication are represented. By precisely recording fundamental data, for example, item use, antagonistic occasions, territories where quality control issues may happen, patients included and areas of unused items, we can track and deal with awful items.

5.11 Connected Information Sharing

In the first place, structure a well-created and coordinated therapeutic system through the sharing of restorative data and records. From one perspective, approved specialist can check the patient's restorative history, history of sickness, treatment, and protection subtleties. Patients will likewise have the opportunity to pick or supplant specialists and emergency clinics. Then again, community and rustic medical clinics can flawlessly interface with focal emergency clinics for data and opportune master guidance, just as orchestrating referrals and get preparing.

5.12 New-Born Anti-Kidnapping System

At a general medical clinic's, obstetrics and gynaecology office or a ladies' and kids' emergency clinic, consolidating mother and kid recognizable proof administration, and baby security the executives will anticipate solo access by pariahs. Specifically, every newly conceived ought to get an RFID anklet that interestingly recognizes the child and has a one of a kind correspondences with the mother's data. To decide whether the family has the right child, the RFID anklet just needs to get checked by a medical caretaker or other staff.

5.13 Alarm System

Through continuous checking and following of clinic medicinal gadgets and patients, the framework will naturally call for assistance in case of patient pain. It will likewise keep patients from leaving the clinic all alone and will avert the harm or robbery of costly gadgets and secure temperature-touchy medications and research center.

6 Conclusion

IoT will definitely change the prosperity of business and the way in which patients are managed. Not only will it advantage pros and various specialists, but also to people who have no passageway to the basic prosperity workplaces. Tending to a couple of issues, for instance, data security will disturb the prosperity of business without breaking the assurance. We should clutch development as a blessing as opposed to a scold and see what happens in the coming years. The usage of IoT is extending bit by bit in each piece of the human administration. In this chapter, we have explored various uses of IoT in different verticals of the restorative business.

The significant passageway of the adaptable applications is changing the way in which patients are given remedial treatment in the crisis centers and watching out for all of the challenges the human administration industry is going up against. Various human administrations have hopped in the social protection industry getting a handle on moved versatility answers for better patient thinking. The more apparent challenges to grasping IoT in social protection are taking care of, managing, and checking data. The nonattendance of EHR compromise is another check to endure. The immovable quality and security issues with data cause nearby interoperability and a nonappearance of planning and establishment among providers in light of the way that despite when data streams energetically, various providers miss the mark on the structure and capacity to get to it. The accompanying degree of the issue lies among the people that can benefit most from IoT – poor web access among frail masses including the more seasoned, those with low preparing levels, lower paid people, commonplace occupants, and minorities.

References

1. https://www.medicaldesignandoutsourcing.com/medical-iot-security-challenges-healthcare/
2. https://breadware.com/blog/iot-technology-healthcare-examples
3. https://www.comparitech.com/internet-providers/iot-statistics/
4. G. Yang et al., A health-IoT platform based on the integration of intelligent packaging, unobtrusive bio-sensor, and intelligent medicine box. IEEE Trans. Industr. Inform. **10**(4), 2180–2191 (2014)
5. L. Catarinucci et al., An IoT-aware architecture for smart healthcare systems. IEEE Internet Things J. **2**(6), 515–526 (2015)
6. A. Al-Fuqaha, M. Guizani, M. Mohammadi, M. Aledhari, M. Ayyash, Internet of things: A survey on enabling technologies, protocols, and applications. IEEE Comm. Surv. Tutorials **17**(4), 2347–2376 (2015)
7. S.B. Baker, W. Xiang, I. Atkinson, Internet of things for smart healthcare: Technologies, challenges, and opportunities. IEEE Access **5**, 26521–26544 (2017)
8. D. Dziak, B. Jachimczyk, W. Kulesza, IoT-based information system for healthcare application: design methodology approach. Appl. Sci. **7**(6), 596 (2017)
9. A.J. Jara, M.A. Zamora-Izquierdo, A.F. Skarmeta, Interconnection framework for mHealth and remote monitoring based on the internet of things. IEEE J. Select. Areas Comm. **31**(9), 47–65 (2013)
10. D. Castro, W. Coral, J. Cabra, J. Colorado, D. Méndez, L. Trujillo, Survey on IoT solutions applied to Healthcare. Dyna **84**(203), 192–200 (2017)
11. D.V. Dimitrov, Medical internet of things and big data in healthcare. Healthcare Inform. Res. **22**(3), 156–163 (2016)

Part III
Safety Issues

Mimicking Biometrics on Smart Devices and Its Application in IoT Security for Health Systems

Dimitrios Myridakis, Georgios Spathoulas, Athanasios Kakarountas, Dimitrios Schoinianakis, and Joachim Lueken

1 Introduction

The Internet of Things (IoT) has introduced many novel applications in various domains such as healthcare, industry, defense, home monitoring, etc. Among these, healthcare holds a key position, since it adopts the research results from various topics (medical technology, sensors, security, privacy, just to name a few). The remote health monitoring, the exercise programs, the chronic illnesses and the care of elderly people are some of the medical applications whose range is enhanced by the power of IoT. Compliance of home treatment with healthcare providers is an additional challenge. In this section, a set of healthcare applications is discussed to pinpoint the importance and the broad scope of security in this research field.

1.1 Healthcare Applications

1.1.1 Healthcare Solutions Using Smartphones

The proliferation of smartphones in our daily life has given access to external smart sensors, highlighting the role of smartphones as a driver for the IoT. Smartphones have become a versatile healthcare device thanks to the design of various software

D. Myridakis (✉) · G. Spathoulas · A. Kakarountas
Computer Science and Biomedical Informatics, University of Thessaly, Lamia, Greece
e-mail: dmyridakis@dib.uth.gr; gspathoulas@dib.uth.gr; kakarountas@uth.gr

D. Schoinianakis · J. Lueken
Cyber Security Munich (CSM), Nokia Bell Labs, Munich, Germany
e-mail: dimitrios.schoinianakis@nokia-bell-labs.com; joachim.lueken@nokia-bell-labs.com

N. Gupta, S. Paiva (eds.), *IoT and ICT for Healthcare Applications*, EAI/Springer Innovations in Communication and Computing,
https://doi.org/10.1007/978-3-030-42934-8_10

and hardware products. In [30] an extensive overview of healthcare applications for smartphones is provided, as well as patient applications and general healthcare applications are discussed. At the same time there is a review on information search applications, training, and medical education (collectively referred to as auxiliary applications). After all, many recent smartphone apps serve similar purposes as the aforementioned applications [2, 20, 29, 34].

1.1.2 Imminent Healthcare Solutions

Despite the abundance of portable medical devices, there is no clear demonstration of their integration with the Internet. It is envisaged that in the near future health devices will be equipped with IoT functions as this is only a matter of time. Applications, devices, and other healthcare cases are increasing in number as the demand for IoT services around the world grows. Some healthcare areas for which integration with IoT is imminent include skin infections, sharp exhalation flow, hemoglobin detection, cancer treatment, abnormal cell growth, ocular disorders, and remote monitoring [15, 33, 36]. Nowadays, the devices beyond their portability and diagnostics are also compatible with most connectivity protocols.

1.1.3 Wheelchair Management

There are quite a few researchers who have worked to develop intelligent wheelchairs, which serve people with disabilities. An example that we should mention is that of the intelligent wheelchair designed by Intel's IoT department [21]. The above development basically shows that standard "things" can evolve into intelligent machines with a degree of automation. This device allows evaluation of a site's accessibility as it collects data from proximity sensors and monitors vital functions of the person sitting on the chair.

1.1.4 Medication Management

One of the most serious threats to public health is the problem of non-compliance with medication. In order to address the above issue, some promising solutions based on IoT are offered. A typical example of Internet-based drug management of a smart drug box packaging method is proposed in [32].

1.1.5 Rehabilitation System

One of the most vital disciplines of medicine is represented by physical medicine and rehabilitation, as they can enhance or even restore the functional ability as well as the quality of life of people with some physical injury or even disability. The health sector suffers from staff shortages, which the IoT is coming to alleviate. Typical examples include an ontology-based automated design method for smart recovery systems based IoT, as suggested in [13]. IoT-based rehabilitation systems such as the smart city medical rehabilitation system [44], migraine rehabilitation education [17], integrated prison application system [4], and childhood education autism [26] highlights the endless possibilities in this area.

1.1.6 Oxygen Saturation Monitoring

Oxygen saturation in the blood can be monitored continuously non-invasively with pulse oximetry. Technology-based healthcare applications are useful to integrate IoT with pulse oximetry. A research into CoAP-based healthcare services looks at the possibilities of pulse oximetry based on IoT [24]. The above device can be used to continuously monitor patient health through an IoT network. For telemedicine applications an integrated pulse oximeter system is described in [25].

1.1.7 Blood Pressure Monitoring

Work on [39] examines how a Bringing BP device works with an Apple mobile computing device, while [22] proposes a device for collecting and transmitting BP data to an IoT network. The combination of a KIT cellphone and a NFC-enabled KIT (BP) blood pressure monitor is being tested at [8] as part of an IoT-based BP monitoring.

1.1.8 Electrocardiogram Monitoring

Electrocardiogram (ECG) monitoring that takes advantage of IoT in addition to being able to be used to its fullest has the potential to provide maximum information [9]. Discussion has been made on some studies on ECG tracking based on IoT [18, 19, 28, 35]. An IoT-based ECG tracking system introduces the innovative [42]. Use a portable monitoring node to collect ECG data, and use Wi-Fi transmitted directly to the IoT cloud. Both HTTP and MQTT protocols in the IoT cloud are used to provide visualization, validation of ECG data to users, as well as detection of abnormal data that may indicate real-time cardiac dysfunction.

1.1.9 Glucose Level Sensing

Diabetes mellitus is a metabolic disease characterized by impaired glucose metabolism and an increase in blood sugar (hyperglycemia), either because of the decreased sensitivity of the body's cells to insulin or as a result of decreased insulin secretion. Gia et al. [16] study the feasibility of an invasive and continuous glucose monitoring (CGM) system using the IoT-based approach. The utility model at [23] examines the m-IoT-based architecture for diabetes management as well as the potential benefits of using m-IoT in non-invasive glucose levels detection.

The enhancing user's experience, the enhancing quality of life, and the reducing costs are to be expected from IoT which are based to healthcare services. From the point of view of healthcare providers, the correct determination of the optimum times for replenishing supplies to their continuous and smooth operation is a possibility provided by IoT. Ensuring the best possible use and service of more patients is achieved by IoT which provides for the efficient planning of scarce resources.

1.2 Security Issues Analysis

Several limitations in IoT devices prevent their full-fledged adoption in healthcare systems. Interoperability and security are especially impacted by such limitations [39]. Since the architecture of the system is not-well defined, data restriction and integrity as well as the robustness of the overall system are not guaranteed. Even though IoT provides the protocols for maintaining the information, various issues are raised. As an example, it is questionable whether Internet technologies and practices such as TCP/IP and OSI protocols or protocols such as IEEE 802.3 or IEEE 802.11 are applicable to IoT. These concerns are due to the fact that IoT devices do not follow the same protocol implementation practices and pertain to specific architectures, computing capabilities, interconnection capabilities, and requirements. As a result, security remains to be effectively addressed.

In order to understand the magnitude of the problem, we need to get a full picture of how the IoT system is encapsulated and used by the protocols and technologies that implement them. OSI Protocol Stack (ISO/IEC 7498-1) is an ideal starting point, since we can place every technology at its level and study it in general, but also in terms of security in particular. For example, technologies based on IEEE 802.15.4 and IEEE 802.15.1 standards cover the Physical Layer and Data Link Layer. Essentially, with the right equipment, one can intercept the data reaching/leaving the IoT device.

With regard to the second level of the Data Link, a very common attack is MAC Spoofing, where the attacker not only gains physical access to the target network but also changes/hides the physical address of his system so that he/she remains covert. However, if the protocol stack used does not take advantage of the MAC addresses, then using a Protocol Analyzer application, we can record the frames that are exchanged. It is a dual-sided attack, since we would be able in principle to access their content if it is not encrypted and at the same time intervene in the protocol by sending similar frames [40].

Attacks at higher levels such as the Network Layer, especially when using the widely used IP, are also quite common: IP Spoofing, Routing (RIP) Attacks, ICMP Flood, ICMP Flood, Ping of Death Attack, Teardrop Attack, and Packet Sniffing [1].

Moving on to the Transport Layer we encounter attacks on protocols such as TCP and UDP: TCP sequential attack, TCP SYN flood, TCP reset attack, UDP flood attack, Smurf attack, etc. [41]. At the Session Layer, we have Session hijacking attacks on existing Transport Level connections such as man-in-the-middle attacks, blind hijacking, etc. [5]. At the Presentation Layer, we have attacks on the way data is delivered to final user applications. At this level, encryption/decryption of data is typically done via the SSL and TLS protocols. Attacks that apply in this category are Session Layer attacks: SSL stripping, STARTTLS command injection attack, beast, padding oracle attacks, compression attacks, and certificates attacks [37].

Finally, at the Application Layer level, the attacks are related to the end user application such as FTP bounce attack on FTP applications, SSH brute force attack on SSH applications, HTTP input attack validation [3], SQL injection attack [14], cross-site request forgery attack [6], and XSS attack (among others).

Secure Shell (SSH) is a protocol for secure remote connection and other secure network services through an insecure network. It consists of three key elements [43]:

1. Transport Layer protocol provides server authentication, data integrity, and transaction confidentiality. Optionally, it can also apply data compression. Typically, it runs over a TCP/IP connection.
2. The User Authentication protocol authenticates the client-user ID on the server. It runs over Transport Layer protocol.
3. The Connection protocol multiplexes the encrypted physical channel into several logical channels and runs over the User Authentication protocol.

The aforementioned analysis applies very well to healthcare systems and allows for intermediate attacks and intruders to access health information which in turn plagues the privacy, security, and reliability of the entire healthcare system in an interconnected framework. More and more healthcare organizations are becoming targets of cyberattacks due to the goldmine of their personal data. Medical practitioners come across the phrase Hippocrates vow, "First, do not harm," and this reflects the utmost importance attached to patient care. A health professional, above all, must consider the well-being of patients. And nowadays, in 2020, concerns about privacy and cybersecurity are on the rise of sophisticated "prosperity."

When discussing these potential problems, we need to think about what IoT applications and what impact an attack may have that will alter or subtract data. As significant as the 2017 WannaCry and NotPeyta ransomware attacks [27] are, the potential real impact of these attacks is just as important. The potential of cyberattackers has increased, resulting in the access, theft, and sale of patient information on the dark web. In addition, another important skill is that they are making effective patient care almost impossible by closing a hospital's access to critical patient systems and records.

A typical example could be a system of devices that monitors a patient's vital health functions (e.g., sugar, pressure, temperature, pulse, etc.) and send these data to a recording/monitoring system within a hospital or a similar system that is installed in a doctor's office. How safe is this transmission? How reliable? If we go a step further, let's consider a corresponding device that controls the drug flow to adjust the sugar intravenously to the patient. What if the right amount is not injected? Which protocols are used and how are we protected from attacks? The use of already known and widely used technologies has the advantage that any known vulnerability may also apply to IoT devices, but there could still be many others related to the specific device. There are also technologies and protocols built exclusively for IoT and the devices that shape it, such as the IEEE 802.15.4 standard protocols or ZigBee's trademark. Extra care to account for the intrinsic vulnerabilities and quirks of those protocols and devices is required.

As the adoption of IoT devices and medical devices increases, the scope for healthcare attacks is increasing. Stagnant cybersecurity budgeting and cybersecurity budget constraints have exacerbated the problem. In order to achieve robust performance in an IoT system, various approaches have been introduced [7]. Developers in information and communication technologies make it easier for researchers and engineers to realize that the system is certified for all applications. In a healthcare system, if we look at the confidentiality and sensitivity of medical data, it must meet advanced access control procedures with stringent security and quality data requirements [38]. The project at [10] aims to integrate into a secure healthcare monitoring system, artificial intelligence technology such as neural networks and the fuzzy system. As a result, the system operates as an autonomous "smart healthcare system" that makes decisions on its own and prioritizes the health parameters collected from the sensor nodes. At [31], the authors propose an access control architecture for limited IoT healthcare resources. The approach is based on a policy that provides detailed access to authorized users of services while protecting valuable resources from unauthorized access, while [11] proposes a hybrid security model to provide diagnostic data on medical images.

The significance of IoT security in general is quite apparent in international standardization and policy-making groups and consortia.An interesting extensive report, titled Key Fun Suggestions for Fun in the Critical Information Infrastructure to Identify Key IoT Cyber Security Recommendations, was published in November 2017 by the European Network and Information Security Agency, formerly called the European Network and Information Security Agency (ENISA) [12].

2 Bio-mimicking on Smart Devices

The concept of Smart BIoTS considers the monitoring of a smart device in terms of functional characteristics, just like humans do to check their health status. Such characteristics on humans are defined by normal values or ranges of biological measurements. In an analogy to the physical characteristics of a living organism, the functional characteristics (e.g., power consumption, noise, temperature, etc.) of a smart device are continuously monitored. Furthermore, the operational characteristics, in an analogy to the behavioral characteristics of a living organism, are also monitored (e.g., motion of a camera, update frequency, etc.). Thus, a smart device may be considered as a simple bio-mimicking system, with well-defined characteristics, which through the study of the system itself have been identified or even identified by the manufacturer. Since the potential characteristics that may be monitored are numerous, the power supply is selected to be considered in this chapter, to prove the concept of this method.

The main idea behind this method is focused on the monitoring of the supply current on smart devices. Given that the functionality and functional characteristics of such a device are limited, it is conceivable that any deviation from normal operation would result in a corresponding deviation for the power consumed. In an analogy to the immunity system of a living organism, that tries to confront any virus attack or to rehabilitate any excessive effort of the body, and the effect is usually expressed by the increase of the body temperature, a smart device is expected to deviate from normal operation characteristics. It can therefore be considered as a sidetrack of the device's unexpected operation, as the energy consumption of a resource is greater than expected in a current deviation.

Furthermore, correlating operation to energy consumption, it may be possible to draw conclusions on the network activity of the medical device. In the presented approach, a device is set up as an intermediate observer that intermediate between the power supply of the device and the device under monitoring. We're talking about a realistic approach, as all devices offer access to their power supply. Reliability issues as well as security-related attacks against the device can eventually be detected using the proposed system. The constructed setup as well as the relative measurement data are presented throughout the rest of the chapter.

Concerning on the proposed setup, we have used a microcontroller, which monitors the supply current of the targeted device. The microcontroller has to be interposed between the power supply and the device that should be monitored:

- At first, the microcontroller is connected to a personal computer (laptop). The code that has been developed for current monitoring purpose is compiled and uploaded to the microcontroller code memory. This is considered a golden application code, hence without any malicious inclusions.
- The monitoring circuit is built on a breadboard. To achieve greater accuracy in calibration, a low resistance of 1 Ohm is used. The above resistance is located between the two inputs of the microcontroller to measure the voltage. According to the following formula, it is possible to calculate the voltage measurement at the two input points.

$$I = \frac{V_2 - V_1}{R} \tag{1}$$

where V_1 and V_2 are the two reference voltages as depicted in Fig. 1 and R is the 1 Ohm resistor.

In this case and in order to be able to prove the concept, we aim to control the discrepancies of normal operation for various installations as well as the network activity and specifically deviations of supply current. Any deviation is a warning

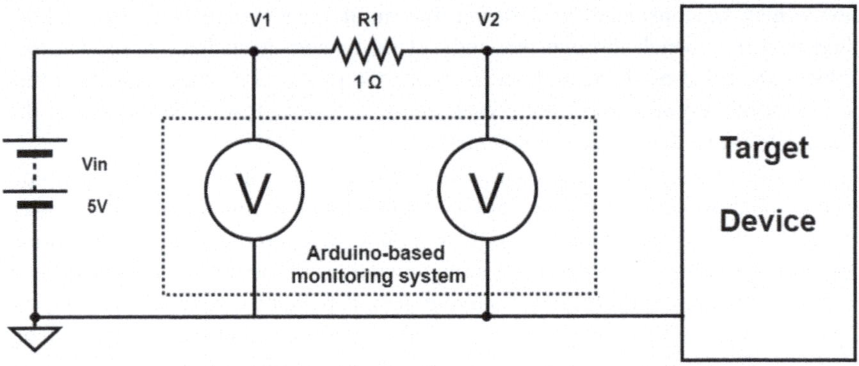

Fig. 1 Circuit of monitoring device

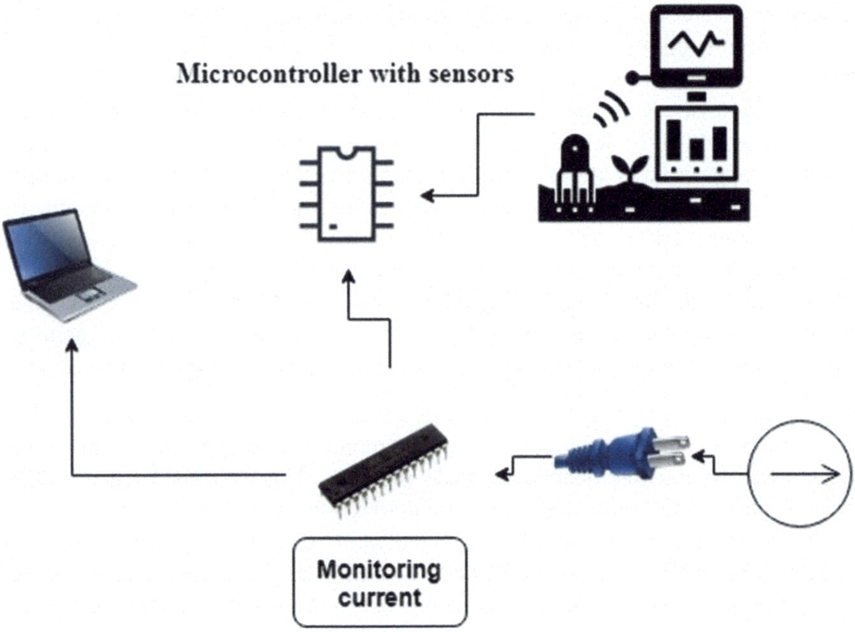

Fig. 2 Microcontroller with sensors

for the presence of an anomaly. Normal operation will be referenced hereinafter as normal profile of the device.

Three applications were considered to demonstrate the validity of this approach.The first application is monitoring humidity and temperature, inside a hospital room, using custom hardware for this purpose, as illustrated in Fig. 2. It includes a microcontroller, along with the required sensors. The second application is based on a commercially available IP camera for monitoring indoor and outdoor

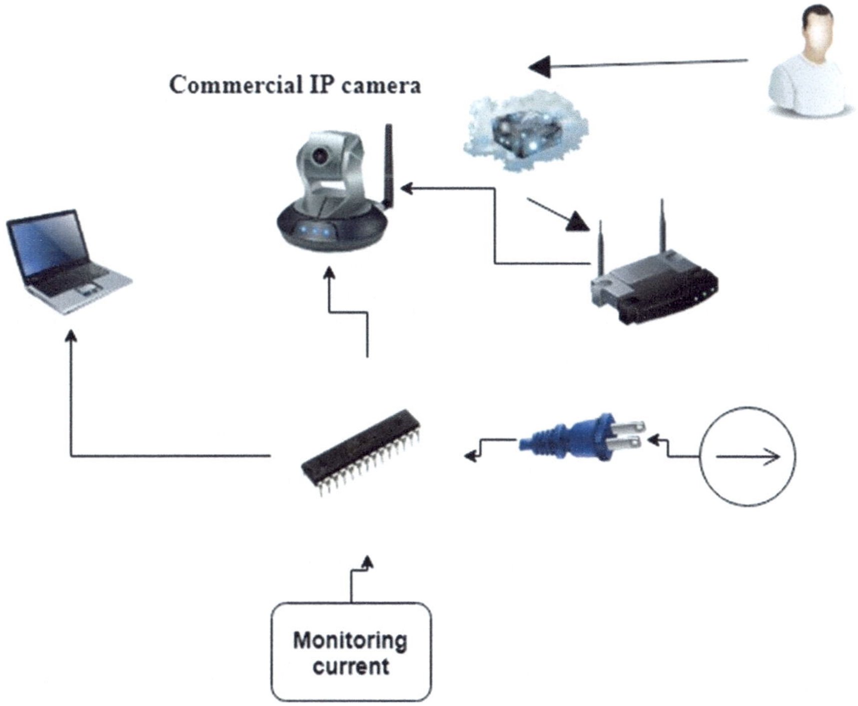

Fig. 3 Commercial IP camera

areas, usually used for identifying falls of older people in their house, depicted in Fig. 3, and the third includes a custom-made IP camera (Fig. 4).

3 Results

In order to prove that the observation of abnormal network activity is feasible via monitoring of supply current, a denial of service (DoS) attack has been mounted. For the case of the microcontroller thermometer, the device performed its normal operation capturing temperature and room humidity. Neither an external environmental factor exceeded normal characteristics nor an attack was performed, in order to have a profile of the power amperage in normal conditions for reference values. The graph of Fig. 5 depicts the measurements under normal operation.

Subsequently, a DoS attack was executed against the microcontroller thermometer device, during its normal operation. The measurements for power amperage, as depicted in Fig. 6, clearly illustrate an increase in power – indicating the increase in the device's network activity.

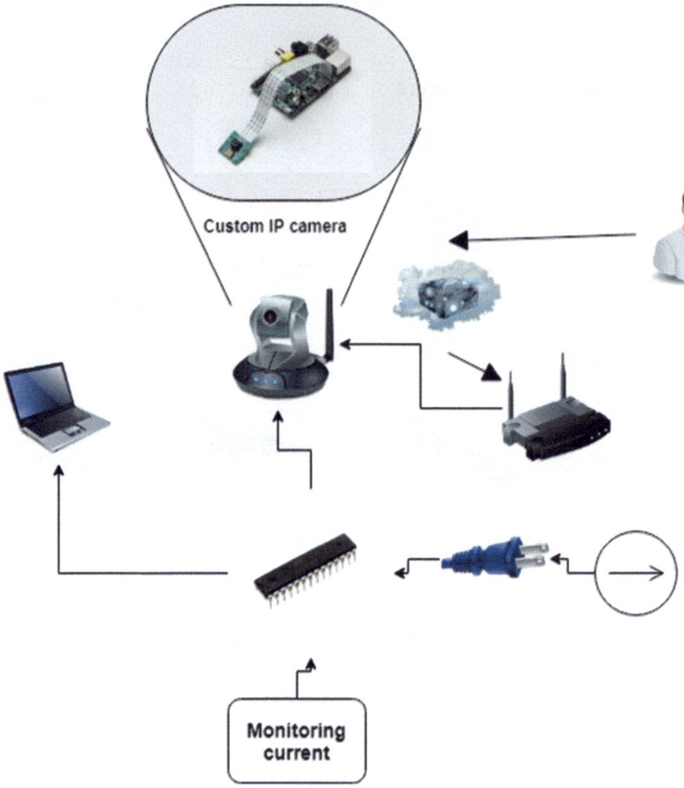

Fig. 4 Custom IP camera

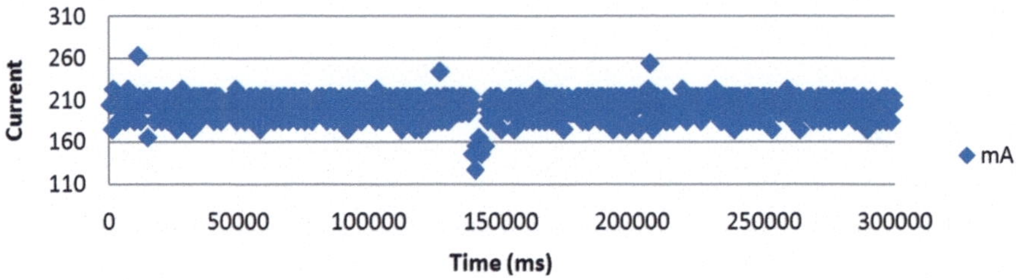

Fig. 5 Thermometer – normal operation

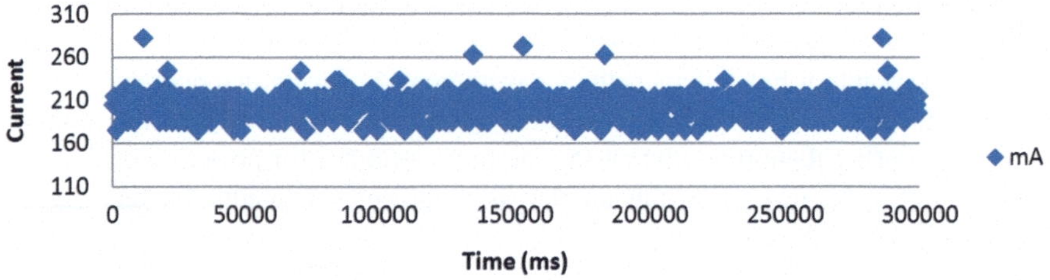

Fig. 6 Thermometer – under attack

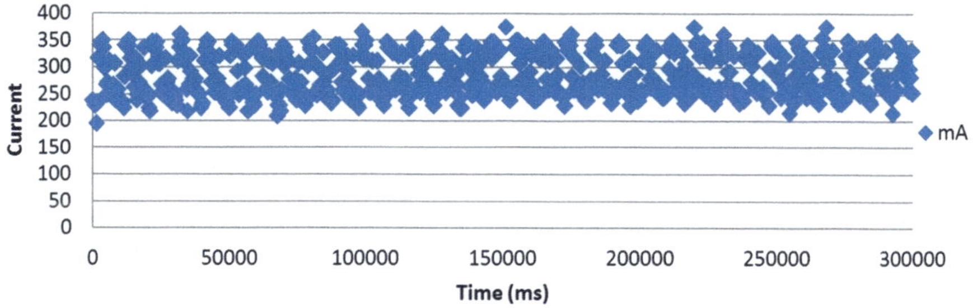

Fig. 7 Commercial camera – normal operation

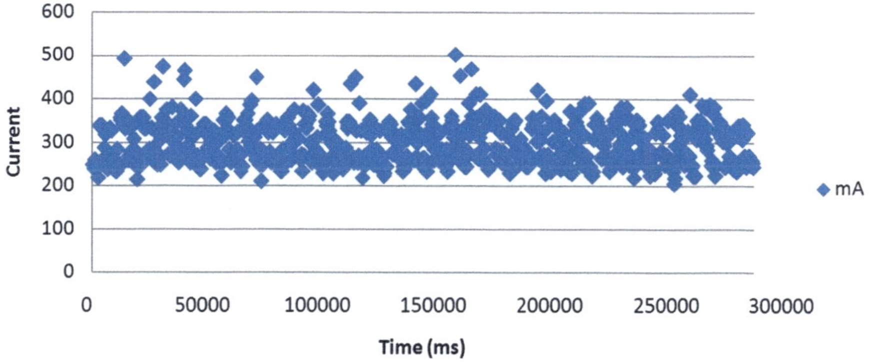

Fig. 8 Commercial camera – under attack

Regarding the commercial IP camera, the scenario involves streaming of a still image without any additional network activity, which corresponds to the typical scenario of a home room monitoring for elders' falls. The measurements graph for the normal behavior is presented in Fig. 7. The respective measurements for when the device was under a DoS attack appear in Fig. 8. As with the thermometer case, the pattern of power amperage increase is clearly visible.

Finally, with regard to the custom-made IP camera, the microelectronic device performed periodical capturing of current values. Neither an external environment factor exceeded normal characteristics nor an attack was performed, in order to construct a baseline profile of the power amperage for comparison reasons. The normal profile figure for this profile is depicted in Fig. 9, where three charts have been included (horizontal axis denotes time in hours): the diagram labeled Power represents the actual current of the device, while the second and third diagram apply moving average filters for better visualization of the results.

As with the other cases, a DoS attack was launched, and the results are summarized in Fig. 10. There is a clear deviation of current supply when a DoS attack is launched. Considering that the features of the DoS attack were realistic, it may be said that even with a simple external circuitry, we were able to sense

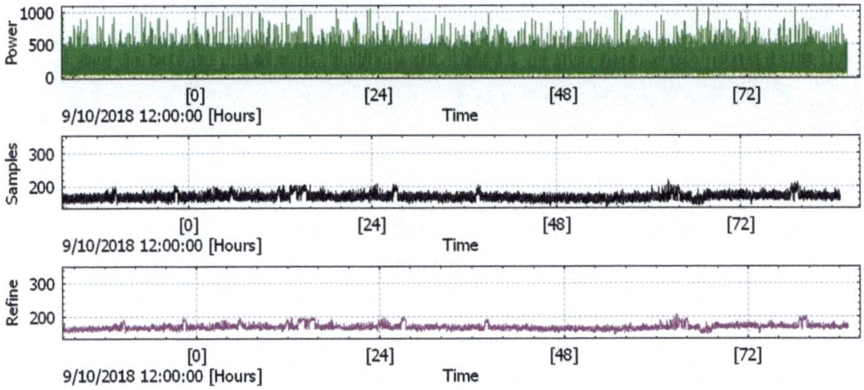

Fig. 9 Custom camera – normal operation

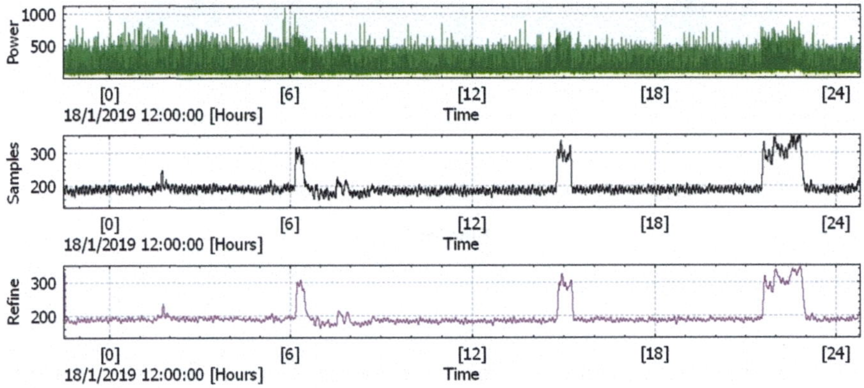

Fig. 10 Custom camera – under attack

deviations and detect the anomaly. It is expected to measure even bigger deviations under a normal Botnet attack and thus trigger a safe mode for the smart device.

4 Conclusions

It is without question that there are still significant steps to be taken toward a safe and secure IoT infrastructure.With regard to the healthcare system, the devices employed offer a wide research spectrum for developing efficient security, privacy, and energy-awareness mechanisms. In this context, we have examined and demonstrated that by monitoring indirect characteristics of IoT devices, it is possible to draw safe results as to whether the device is under attack and trigger respective protection mechanisms.

Although in our experiments we considered simple scenarios regarding health applications (monitoring conditions in a hospital room and monitoring elders for falls in their house) for showcasing the effectiveness of the method, we consider the presented methodology to be rather generic, and it is expected to behave equivalently in most IoT devices.

Ultimately, Smart BIoTS attempts to address both reliability and security issues based on the bioequivalence of IoT's devices and to safeguard healthcare devices by examining the "health" of the device itself.

The extension of monitored parameters for security issues confronted by intrusion detection systems is expected to be integrated soon in major systems. The presented method is the basis for this kind of parameter extension that will provide additional measurements for protecting critical IoT infrastructures. The handling of the collected data in an IDS, and the exploitation of machine learning algorithms, in order to identify promptly new kinds of attacks, is a challenge for future research.

Acknowledgments This work was supported by NOKIA Bell Labs through global donation program for "Research on IoT Security based on Bio-informatics."

References

1. A.H. Alqahtani, M. Iftikhar, TCP/IP attacks, defenses and security tools. Int. J. Sci. Modern Eng. (IJISME) **1**(10) (2013). ISSN: 2319–6386
2. Apple, Apple. https://www.apple.com/itunes/charts/ (2019). Online; Accessed 10 June 2019
3. H. Beitollahi, G. Deconinck, Tackling application-layer DDoS attacks. Proc. Comput. Sci. **10**, 432–441 (2012)
4. M.I. Bhat, S. Ahmad, A. Amin, S. Ashraf, e-health with internet of things. Int. J. Comput. Sci. Mob. Comput. **6**(6), 357–362 (2017)
5. I. Dacosta, S. Chakradeo, M. Ahamad, P. Traynor, One-time cookies: preventing session hijacking attacks with stateless authentication tokens. ACM Trans. Internet Technol. (TOIT) **12**(1), 1 (2012)
6. P. De Ryck, L. Desmet, W. Joosen, F. Piessens, Automatic and precise client-side protection against CSRF attacks, in *European Symposium on Research in Computer Security* (Springer, 2011), pp. 100–116
7. M.M. Dhanvijay, S.C. Patil, Internet of things: a survey of enabling technologies in healthcare and its applications. Comput. Netw. **153**, 113–131 (2019)
8. A. Dohr, R. Modre-Opsrian, M. Drobics, D. Hayn, G. Schreier, The internet of things for ambient assisted living, in *2010 Seventh International Conference on Information Technology: New Generations* (IEEE, 2010), pp. 804–809
9. B.J. Drew, R.M. Califf, M. Funk, E.S. Kaufman, M.W. Krucoff, M.M. Laks, P.W. Macfarlane, C. Sommargren, S. Swiryn, G.F. Van Hare, Practice standards for electrocardiographic monitoring in hospital settings: an American heart association scientific statement from the councils on cardiovascular nursing, clinical cardiology, and cardiovascular disease in the young: endorsed by the international society of computerized electrocardiology and the American association of critical-care nurses. Circulation **110**(17), 2721–2746 (2004)
10. H.A. El Zouka, M.M. Hosni, Secure IoT communications for smart healthcare monitoring system. Internet Things, In press, corrected proof, Available online 24 January 2019 Article 100036 (2019)

11. M. Elhoseny, G. Ramírez-González, O.M. Abu-Elnasr, S.A. Shawkat, N. Arunkumar, A. Farouk, Secure medical data transmission model for iot-based healthcare systems. IEEE Access **6**, 20596–20608 (2018)
12. Enisa, Enisa. https://www.enisa.europa.eu/publications/baseline-security-recommendations-for-iot/ (2019). Online; Accessed 10 June 2019
13. Y.J. Fan, Y.H. Yin, L. Da Xu, Y. Zeng, F. Wu, IoT-based smart rehabilitation system. IEEE Trans. Ind. Inf. **10**(2), 1568–1577 (2014)
14. J. Fonseca, M. Vieira, H. Madeira, Testing and comparing web vulnerability scanning tools for SQL injection and XSS attacks, in *13th Pacific Rim International Symposium on Dependable Computing (PRDC 2007)* (IEEE, 2007), pp. 365–372
15. E. Gayat, A. Bodin, C. Sportiello, M. Boisson, J.F. Dreyfus, E. Mathieu, M. Fischler, Performance evaluation of a noninvasive hemoglobin monitoring device. Ann. Emerg. Med. **57**(4), 330–333 (2011)
16. T.N. Gia, M. Ali, I.B. Dhaou, A.M. Rahmani, T. Westerlund, P. Liljeberg, H. Tenhunen, IoT-based continuous glucose monitoring system: a feasibility study. Proc. Comput. Sci. **109**, 327–334 (2017)
17. Z. Guangnan, L. Penghui, IoT (internet of things) control system facing rehabilitation training of hemiplegic patients. Chin. Patent **202**(587), 045 (2012)
18. M. Hassanalieragh, A. Page, T. Soyata, G. Sharma, M. Aktas, G. Mateos, B. Kantarci, S. Andreescu, Health monitoring and management using internet-of-things (IoT) sensing with cloud-based processing: opportunities and challenges, in *2015 IEEE International Conference on Services Computing* (IEEE, 2015), pp. 285–292
19. M.S. Hossain, G. Muhammad, Cloud-assisted industrial internet of things (IIoT)–enabled framework for health monitoring. Comput. Netw. **101**, 192–202 (2016)
20. Imedicalapps, Imedicalapps. http://www.imedicalapps.com/2014/01/diagnose-app-evidence-based-clinical-decision/ (2019). Online; Accessed 10 June 2019
21. Intel, Intel. http://www.intel.co.kr/content/www/kr/ko/internet-of-things/videos/dr-hawkings-connected-wheelchair-video.html/ (2019). Online; Accessed 10 June 2019
22. S.R. Islam, D. Kwak, M.H. Kabir, M. Hossain, K.S. Kwak, The internet of things for health care: a comprehensive survey. IEEE Access **3**, 678–708 (2015)
23. R.S. Istepanian, S. Hu, N.Y. Philip, A. Sungoor, The potential of internet of m-health things (m-IoT) for non-invasive glucose level sensing, in *2011 Annual International Conference of the IEEE Engineering in Medicine and Biology Society* (IEEE, 2011), pp. 5264–5266
24. H.A. Khattak, M. Ruta, E. Di Sciascio, Coap-based healthcare sensor networks: a survey, in *Proceedings of 11th International Bhurban Conference on Applied Sciences and Technology (IBCAST)*, 2014, pp. 499–503
25. E.C. Larson, M. Goel, M. Redfield, G. Boriello, M. Rosenfeld, S.N. Patel, Tracking lung function on any phone, in *Proceedings of the 3rd ACM Symposium on Computing for Development* (ACM, 2013), p. 29
26. S. Liang, Y. Zilong, S. Hai, M. Trinidad, Childhood autism language training system and internet-of-things-based centralized training center. Chin. Patent **102**(184), 661 (2011)
27. S. Mansfield-Devine, Ransomware: the most popular form of attack. Comput. Fraud Secur. **2017**(10), 15–20 (2017)
28. M. Marudhapandi, D. Ramkumar, R. Ramkumar, S. Jeevanandham, N. Suguna, Wearable ecg monitoring system and data analysis. Int. J. Sci. Res. Comput. Sci. Eng. Inf. Technol. **5**, 667–673 (2019)
29. Medicaljoyworks, Medicaljoyworks. http://www.medicaljoyworks.com/ (2019). Online; Accessed 10 June 2019
30. A.S.M. Mosa, I. Yoo, L. Sheets, A systematic review of healthcare applications for smartphones. BMC Med. Inform. Decis. Mak. **12**(1), 67 (2012)
31. S. Pal, M. Hitchens, V. Varadharajan, T. Rabehaja, Policy-based access control for constrained healthcare resources, in *2018 IEEE 19th International Symposium on "A World of Wireless, Mobile and Multimedia Networks" (WoWMoM)* (IEEE, 2018), pp. 588–599

32. Z. Pang, J. Tian, Q. Chen, Intelligent packaging and intelligent medicine box for medication management towards the internet-of-things, in *16th International Conference on Advanced Communication Technology* (IEEE, 2014), pp. 352–360
33. M. Pesta, J. Fichtl, V. Kulda, O. Topolcan, V. Treska, Monitoring of circulating tumor cells in patients undergoing surgery for hepatic metastases from colorectal cancer. Anticancer Res. **33**(5), 2239–2243 (2013)
34. Prognosisapp, Apple. http://www.prognosisapp.com/ (2019). Online; Accessed 10 June 2019
35. A. Rahman, T. Rahman, N.H. Ghani, S. Hossain, J. Uddin, IoT based patient monitoring system using ecg sensor, in *2019 International Conference on Robotics, Electrical and Signal Processing Techniques (ICREST)* (IEEE, 2019), pp. 378–382
36. V.M. Rohokale, N.R. Prasad, R. Prasad, A cooperative internet of things (IoT) for rural healthcare monitoring and control, in *2011 2nd International Conference on Wireless Communication, Vehicular Technology, Information Theory and Aerospace & Electronic Systems Technology (Wireless VITAE)* (IEEE, 2011), pp. 1–6
37. Y. Sheffer, R. Holz, P. Saint-Andre, Summarizing known attacks on transport layer security (TLS) and datagram TLS (DTLS). Technical report, Technische Universitaet Muenchen, 2015
38. S. Sicari, A. Rizzardi, L. Grieco, G. Piro, A. Coen-Porisini, A policy enforcement framework for internet of things applications in the smart health. Smart Health **3**, 39–74 (2017)
39. L.M.R. Tarouco, L.M. Bertholdo, L.Z. Granville, L.M.R. Arbiza, F. Carbone, M. Marotta, J.J.C. De Santanna, Internet of things in healthcare: interoperatibility and security issues, in *2012 IEEE International Conference on Communications (ICC)* (IEEE, 2012), pp. 6121–6125
40. W. Xu, K. Ma, W. Trappe, Y. Zhang, Jamming sensor networks: attack and defense strategies. IEEE Netw. **20**(3), 41–47 (2006)
41. G. Yang, M. Gerla, M. Sanadidi, Defense against low-rate TCP-targeted denial-of-service attacks, in *Proceedings. ISCC 2004. Ninth International Symposium on Computers and Communications (IEEE Cat. No. 04TH8769)*, vol. 1 (IEEE, 2004), pp. 345–350
42. Z. Yang, Q. Zhou, L. Lei, K. Zheng, W. Xiang, An IoT-cloud based wearable ecg monitoring system for smart healthcare. J. Med. Syst. **40**(12), 286 (2016)
43. T. Ylonen, C. Lonvick, The secure shell (SSH) protocol architecture. Technical report, The Internet Society, 2006
44. Y. Yue-Hong, F. Wu, F.Y. Jie, L. Jian, X. Chao, Z. Yi, Remote medical rehabilitation system in smart city. Chin. Patent **103**(488), 880 (2014)

Data Privacy and Security for IoMWT (Internet of Medical Wearable Things) Cloud

Raluca Maria Aileni, George Suciu, Maheswar Rajagopal, Sever Pasca, and Carlos Alberto Valderrama Sukuyama

1 Introduction

The latest technology developments have produced a sudden release of intelligent and deeply associated computing devices that are in a Small Form Factor (SFF). Communication devices interact with people these days, in different domains, such as lifestyle, well-being, and entertainment [1]. Scaling down the hardware equipment carries out a crucial role in this improvement; the Internet was a less known part [1, 2]. The Internet performed by far the most critical role in the ascension of the technology, and it is fundamental because of the transformation it has gone through since the 1990s. During the time, the Internet was utilized mainly only for correspondence purposes, such as emails. Nonetheless, it has been observed an evolution during the years, when in the 2000s, the mobile phone was the revolution in the field of wireless technologies [2]. Nowadays, there are many numbers of customers associated with the Internet lead to the Internet

R. M. Aileni (✉) · S. Pasca
Politehnica University of Bucharest, Faculty of Electronics, Telecommunication and Information Technology, Bucharest, Romania

G. Suciu
Politehnica University of Bucharest, Faculty of Electronics, Telecommunication and Information Technology, Bucharest, Romania

Beia Consult International, Bucharest, Romania

M. Rajagopal
Sri Krishna College of Technology, Coimbatore, India

C. A. Valderrama Sukuyama
Mons University, Faculty of Engineering, Department of Electronics and Microelectronics, Mons, Belgium

N. Gupta, S. Paiva (eds.), *IoT and ICT for Healthcare Applications*, EAI/Springer Innovations in Communication and Computing,
https://doi.org/10.1007/978-3-030-42934-8_11

of Things (IoT). IoT comprises several items, for example, robots, sensors, and actuators to the Internet [2]. When these kinds of items, usually called wearable devices, are attached to humans' body, it can be monitored individual's health parameters and safety. Considering suitable specifications, IoT can be defined as a system of physical objects supported by sensors and introduced innovations for data communication that supplies interchange with the environment [3]. Wearable devices have the advantage that they can be worn without any difficulty and can monitor a person's physical activity easy [3, 4]. Such wearable devices, tri-axial accelerometers, magnetometers, altimeters, and gyroscopes shape an automatic virtual environment [3, 5].

Many essential benefits are encouraging healthcare organizations to embrace a connected future. Primary among them is the possibility to enhance patient outcomes when data are shared in real time. The IoT permits healthcare specialists to extract data from medical devices, mobile apps, and chips inserted in our bodies to help diagnose patient's health more quickly [6].

Wearable devices and Medical IoT Interoperability & Intelligence, empowered by the Internet of Medical Things (IoMT), is a fast-growing field. Numerous patients are wearing IoMT devices—from connected health and wellness devices to connected insulin pumps and embedded pacemakers. Leaders in the department estimate that a large number of health organizations—up to 87%—plan to utilize the Internet of Things (IoT) technology by 2019 [7].

Wearable devices and tracking devices have grown as part of standard healthcare methods, intertwined with an evolving healthcare transmission model. According to recent studies, the wearable technology market is assumed to rise from $20 billion in 2015 to $70 billion in 2025. Medical devices are implemented with Wi-Fi or Near-field communication technology in order to allow machine-to-machine (M2M) communication that is the base of IoMT [8].

Generally, the IoMWT construction relies on three layers: the perception layer, the network layer, and the application layer, as shown in Fig. 1. The primary responsibility of the perception layer is to gather healthcare data with a type of devices. The network layer is composed of the wired and wireless system and middleware and performs processing and transmission of the input collected by the perception layer. Well-designed transport protocols should improve transmission efficiency, reduce the energy consumption, and ensure the security and privacy [9].

This book chapter, structured on seven chapters, presents several aspects concerning sensitive data management, cloud security and patients' data anonymization using ARX application.

In this chapter, we present aspects concerning massive medical data processing (e.g., data collected using the electroencephalograph (EEG) for patients suffering with epilepsy, Parkinson, or Alzheimer) using Hadoop, health predictive modeling of the massive data, and an edge computing solution for patient data anonymization. Our proposed solutions for security and privacy in medical area address the security aspects on different levels of account type and data anonymization by differential privacy model using ARX application and simulating several reidentification attacks

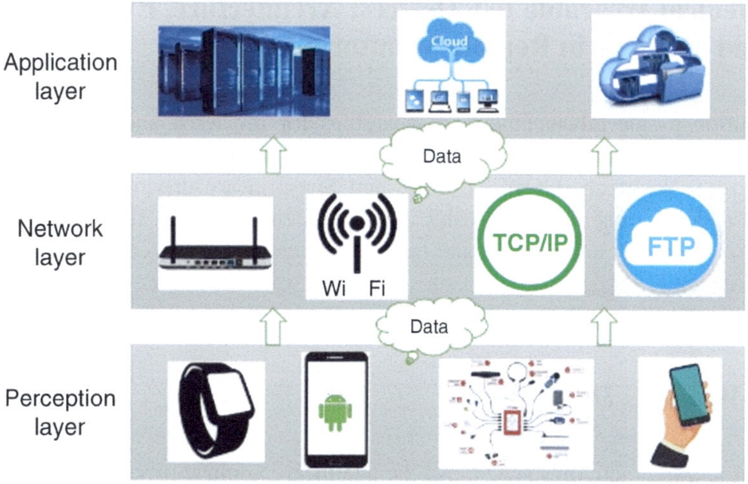

Fig. 1 Structure of medical internet of things

(persecutor, journalist, and marketer attack) analysis in order to provide the risks of data reidentification.

An essential application sector of the IoT is the healthcare sector. The IoT has acted a vital role in this area by intensifying service quality while decreasing costs. It is understandable to track health parameters, such as blood pressure, blood glucose, body temperature, and many more in real time by using wireless sensors. The evolution of sensors, better data processing technologies, and advanced technologies for wireless communication has driven to the expanding implementation of the IoT in the healthcare sector. The development of WBSNs (Wearable Body Sensor Networks) to frequently monitor patients' movements is an added milestone for the implementation of the IoT. Medical devices have endured severe changes, from the conventional unconnected devices to wireless modified devices. These improvements comprise the evolution of medical IoT systems that connect to cell phones. The medical IoT is a system involving mainly of health-monitoring devices. A back-end system remotely reads Patients' health parameters and eventually, investigates the collected data and presents suitable feedback to the clinical team [10, 11].

The medical technology business designs and manufactures a broad range of medical products that aid to diagnose, monitor, and treat diseases and health conditions.

The advancement of the IoMT is being serviced by an increase in the quantity of connected medical devices that can generate, collect, analyze, or transmit health data or images and connect to the healthcare provider networks, carrying data to either a cloud repository or internal servers [12].

2 Medical Wearable IoT Devices: Architecture, Evolution, and Methods in Remote Monitoring

In addition to the advancement in wearables for the wrist and eyewear, smart clothing ambitions are also on the rise. NanoSonic, Textronics (which manufactures NuMetrex), Weartech (GOW Trainer), and Sensoria are driving the application of those developing IoT in textiles. Specific items such as shirts can directly measure a person's temperature and heartbeat; therefore, socks can provide impact measurement [13].

2.1 Wearable Devices

Several research activities specific to the healthcare applicability domain currently performed worldwide have as primary interest sensor devices. Many projects were developing, and there are other projects already existing on the market. Several current projects have focused on wearable health devices [14]. The goal of this book chapter is to set the optimal safety for using wearable biosensors for patients according to the condition and physical state monitoring. The goal in IoMWT is to minimize the security risks by using the adequate methodology for working with personal sensitive data, anonymization, and data analytics in the cloud. The anonymization of the patients' data is crucial because it allows the legal use of data for predictive modeling, the study of the evolution of the disease, and correlations between the various symptoms.

The following are the most impressive indoor/outdoor health monitoring applications that operate on real-time and non-real-time modes.

(a) Health Gear – represents a wearable real-time system produced by Microsoft Research; it monitors and analyzes physiological signal from the sensors connected by Bluetooth to a mobile phone [15].
(b) MobiHealth – health project funded by the European Commission; enables patients to perform any kind of physical activity while continuously monitoring humans' parameters using UMTS and GPRS networks [16].
(c) Ubimon – developed by the Department of Computing from the Imperial College London, Ubimon has the purpose to identify, show, and express the most important problems when talking about the usage of wearable and implantable sensors for shared mobile monitoring [17].
(d) CodeBlue – represents a research working program developed by Harvard University, USA, and incorporates sensor node and wireless devices into a disaster response setting. It can work with a large number of wireless devices [18].
(e) eWatch – a wearable sensor device that makes available a platform for context-aware computing research. eWatch is developed as a wristwatch and has

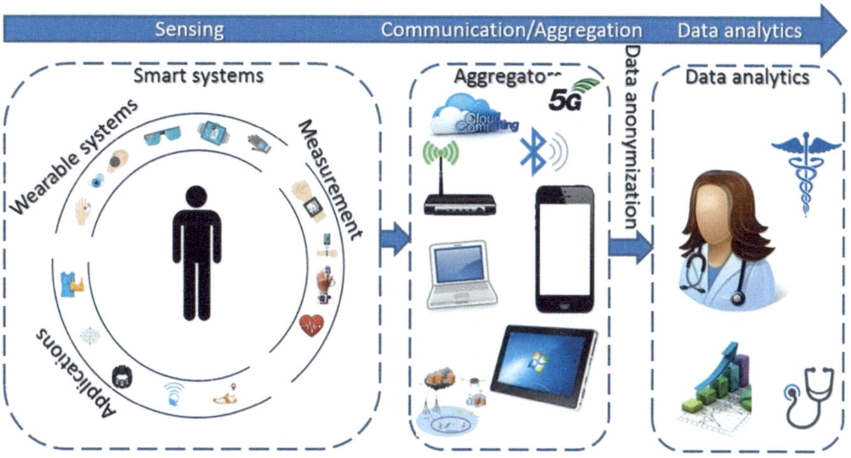

Fig. 2 Wearable internet of things application

different features: notifies when it is recorded light, motion, sound, abnormal temperatures, etc., and it is easy to use and view all the information [19].

(f) The vital jacket is based on mobile computing device that allow heart rate monitoring and can be used for medical monitoring, high performance sport and fitness application that transmit data via Bluetooth to PDA and store it in a memory card in a split second [20].

Figure 2 shows the phenomena of the transition of information from the patient to physicians.

The embedded sensors for a medical purpose will allow remote measurement for blood pressure, temperature, skin moisture, glycemia, or stress and will include actuators to turn on and off devices or adjust in real time. Patient vital sign data could be monitored for predictive analytics. Figure 3 describes the entire case scenario, why do we need to advance technologically in the field of medicine, when aged patients are suffering from diseases and medical care center are not enough as required. IoT wearable devices can monitor the patients 24/7 at home, and in case of an emergency, alerts are sent to the caretakers, as well as to the medical department to take the measures as it is described in Fig. 3. Readings are gathered through the biomedical sensors and sent to the wearable autonomous devices for monitoring, which generates the alerts according to the situation.

The process of using data is to filter the information in the private cloud and to anonymize data analyses in public cloud [20]. This process would help in keeping the privacy of the patient secure and authentic. The solution proposed in this chapter will use different security levels for different types of accounts (medical staff and patients) in order to provide security and privacy [21]. The second purpose of

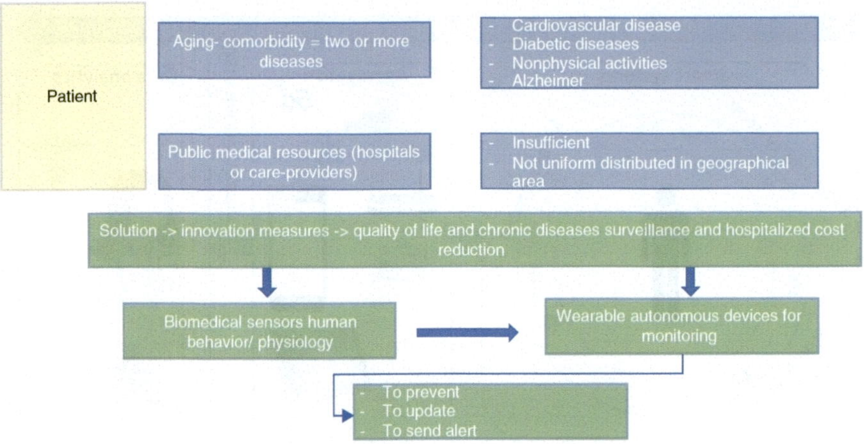

Fig. 3 Information privacy for medical records

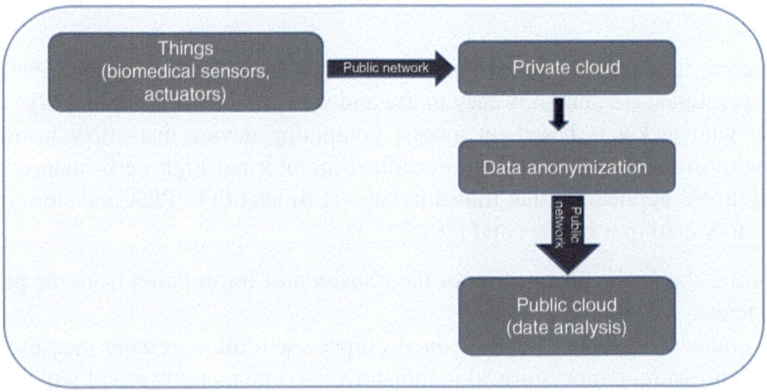

Fig. 4 Medical monitoring

this solution is to perform predictive analyses modeling for a better understanding and anticipation of patients' possible diseases or preventive actions. This is the reason for the need for anonymization of the patients' data performed in a cloud environment. Figure 4 illustrates the data privacy of the patient's data, as seen below.

For the process of anonymization [22, 23], several techniques can be used, such as the following:

- Kasking – protect X data by converting it in Y data (non-perturbative masking methods, perturbative masking methods, and synthetic data generation).
- Synthetic data – represented by protected data X, which consists of randomly generated records that do not directly derive from the records in Y (fully synthetic data; partially synthetic data, and hybrid data).

In this case, we choose to use the non-perturbative masking method and synthetic data method with partially artificial and hybrid data types. The steps for generating synthetic data are as follows:

- Set the model for the population
- Adjust the model to the original data set X
- Generate the synthetic data Y from the model

The challenges for IoT security for healthcare ecosystem, at the physical and logical level, are cryptographic algorithms development, authentication protocols, access control, and privacy.

3 Internet of Wearable Medical Things (IoMWT): Security Risks and Vulnerabilities

The massive amount of information that IoT and wearable technologies can gather, the privacy and security-related concerns become more and more critical as the number of these devices increases rapidly [24]. Users benefit from the personalization and customization that IoT and wearable technologies provide, yet those same capabilities have significant demand and exacerbate digital privacy and data security risks that are already in the market, taken in the category of traditional online services and technologies [25]. These privacy and security-related concerns can be observed in order to gain:

- Access to the device itself (i.e., if it is lost or stolen, makes a different kind of scenario)
- Access to the information that a device shares with nearby devices or systems (i.e., data transferred over Wi-Fi or other wireless networks)
- Access to data transmitted to the cloud or any remote storage system [26].

3.1 Related Work in Security and Privacy

A matter concerning security and privacy has been under discussion and hot topic of the modern era. There have been numerous authors [27] who illustrated the problems regarding the eHealth sector. The authors in [28] have taken into account some of the difficulties found in the private health sector for monitoring applications. This study assesses the security breaches for sensor network applications in wearable medical devices. The sensor devices must be foolproof since they are being used by nonexpert patients in case of medical application. The setup and control process of the data security mechanisms are node related, and they involve a little and intuitive human interaction. For the rest of the applications, to bootstrap initial secure communication between all the nodes in a BAN for secure data communication, device pairing techniques are approved, which is not easy for utilization. If we ignore the manual steps for usability increment, then it will sacrifice security [29, 30]. From all the above mentioned, the main conclusion is that between security issues and privacy issues, the first ones are the most important and most analyzed.

3.2 Security Risks and Vulnerabilities

Security represents one of the most significant aspects, which, in general terms, is seen as a concept for defining safety. According to the site of the US Department of Commerce [29], cautions against security breaches that might influence the personal data of wearable medical devices have to be taken care of. Healthcare systems mostly use sensor networks that face many downfalls in terms of security threats and cyberattacks. These can cause severe problems in the social life of people wearing wireless sensors devices because the gathered data can be utilized to harm the individual. As part of the active research, security issues regarding sensor networks, for example, eavesdropping on medical data, modification of medical data, forging of alarms on medical data, denial of service, and location tracking of users [31, 32]. Similarly, many people have specifically addressed security issues concerning healthcare applications [33]. In the following section, we will highlight and discuss some threats, attacks, and possible countermeasures.

Security breaches in healthcare applications within end devices networks are a significant concern which must be addressed. Applications in the healthcare sector based on sensor networks have a similarity to WSN application environments. Security problems can be categorized into two primary levels, one of them being the system security and the other one the information security. Threats and attacks [30] are divided into two major sections: active and passive. Passive attacks can take place during the routing phase of the data packages in a system. The hackers have the possibility to change the route of the packets or can cause the transmission to be unpredictable. The hackers can cheat medical data by snooping to the wireless network. Current attacks are more dangerous than passive ones. Attacks may find the location of the user by eavesdropping, which may lead to life-threatening scenarios [34]. The prevailing trend of sensor device designs is that they have little external security characteristics and hence are prone to physical tampering. This aspect enhances the vulnerability of the devices and implies difficult security challenges. Likewise, essential data transmission from WBAN networks through GPRS or any other similar network systems can also be stolen using eavesdropping.

Attacks in health monitoring are in detailed manners regarding the eavesdropping and modification of medical data, falsification of alarms on medical data, DoS (denial of service), location and users' activity tracking, physical tampering with devices, and jamming attacks. People with malicious purpose may use stolen information to conduct harmful activities. Attacks, which can take place in healthcare systems, performed using WSN are presented in Table 1.

Table 1 The security risk to WBAN and corresponding security requirements

Attack assumption	The risk to WBAN	Security requirements
Computation capabilities	Data modificationImplementation	Data integrityAuthentication
Listening capabilities	Eavesdropping	Encryption
Broadcast capabilities	Replaying	Freshness protection

1. Modification in medical data—when attackers can alter the personal medical data during the gathering, transmission, or storing of it, it can result in corrupted medical records of the patients. This can lead to fake alerts, e.g., resulting in futile rescue missions. Or in a worse case, the false negatives (i.e., changing alerting data into natural result) can disguise real alerts or emergencies. The hackers have the possibility of deleting or replacing a section of eavesdropped information or the complete data that were spoofed and can send the alerting data back to the original receiver for malicious purposes. Medical data are essential, and the modifications brought to it may lead to hazardous resulting, taking into account the well-being of a patient or a healthy person even.
2. Impersonation attack—If a hacker spies on the identity of any wireless sensor node can tamper the other nodes.
3. Eavesdropping—Regarding the open characteristics used by the sensor networks of the wireless channel, any enemy may capture radio transmissions among wireless nodes without any effort. Stolen information can be utilized for mischievous activities.
4. Replaying—A piece of the accessible data can be eavesdropped and resent to the first receiver to accomplish a similar reason in an alternate case.
5. Threats and attacker can be categorized in internal or external. Because of the fact that external attackers are not part of the system, their malicious intentions cannot be prevented. Because external attackers are not part of the system, their harmful activities cannot be stopped. The primary intention of these attacks is to abduct sensitive personal data. For the reason that wireless connections are more vulnerable than wired connections, the attackers can find them more comfortable. When they know the personal health data value, they may try to steal it by using both internal and external attacks.
6. Eavesdropping on medical data—during the gathering, transmission, and storing processes of the medical data, the attacker can take advantage and attempt to access that information illegally. An example is represented by unapproved spying radio transmissions between nodes. Medical data are highly sensitive to abuses and require to be protected against eventual attacks.
7. Forging of alarms on medical data—Attackers can generate fake messages instead of modifying the regular ones. This action performed by the attackers can lead to inaccurate data records or false system reactions.
8. Tracking users' location—A PEMS (Portable Emissions Measurements System) system user permits a constant track of the messages that are sent, and since the

system supports the localization of people, data can be obtained, reunited, and analyzed to result precise location profiles.

9. Tracking users' activity—This is a very common attack when it comes to eHealth systems. Considering the data records, patients' activities can be analyzed. For example, we can identify how much time a person spends during workouts and monitor the heart rate, oxygen saturation, etc.

4 Edge IoT Software- and Hardware-Level Security

IoT edge computing offers the possibility of gathering computational and analytics capabilities that are connected to the concept of data generation. IoT deployments became secure since some processes will occur in a specific location. Even though the IoT systems are based on different types of architectures, collaboration and edge intelligence are two characteristics that are common to all of them. One essential thing about this technology is that it uses the advantages of the fact that IoT devices are interconnected and of the gateways that will ensure data processing and device management.

IoT edge computing provides a particularly extensive protocol that will maintain data ingestion. Nowadays, enterprises need a platform that will be used to collect machine data and display them to other IoT systems. Moreover, there are a couple of accepted standards regarding enterprise applications that will be used. IoT platforms must support several devices that follow specific data ingestion protocols. The platform must be modular, and it should allow new forms of communication. Finally, several functionalities such as encryption or data protection will be provided to ensure essential security operations.

Moreover, IoT edge computing must provide the ability to work offline as this measurement will reduce the costs and will offer higher performance; currently, most companies will work with a large amount of data. First, this type of system must establish a secure connection between the cloud and the edge. An engine that will process the information gathered and analyzed by different learning tools that will allow the possibility of sending alerts in case of problems. Furthermore, the IoT edge platform must enhance operational processes that already exist. This facility will ensure that people have access to all types of data in real time.

When choosing the optimum IoT platform, each need must be taken into consideration. When building an application, its progress is strictly monitored. Although edge computing is still developing in terms of facilities, selecting the best IoT platform is something that needs to be considered in the long run as current and future needs are essential.

The hardware architecture of these systems needs to support deployments at multiple scales. Therefore, many gateways are used within the IoT platforms. Usually, ARM, x86, and MIPS are used along with containerization technologies that will facilitate the implementation of the same set of functionalities within the same IoT hardware without the need of making any further changes. It lowers the costs, and it reduces the number of employees hired to maintain different versions of production hardware and software. Frequently, platforms that can exchange resources with cloud are preferred, as they will anticipate unexpected requests coming from applications [35].

When talking about hardware layer in the context of edge computing, security needs to be taken into consideration while making the design of the device. Cryptographic keys are integrated within the chips and can be used to authenticate the user. However, this kind of system is still vulnerable as the keys are shared on a universal bus but can be solved if the keys are stocked at a different level and not through sharing keys.

For the communication layer, edge gateways are needed to ensure security by encrypting the data. If the network has a low bandwidth, MQTT (Messaging Queuing Telemetry Transport) can also be used. As for the cloud part, sensitive data need to be moved to the cloud. Furthermore, certificates will help within the authentication process. Moreover, the update devices need to be updated remotely so that attacks will be avoided [36].

There are a couple of IoT attacks that were taken into consideration in the last couple of years. The first category is related to the OSI (Open Systems Interconnection) physical layer that requires unauthorized access to physical systems. There is another category of attacks related to the software part, which includes viruses. DDoS (distributed denial of service) are can also be included in this section, although they can appear at lower levels within the OSI Model. The risk related to this section is that some warning might not be noticed.

Network attacks can also occur, and they represent one major vulnerability of IoT devices due to their wireless connectivity. A cryptanalysis attack occurs when someone tries to recover a message without having an encryption key. This situation also includes the case when the hacker tries all the possible combination for a password. The side-channel attack is related to the encryption that was used to gain access to the information [37].

5 Data Security in Private, Public, and Hybrid Cloud

Cloud's most significant advantage is that it makes the information more accessible when an Internet connection is available. However, there are many.

The public cloud is a free service such as Office 365, Dropbox, and Google apps that can be accessed via the web. On the other hand, the private cloud provides the same function but uses a firewall that will block the access of any unknown users. While giving this type of service, some problems such as the loss of resources or the vulnerability of the information might arise [38].

However, there are still some advantages regarding the use of a public cloud such as the inability of locating the exact information, the lack of hardware failures, and the use of degrees of physical security. On the other hand, the public cloud allows access to be granted from every location. Also, international issues might become a critical problem and can lead to criminal misconduct.

The private cloud aims to grant access to physical servers and to keep the information secured by a firewall. Moreover, the design of the architecture can be changed according to user's preferences, and this allows the data infrastructure to be isolated. Still, it can be an essential disadvantage if users have access to physical access. In addition, the owner is responsible for the security of the cloud [39]. A private cloud offers precise control over the security parameters, as all the measurements are taken inside the system and redistributed to the security provider [40]. Therefore, private cloud implementations provide multiple security advantages, although it still requires maximum attention from the organizations that will use it [41].

On the market, there is a third option called a hybrid cloud. This offers a mixture between the best elements of both public and private cloud. It gives the possibility of moving between the two of them, providing more flexibility and multiple options regarding the deployment of the data. It is essential to have a cloud team that will plan and organize the entire cloud strategy as they should decide whether an application should be moved to the cloud or not and keep track of the policies and costs. It is essential to have fluency during the process of handling data, as the speed is an essential key within this type of service [42].

Hybrid cloud has a couple of requirements such as the presence of public infrastructures like Microsoft Azure or a WAN (wide area network) established between the public and the private cloud. To use a hybrid cloud, servers, a LAN network (local area network), and a load balancer must be implemented. A private cloud software layer can be installed to create the virtualization layer or a hypervisor that will be used to sustain VMs (virtual machines) or containers.

The maximum interoperability between API (application programming interfaces) and services must be ensured. Therefore, cloud layers that are compatible with the targeted public cloud must be selected. Hybrid clouds will allow the enterprise to implement a local private cloud that contains critical data and resources [43].

6 Personal Data Management and Anonymization in the Cloud

The start of the last decade manifested that the speed and the volume of data generated are surpassing the current memory of institutions' data management [44]. Cloud-based data management is, indeed, stimulating to recognize the potential of large-scale data management solutions by providing adequate scaling of resources. In a cloud-based data management situation, institutions or organizations pay storage and computing power to execute the data management applications to work preferably. Data management is one of the most significant research domains in cloud computing. Many cloud-based data management systems are in service now, such as Bigtable in Google, Cassandra, and Hive in Facebook, HBase in Stream, PNUTS in Yahoo!, and several other systems. Cloud computing has grown into significant influence on data management research and performs as a critical role [45].

The Registration and Unique Identifier generation module is necessary for creating a PHR (Patient Health Record) tool. The module is used to provide information for all patients with a unique ID value. All hospital employees can fetch the data of the patient by using only the ID from the database. Additionally, this module can reduce consumption. To maintain or create one patient medical record, every hospital needs to have a unique identifier value for the patient in an organization. For generating an ID value for each user, the Md5 hash algorithm is used. That takes a password as input and 32-bit id as an output. MongoDB application preparation is used to store patients' records. Every data, along with this metadata, are to be stored in a single place; it will simplify the access time of data and minimize the use of joining the robust modules [46].

MD5 hash algorithm presents a processing 512-bit block as input and produces a 128-bit (16 bytes) blocks, often expressed as a 32-digit hexadecimal value. After getting all the medical records from the patients, it is needed to store in an encrypted format both rest and the transformation. All patients' sensitive medical information is changed into critical values by using an Md5 hash algorithm only for the limit. After hashing, the encrypted data are separately stored on a different cluster. Transferred medical records of the patients are encrypted, so whenever we want to transfer medical records for patient's treatment, we need to decrypt records in the opposite side. Hadoop MapReduce algorithm for anonymizing data is used to help to deal with the massive amount of the daily updatable patient's records. Hadoop MapReduce has several components, such as HDFS and MapReduce (Fig. 5). *HDFS (Hadoop Distributed File System)* was entirely used for data storage and contained the data node and the name of the node. MapReduce is used for operating by Mapper and Reducer [46].

Several records of the human body electrical activity produced by heart, brain, or muscles can be used in medical diagnosis.

These records must be anonymized when are used for statistics in order to protect the sensitive data of the patients.

- In case of medical bioheat transfer, MapReduce can be used to process data and to select only the appropriate data based on the default limits (by Reducer – Fig. 6):
- Set the temperature limits [47]:

```
INSERT INTO TABLE temperature
Select
*,
high_limit_temperature-actual_temperatureas
critical_temperature,
IF ((high_limit_temperature-actual_temperature)>1.2, 'Low',
IF ((actual_temperature-high_limit_temperature)>1, 'High'))
AS temperature_limit
```

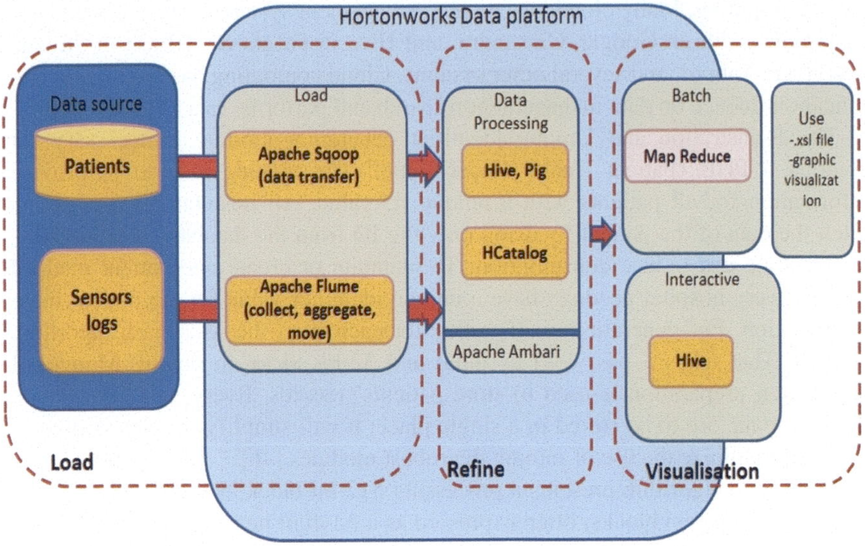

Fig. 5 Big data predictive analytics for bioheat transfer modeling [46, 47]

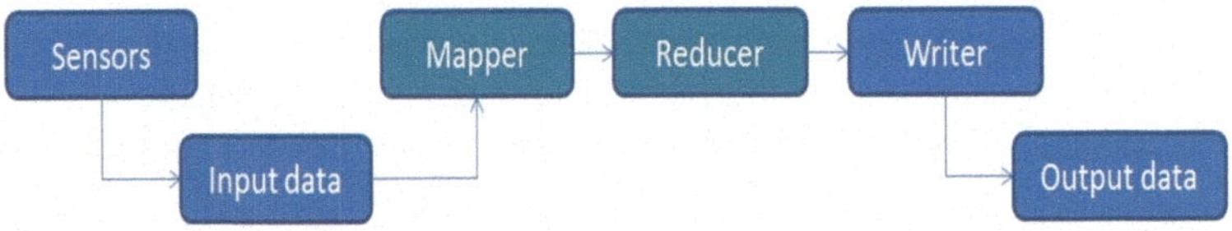

Fig. 6 Data processing by MapReduce [46, 47]

```
IF ((high_limit_temperature-actual_temperature)<1.2, 'Normal'
IF ((high_limit_temperature-actual_temperature)<1, 'Normal'))
AS temperature_normal from temperature_data;
```

- In case of the diagnosis or disease evolution studies that involve patients with epilepsy, Parkinson, or Alzheimer, the edge computing solution is necessary in order to process the massive volume of data collected by EEG (Figs. 7 and 8).

The EEG method based on electrodes is used for study neurological disorders such as the following:

- *Epilepsy* represents a chronic disease of the brain, demonstrated by the crises convulsive partial (focal lengths) or preferences, due to spontaneous electrical discharges that occur at the level of the brain.
- *Parkinson* represents a progressive neurological illness characterized by such trembling of the extremities, hardness of the muscles).
- *Alzheimer* involves progressive nerve degeneration due to the reduction of the number of neurons, brain atrophy, and the presence of the "caterpillar" plates, indicating the loss of memory and disorientation in time and space.

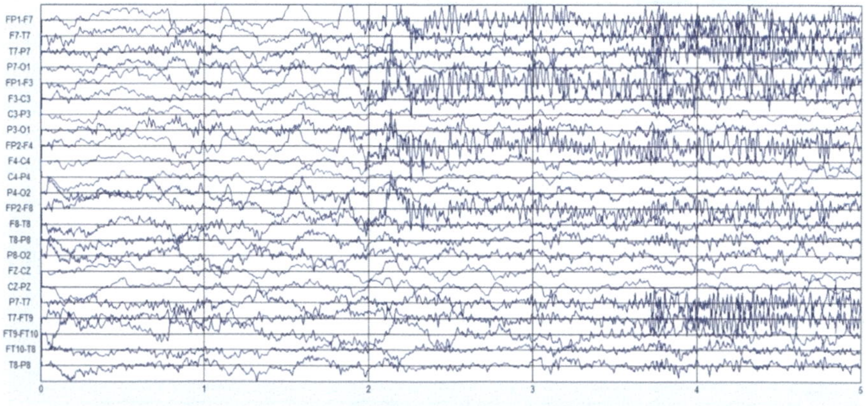

Fig. 7 EEG-23 channels, patient with epilepsy [48]

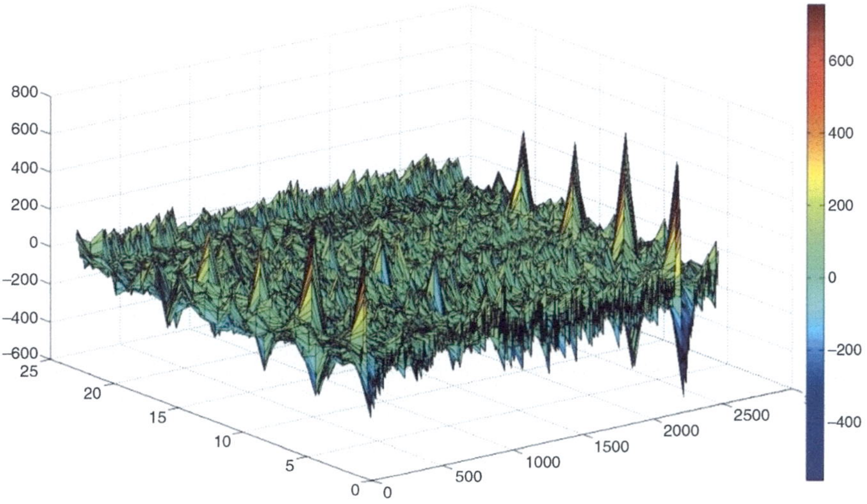

Fig. 8 EEG-3D view seizure, patient with epilepsy [48]

In an analysis of the EEG signals, we used discrete wavelet transform (DWT) and complementary filters in order to obtain the doubling of the data (Figs. 9 and 10) and reduction of the samples [48].

$$s = a_{12} + \sum_{i=1}^{12} d_i \tag{1}$$

The proposed solution for wearable EEG and data analysis, in correlation with other body parameters such as ECG, PPG, temperature, and skin moisture, is presented in Fig. 11.

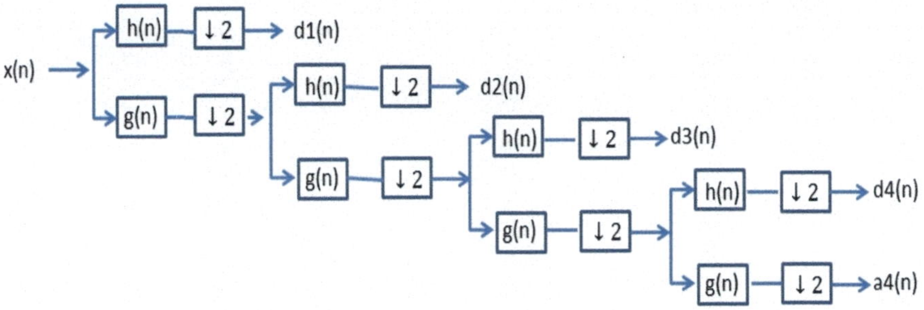

Fig. 9 Signal decomposition in details and approximations [48]

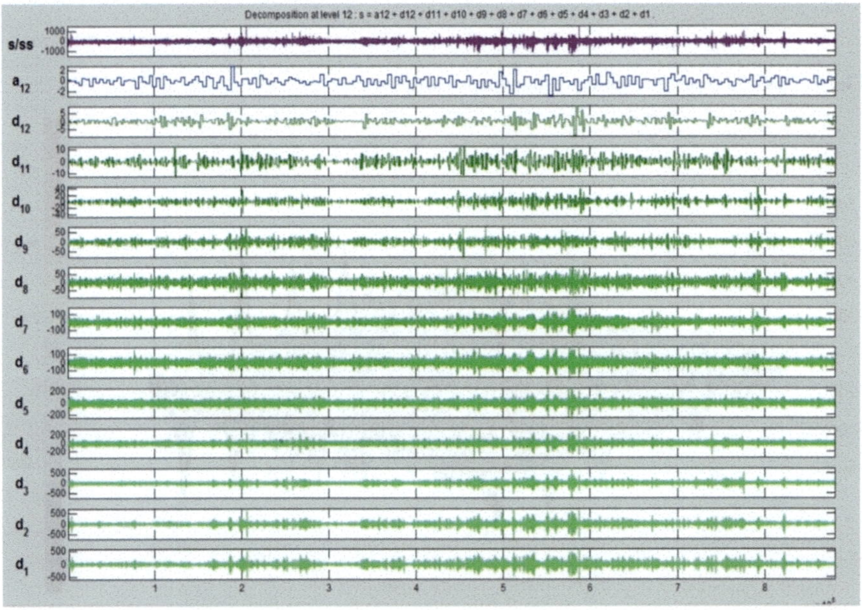

Fig. 10 Wavelet decomposition on 12 levels (1) [48]

The applications dedicated for data anonymization are open-source (ARX, Amnezia, μ-ARGUS, sdcMicro) or professional anonymization software (Aircloak Insights) GDPR-compliant. The Aircloak Insights application allows instant data anonymization, control, and the right balance between costs and efficiency. Also, this application can be created as a data-based business model by using anonymized data.

The anonymized patient records are generated by using the technique of generalizing and specification. In this technique, the critical attribute of quasi domains value is changed into a more general form. For example, if the native patient residency is Chennai, after the anonymization algorithm, the output will be India.

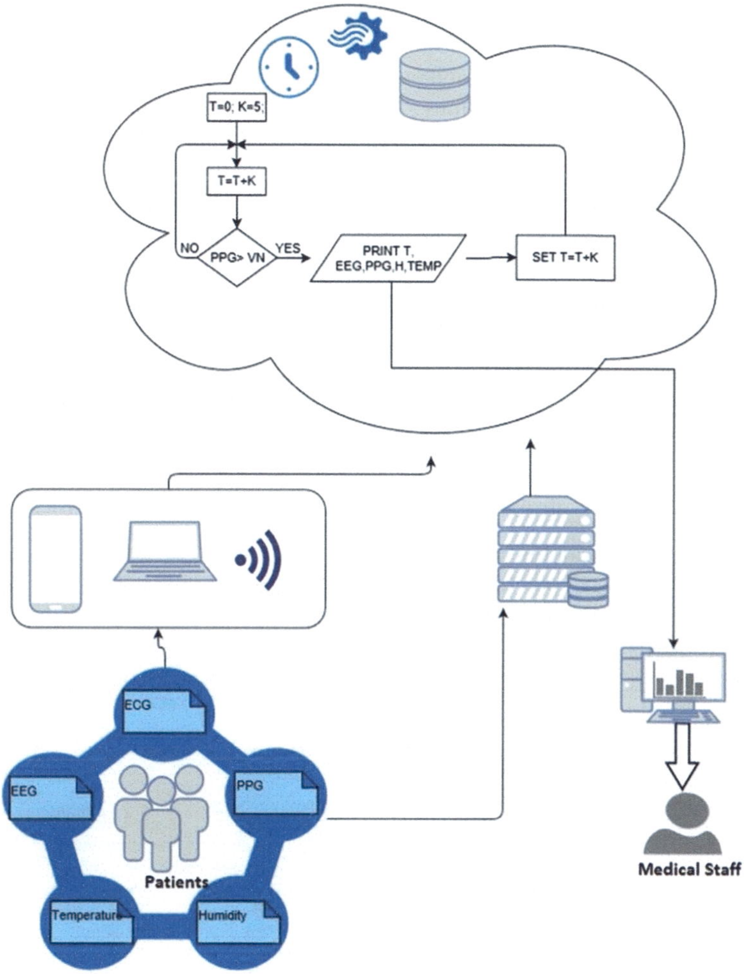

Fig. 11 WearEEG apps – block diagram [48]

This technique hides the values by symbols (for example, if the PIN code number has six digits, changing values of the number into symbols will be done only for the last three digits. By using a novel AAPM (Authorized Accessible Privacy Model), only the authorized personnel will be able to view and edit the patient's records. The authorized person has full rights over the data. The indirectly authorized person has only rights to insert data rather than edit or view.

Anonymization is the best way to provide privacy over the microdata, which is released by the data publisher. Most of the existing systems are using several anonymization algorithms. A large number of top-down specification and bottom-up generalization techniques are developed separately for many organizations. Security and privacy over the patient's sensitive information must be provided. Specific attributes are also available in the data sets. For categorical attributes, the hybrid anonymization algorithm is used and includes both TDS (Top-Down Specialization) and BUG (bottom-up generalization).

The term [49] anonymization in cloud computing refers to anonymizing data while hiding any data which is public or sensitive to the original data. Data anonymization is widely used for securing sensitive data, and the techniques used in this field are focused on the public cloud. Mainly speaking, this feature enables the preservation of data to be private while just exposing a few critical data.

One of the algorithms used in anonymization is the k-Anonymity, an algorithm used for making each record alike from k-1 records. K-anonymity has three attributes, such as the following:

- Attributes having the propriety of recognizing an individual straight forwardly
- Quasi-identifier: linked attributes to external information that identifies each particular individual
- Sensitive attributes: attributes that should not be revealed to third parties.

We generated a model in ARX, we used as input data (Fig. 8) a set of sensitive (age, name, address), and insensitive data (disease, wearable devices), and we applied the models 2-Anonymity and differential privacy. The results are presented in Figs. 15 and 16 that represent the reidentification attack risks before and after data anonymization. In Figs. 12 and 13 the input microdata set, before the attack, and the output microdata after the attack are presented.

	ID	Name	Country	Town	Zip code	Age	Gender	Disease	Wearable devices
1	1	Marlon	UK	London	112034	67	M	Diabetes	Glucose monitor
2	2	Glenn	France	Lyon	112432	75	M	Emphysema	Pulse oximeter, temperature and skin moisture
3	3	Diana	Romania	Bucharest	203412	58	F	Cardiac Insufficiency	Peacemaker
4	4	Flavius	Lithuania	Vilnius	127734	66	M	Arthritis	Pedometer, accelerometer
5	5	Clara	Poland	Varsovia	112732	77	F	Angina	Wearable ECG
6	6	Mihai	Romania	Cluj	231298	72	M	Diabetes	Glucose monitor
7	7	Dane	Ireland	Dublin	342232	44	M	Angina	Wearable ECG
8	8	Carla	Austria	Viena	331245	42	F	Diabetes	Glucose monitor
9	9	Anca	Romania	Brasov	129867	59	F	Aneurysm	Wearable ECG
10	10	Ovidiu	Romania	Bucharest	324578	66	M	Diabetes	Glucose monitor
11	11	Klaus	Germany	Munchen	226743	56	M	Diabetes	Glucose monitor
12	12	Flavius	Italy	Napoli	334897	64	M	Arthritis	pedometer, accelerometer
13	13	Liviu	Romania	Craiova	376521	73	M	Cardiac Insufficiency	Peacemaker
14	14	Eugen	Romania	Brasov	129854	68	M	CAD	Wearable ECG
15	15	Carla	Belgium	Leuven	226436	56	F	Diabetes	Glucose monitor
16	16	Bianca	Italy	Rome	296641	67	F	Diabetes	Glucose monitor
17	17	Darius	Italy	Calabria	129844	56	M	Atherosclerosis	Pedometer, accelerometer
18	18	Anda	Romania	Bucharest	316345	66	F	Stroke	Wearable ECG
19	19	Glenn	UK	London	197745	56	M	Arrhythmia	Peacemaker
20	20	Alina	Romania	Brasov	145322	69	F	Asthma	Pulse oxymeter
21	21	Eusebiu	Romania	Bucharest	236754	72	M	Arrhythmia	Wearable ECG
22	22	Jan	Germany		342312	66	M	Epilepsy	Wearable EEG
23	23	Michele	France	Rouen	229017	75	F	Alzheimer	pedometer, accelerometer
24	24	Monica	France	Nantes	235467	73	F	Parkinson	pedometer, accelerometer
25	25	Adrian	Italy	Pisa	224312	70	M	CAD	Peacemaker

Fig. 12 Input data –ARX

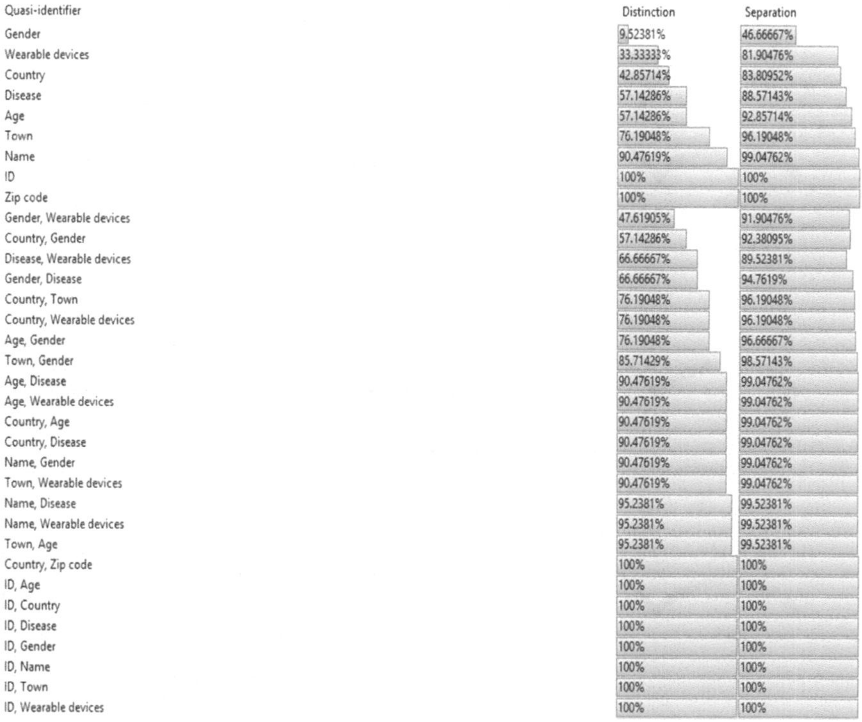

Fig. 13 Risk analysis based on quasi-identifiers

For quasi-identifier (Fig. 13), we used the zip code, and names which could be linked to external data to reidentify individual record owners, and that was removed from output data.

The models used for reidentification attacks are persecutor, journalist, and marketer attack.

Using the reidentification risk analysis implemented, we obtained the estimated risk provided for three different attacker models:

- The prosecutor scenario
- The journalist scenario
- The marketer scenario

In the prosecutor model, the attacker already knows that the data for an individual patient is contained in the data set. In the journalist model, the attacker does not know about data sets content. In the marketer model, the attacker is not interested in reidentifying a specific individual but in attacking a more significant number of individuals' records.

In Figs. 14 and 15 the risk analysis for prosecutor attacker model, journalist attacker model, and marketer attacker model, before and after anonymization by differential privacy model are presented.

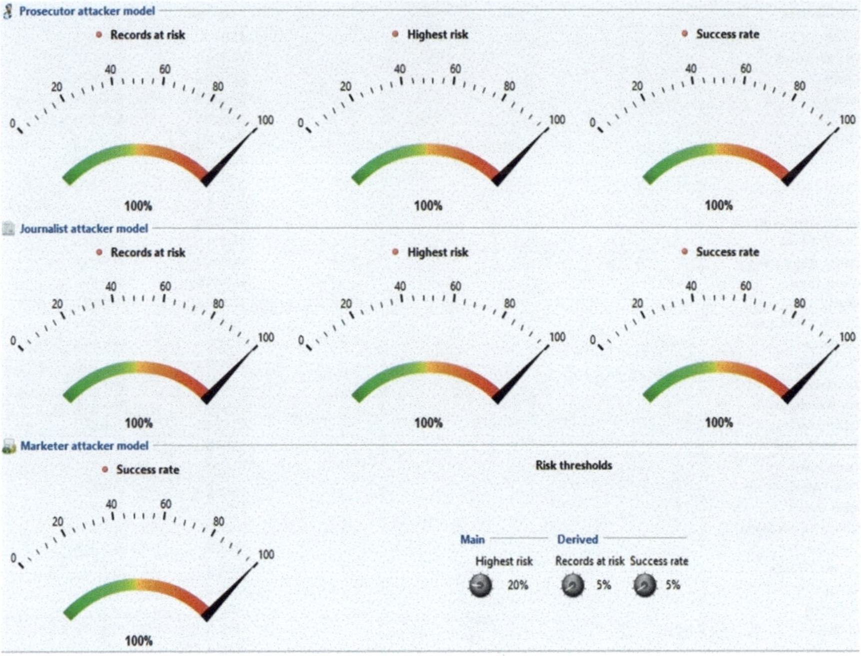

Fig. 14 Reidentification attack risks before data anonymization

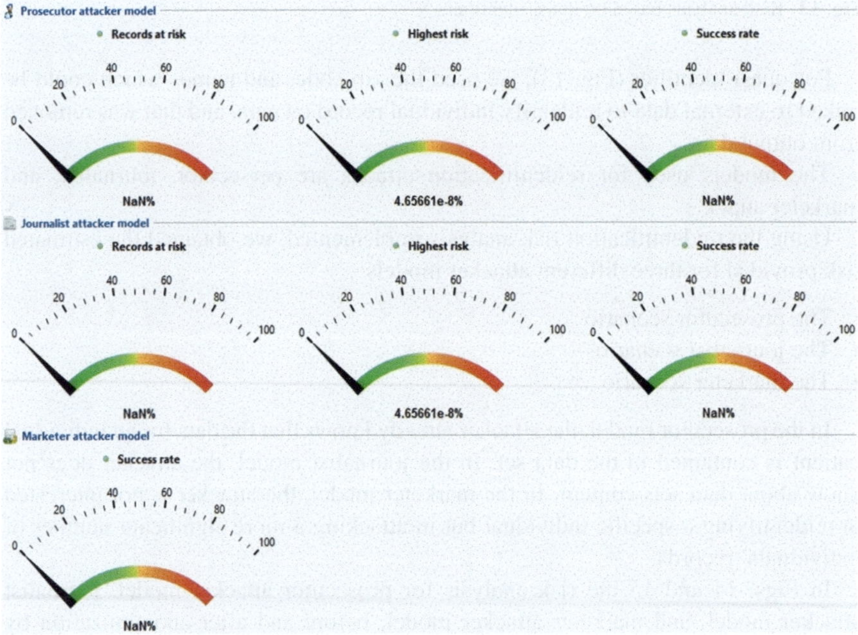

Fig. 15 Reidentification attack risks after data anonymization

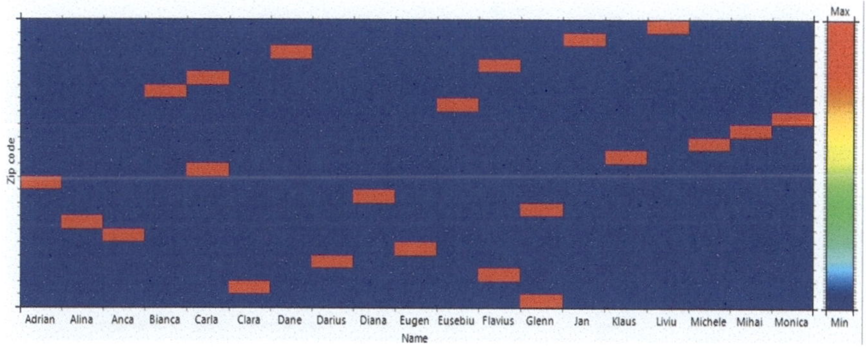

Fig. 16 Contingency before anonymization

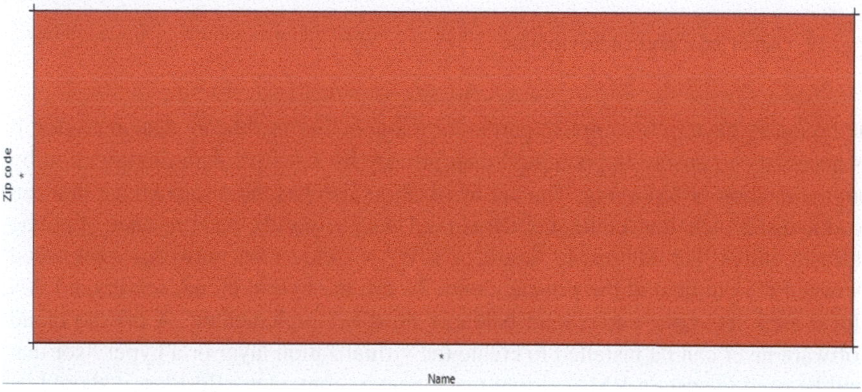

Fig. 17 Contingency after anonymization

In Figs. 16 and 17 the contingency before and after anonymization are presented. The contingency Name = f(zip code) represents the bivariate frequency distribution of the variables (zip code and name).

In Fig. 18, it can be observed that the patient cannot be identified by zip code or name after the anonymization.

7 Conclusions

Sensor network applications in health care are the hot topic of the modern age, which is being researched and deployed all over the world. With the advancement of these applications, effects will arise as well. In this chapter, we tried to highlight the concerns of significant social implications like privacy and security. In order to meet a set of specific security requirements, eHealth monitoring systems confidentiality

	ID	Name	Country	Town	Zip code	Age	Gender	Disease	Wearable devices
1	*	*	UK	London	*	67	M	Diabetes	Glucose monitor
2	*	*	France	Lyon	*	75	M	Emphysema	Pulse oximeter, temperature and skin moist
3	*	*	Lithuania	Vilnius	*	66	M	Arthritis	Pedometer, accelerometer
4	*	*	Poland	Varsovia	*	77	F	Angina	Wearable ECG
5	*	*	Romania	Cluj	*	72	M	Diabetes	Glucose monitor
6	*	*	Ireland	Dublin	*	44	M	Angina	Wearable ECG
7	*	*	Austria	Viena	*	42	F	Diabetes	Glucose monitor
8	*	*	Romania	Brasov	*	59	F	Aneurysm	Wearable ECG
9	*	*	Romania	Bucharest	*	66	M	Diabetes	Glucose monitor
10	*	*	Germany	Munchen	*	56	M	Diabetes	Glucose monitor
11	*	*	Romania	Craiova	*	73	M	Cardiac Insufficiency	Peacemaker
12	*	*	Romania	Brasov	*	68	M	CAD	Wearable ECG
13	*	*	Belgium	Leuven	*	56	F	Diabetes	Glucose monitor
14	*	*	Italy	Rome	*	67	F	Diabetes	Glucose monitor
15	*	*	Italy	Calabria	*	56	M	Atherosclerosis	Pedometer, accelerometer
16	*	*	UK	London	*	56	M	Arrhythmia	Peacemaker
17	*	*	Romania	Bucharest	*	72	M	Arrhythmia	Wearable ECG
18	*	*	Germany		*	66	M	Epilepsy	Wearable EEG
19	*	*	France	Rouen	*	75	F	Alzheimer	pedometer, accelerometer
20	*	*	France	Nantes	*	73	F	Parkinson	pedometer, accelerometer

Fig. 18 Output data: after anonymization

and security have to accomplish particular features. For healthcare data analytics, it is necessary to ensure the patients' data privacy by sensitive data modification by anonymization or removing. The anonymization also has the disadvantage that can lead to insufficient data content or data loss. Using a hybrid cloud presence of public infrastructures like Microsoft Azure or a WAN (wide area network) established between the public and the private cloud. To use the hybrid cloud, servers, a LAN (local area network), and a load balancer must be implemented. A private cloud software layer can be installed to create the virtualization layer or a hypervisor that will be used to sustain VMs (virtual machines) or containers. Besides, it should be mentioned and understood that there is not enough focus on the sensor network, but an overall discussion is needed on the whole system, including backends. Another critical aspect to consider is national security and data protection laws. It is essential to discuss legal and organizational questions, as well as to extend existing PEMS prototypes through security mechanisms. The general public needs to know the benefits and what this implies to be prepared at any time. Rules and regulations of cyber laws, as well as existing health regulations, need to be updated and formalized.

As future work, we envision implementing a cloud platform for managing data privacy for wearable medical devices.

Acknowledgments This work has been supported in part by UEFISCDI Romania and MCI through projects PARFAIT, ESTABLISH, and WINS@HI, funded in part by European Union's Horizon 2020 research and innovation program under grant agreement No. 787002 (SAFECARE) and No. 813278 (A-WEAR).

References

1. S. Ray, J. Park, S. Bhunia, Wearables, implants, and internet of things: the technology needs in the evolving landscape. IEEE Trans. Multi-Scale Comput. Syst. **2**(2), 123–128 (2016)
2. A. Ometov, S.V. Bezzateev, J. Kannisto, J. Harju, S. Andreev, Y. Koucheryavy, Facilitating the delegation of use for private devices in the era of the internet of wearable things. IEEE Internet Things J. **4**(4), 843–854 (2017)
3. M. Haghi, K. Thurow, R. Stoll, Wearable devices in medical internet of things: scientific research and commercially available devices. Healthc. Inform. Res. **23**, 4–15 (2017)
4. J. Wei, How wearables intersect with the cloud and the Internet of Things: considerations for the developers of wearables. IEEE Consum. Electron. Mag **3**, 53–56 (2014)
5. Y. Athavale, S. Krishnan, Biosignal monitoring using wearables: observations and opportunities. Biomed. Signal Process. Control **38**, 22–33 (2017)
6. https://www.carestream.com/blog/2019/01/08/the-iot-in-healthcare-in-2019/
7. Davis J., 87 percent of health organizations plan to adopt IoT technology by 2019, study shows, Healthcare IT News. (2017). https://www.healthcareitnews.com/news%20/87-percent-health-organizations-plan-adopt-iot-technology-2019-study-shows
8. Internet of Medical Things (IoMT) Connecting Healthcare for a Better Tomorrow, UST Global INc. (2017). https://www.ust-global.com/sites/default/files/internet_of_medical_things_iomt.pdf
9. W. Sun, Z. Cai, Y. Li, F. Liu, S. Fang, G. Wang, Security and privacy in the medical internet of things: A review. Security and Communication Networks, in Journal Security and Communication Networks **2018**, 1939–0114 (2018). https://doi.org/10.1155/2018/5978636
10. P.A. Williams, A.J. Woodward, Cybersecurity vulnerabilities in medical devices: a complex environment and multifaceted problem. Medical Devices (Auckland) **8**, 305–316 (2015)
11. D. Miorandi, S. Sicari, F. De Pellegrini, I. Chlamtac, Internet of things: vision, applications and research challenges. Ad Hoc Netw. **10**, 1497–1516 (2012)
12. Deloitte Center for Health Solutions, Medtech and the Internet of Medical Things – How connected medical devices are transforming healthcare (2018). https://www2.deloitte.com/content/dam/Deloitte/global/Documents/Life-Sciences-Health-Care/gx-lshc-medtech-iomt-brochure.pdf
13. The Internet of Me: How Wearable Tech is Changing the Internet of Things, the Application Developers Alliance Emerging Technology Working Group, https://static1.squarespace.com/static/5386471 8e4b07a1635424cdd/t/54b6aacae4b0b6572f7175e5/1421257418695/ADA+-+IoT+Wearables_Final.pdf
14. L. Wolf, S. Saadaoui, Architecture concept of a wireless body area sensor network for health monitoring of elderly people. Consumer Communications and Networking Conference, 2007. CCNC 2007. 4th IEEE, (2007), pp. 722–726
15. N. Oliver, F. Flores-Mangas, HealthGear: a real-time wearable system for monitoring and analyzing physiological signals. *International Workshop on Wearable and Implantable Body Sensor Networks* (2006) 4
16. W. Yong, G. Attebury, B. Ramamurthy, A survey of security issues in wireless sensor networks. IEEE Commun. Surv. Tutor. **2**(8), 2–23 (2006)
17. T. Zia, A. Zomaya: Security issues in wireless sensor networks. *In Proceedings of International Conference on Systems and Networks Communications, 2006. ICSNC '06*, (2006), pp. 40–40
18. M. Welsh, D. Malan, B. Duncan, T. Fulford-Jones, S. Moulton: Wireless sensor networks for emergency medical care. GE Global Research Conference, Boston (2004)
19. U. Maurer, A. Rowe, A. Smailagic, D.P. Siewiorek, eWatch: a wearable sensor and notification platform. *International Workshop on BSN, Wearable and Implantable Body Sensor Networks*. (2006), pp. 4–145
20. http://fiji.eecs.harvard.edu/CodeBlue

21. S. Vaudenay, *A Classical Introduction to Cryptography: Applications for Communications Security, in Applications for Communications Security.*, Springer US (2006). https://doi.org/10.1007/b136373
22. http://www.cms.hhs.gov/
23. W. Yong, G. Attebury, B. Ramamurthy, A survey of security issues in wireless sensor networks. IEEE Commun. Surv. Tutor. **2**(8), 2–23 (2006)
24. A. Milenkovic, C. Otto, E. Jovanov, Wireless sensor network for personal health monitoring: issues and an implementation. Comput. Commun. **29**, 2521–2533 (2006)
25. P. Thibodeau, The Internet of Things could encroach on personal privacy. *Computerworld* (2014)
26. J. Singh, J. Powles, The Internet of Things—the next big challenge to our privacy. *Guardian* (July 28, 2014
27. A. Sacco, Fitness trackers are changing online privacy—and it's time to pay attention, CIO (2014)
28. F. Kargl, E. Lawrence, M. Fischer, Y.Y. Lim, Security, privacy and legal issues in pervasive eHealth monitoring systems. 7th International Conference on Mobile Business ICMB, (2008), pp. 296–304
29. A. Ashraf, A. Rajput, M. Mussadiq, B.S. Chowdhry, M. Hashmani, SNR based digital estimation of security in wireless sensor networks. In *Communications Infrastructure. Systems and Applications in Europe*, vol. 16, (2009), pp. 35–45
30. D. Kouvatsos, G. Min, B. Qureshi: Performance issues in a secure health monitoring wireless sensor network. In *Proceedings of 4th Int. Conference on Performance Modelling and Evaluation of Heterogeneous Networks (HET-NETs'2006), British Computer Society (BCS)*, IEE, Ilkley, UK, September 11–13, (2006) WP01(1–6)
31. H.S. Ng, M.L. Sim, C.M. Tan, Security issues of wireless sensor networks in healthcare applications. BT Technol. J. **24**(2), 138–144 (2006)
32. http://www.its.bldrdoc.gov/
33. T. Zia, A. Zomaya, Security issues in wireless sensor networks. In *Proceedings of International Conference on Systems and Networks Communications, 2006. ICSNC '06*, (2006), pp. 40–40
34. M. Meingast, T. Roosta, S. Sastry, Security and privacy issues with healthcare information technology. 28th Annual International Conference of the IEEE Engineering in Medicine 100 J Med Syst (2012) 36:93–101 and Biology Society, EMBS '06, (2006), pp. 5453–5458
35. https://internetofthingsagenda.techtarget.com/blog/IoT-Agenda/Five-requirements-of-a-leading-IoT-edge-platform
36. https://www.networkcomputing.com/network-security/role-security-edge-computing
37. https://www.iotworldtoday.com/2017/09/23/iot-device-security-comprehensive-look-edge-cloud/
38. G. Suciu, E.G. Ularu, R. Craciunescu, Public versus private cloud adoption—A case study based on open source cloud platforms. In *IEEE 20th Telecommunications Forum (TELFOR)*, (2012) 494–497
39. https://www.networkworld.com/article/3158424/cloud-computing/public-vs-private-cloud-why-the-public-cloud-is-a-real-threat-to-security.html
40. Z. Xiao, X. Fu, R.S.M. Goh, Data privacy-preserving automation architecture for industrial data exchange in smart cities. IEEE Trans. Industr. Inform. **14**(6), 2780–2791 (2017)
41. https://searchcloudsecurity.techtarget.com/tip/Private-cloud-computing-security-issues
42. https://www.forbes.com/sites/forbestechcouncil/2018/06/29/four-major-considerations-for-hybrid-cloud-security/#2c7dad4544a5
43. https://searchsecurity.techtarget.com/magazineContent/Challenges-with-data-protection-in-the-cloud
44. G. Suciu, M. Anwar, I. Rogojanu, A. Pasat, A. Stanoiu, Big data technology for scientific applications. In *IEEE Conference Grid, Cloud & High Performance Computing in Science (ROLCG)*, (2018), pp. 1–4
45. A.A. Vandurke, Data management in the cloud computing IJSRD. Int. J. Sci. Res. Dev. **5**(12) (2018). ISSN (online): 2321–0613

46. R.M. Aileni, S. Pasca, R. Strungaru, Big data predictive analytics for bioheat transfer modeling, *ROLCG Conference*, (2016) ISBN 978-973-0-22868-7
47. R.M. Aileni, S. Pasca and C.A. Valderrama Sukuyama, Wearable health care: Technology evolution, algorithms and needs, in *Book Enhanced Living Environments: From Models to Technologies*, IET Digital Library, pp. 315–343, (2017), doi:https://doi.org/10.1049/PBHE010E_ch13
48. R.M. Aileni, B. Hurezeanu, S. Pasca, Wavelet transform for seizures detection in EEG Records. In *2018 10th International Conference on Electronics, Computers and Artificial Intelligence (ECAI)*, IEEE (2018), pp. 1–6
49. A. Raj, R.G.L.D. Souza, Big data anonymization in cloud using k-anonymity algorithm using map reduce framework. Int. J. Sci. Res. Comput. Sci. Eng. Inform. Technol. **5**(1), 50–56 (2019)

Part IV
Case Studies

Detection of Squamous Cell Carcinoma Through Smart Technologies

S. M. Sagari and S. Venkatesan

1 Introduction

Cancer occurs due to abnormal growth of cells invading on healthy cells, thus changing the tissue texture. This can also be termed as carcinoma in situ. There are different types of cancer [12], which should be detected and diagnosed at its early stage. Oral cancer is one such type of cancer that occurs most common among rural and older population who have less knowledge of their habit of chewing gutka, tobacco, excessive use of bidi, cigarette and overconsumption of alcohol. Generally, due to the unawareness and improper care for health, the tumours are left unnoticed, which can grow to an extent where the treatment becomes ineffective to overcome the disease. So in order to avoid morbidity and mortality at later stages, it is better to get detected and diagnosed at its early stages. Squamous cell carcinoma is the most common type of oral cancer that occurs in oral mucosa lining [1].Various modalities are used to detect the changes of tissue in affected regions [2, 3]. Nowadays, various research are carried out to track object in RGB images by locating the colour changes [4, 7]. Locating the colour difference in oral region is the best approach to diagnose the disease at its early stage. In [5, 6], tongue colour is evaluated with differences between tongue and tongue coating.

S. M. Sagari (✉)
Department of CS&E, Dayananda Sagar College of Engineering, Bengaluru, India

S. Venkatesan
Department of Computer Science, Dayananda Sagar College of Engineering, Bengaluru, India

N. Gupta, S. Paiva (eds.), *IoT and ICT for Healthcare Applications*, EAI/Springer Innovations in Communication and Computing,
https://doi.org/10.1007/978-3-030-42934-8_12

The proposed work focuses on the clinical appearance of colour feature of tumour or lesion to locate ROI in a more precise way. As cancer lesions grow, they change their colour from its surrounding region forming into white or red patches at their nascent stage; in this stage, they can be benign or malignant. When they are left untreated, they may get converted into malignant tumours forming into lumps. The algorithm focuses on segmenting an ROI from its surrounding region, by creating a colour mask of constituent region, thus separating from its surrounding regions based on colour. This is implemented by varying pixel intensity using track bars. In order to achieve accuracy of the results, algorithm implements active contours on ROI based on its colour intensity. The brief outline of the paper is as follows. Section II explains about literature survey. Section III describes about methodology. Section IV on results and discussion and Sect. V is about conclusion and future work.

2 Literature Survey

The research based on detection and localisation of oral cancer plays a major role in medical field, as more and more population are unaware of the impact on their health due to their addiction habits that are not good for health. Various research have been carried out based on clinical perspective, as well as on technical perspective. In [1], the study explains the clinical perspective of squamous cell carcinoma and different types of it. Since oral cavity comprises of different regions such as hard palate, soft palate, tongue, tooth and gums, lesions or colour changes occur in different regions of oral cavity have different appearance of colour and shape, so they are location variants. Before detecting a tumour, it is good to research a premalignant appearance of the lesions [8] as it gives histopathological results, that which gives accuracy to results of the present study. Sometimes ulcers can be predicted as cancer lesion, the study [9] is about the assessment wounds based on colour. In [10], the study describes about future work in oral cancer that can extract patches in the region. Lesion characteristics in oral cavity [11] explains about pigmented lesion that may convert into malignant neoplasm and understands the cause of mucosal pigmentation. Conversion of RGB colour space to HSV colour space [13, 14] plays a major role in detecting the different skin tone. HSV colour space is the best approach to detect the colour and texture [15]. HSV plays a major role in object as in [16]. In [17], HSV validates the pixel-based object detection as in proposed work. Segmenting the image is applied to separate ROI from its surrounding regions [18]. Applying contours are generally used for analysis of pixel intensities having continuous points. Contours draw a line on the ROI, with continuous value points. Here, in this proposed work, they are used as a marker of same intensity. This can be called as the best approach to mark areas in the image which have same colour intensity. This process can also be implemented extensively for shape analysis [19].

3 Methodology

Methods in the proposed work implements a simple but yet effective techniques applied on colour images. The methods implemented using opencv, python. Opencv runs by importing its library like cv2, numpy and marplotlib to perform each of these steps. The methodology implemented in the proposed system is described as a step-by-step procedure and corresponding snippet used in opencv is shown below:

Step 1: Input RGB colour image is taken
A simple RGB image is captured in opencv: RGB image is imported as:

```
cv2.imread("Image in filename.jpg")
```

Step 2: Input RGB image imported into opencv
Once the RGB image is captured, the image is imported by opencv, imread function in opencv stores image as BGR instead of RGB; conversion is default in opencv library.

```
cv2.imread("Image in filename.jpg")
```

Step 3: BGR image is converted into HSV
In this step, BGR is converted to HSV; for object detection here in the proposed work, it is oral lesion detection.

```
cv2.cvtColor(img,cv2.COLOR_BGR2HSV)
```

Step 4: Track bars are created to adjust the parameters
Track bars are been created in this phase to adjust the hue, saturation and value of the image to create a mask. If any indication of noise, it can be removed by applying Gaussian mask with blur to avoid noise in extra regions.

```
cv2.createTrackbar("L-H","Trackbars",0,179,nothing)
cv2.createTrackbar("L-S","Trackbars",0,255,nothing)
cv2.createTrackbar("L-V","Trackbars",0,255,nothing)
cv2.createTrackbar("U-H","Trackbars",180,179,nothing)
cv2.createTrackbar("U-S","Trackbars",255,255,nothing)
cv2.createTrackbar("U-V","Trackbars",255,255,nothing)
```

Step 5: ROI segmented into mask
By using track bars, the ROI is localised using binary mask. Segmenting an image into its corresponding mask makes a clear view of image. It is an effective approach for high contrast images, thus separating foreground and background regions.

```
lower_red=np.array([l_h,l_s,l_v])
upper_red=np.array([u_h,u_s,u_v])
mask=cv2.inRange(img1,lower_red,upper_red)
```

Step 6: Adding contours into ROI of an image
Contours are generally used for shape analysis by joining the continuous points having same intensity. Here in this proposed work, they are used as a marker of same intensity.

```
cv2.drawContours(img,contours,-1,(0,255,0),3)
```

4 Results and Discussion

This section includes results of the methods discussed above, experimented on a number of colour RGB oral images. Various intensities are been initialised to get a segmented ROI of an image. The different steps to show experimental results of the proposed work are discussed below:

Step 1: Input RGB colour image is taken
Four RGB oral images are taken for experimentation, each of which is of different subjects: normal oral cavity, lesion found on above gum region, tissue changes, in the soft region of the oral cavity, and red lesion, as shown below:

- The image (a) is taken in 12.6 MP mobile camera.
- Image (b) and (c) oral cavity is taken in 12MP mobile camera.
- And image (d) is downloaded from Google images (Images 1).

Step 2: Input RGB image imported into opencv
In this step, since opencv library is used, it reads an image as BGR instead of RGB, as shown below, for example, lesion on gum, and histogram of each image is shown below (Images 2 and 3).

We can notice in the above histogram that how pixels having RGB and BGR colour intensities are distributed.

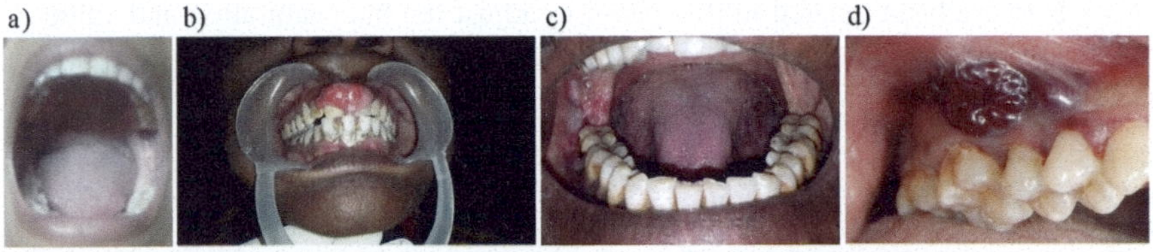

Images 1 (**a**) Healthy oral cavity, (**b**) lesion on gum, (**c**) tissue changes, (**d**) red lesion

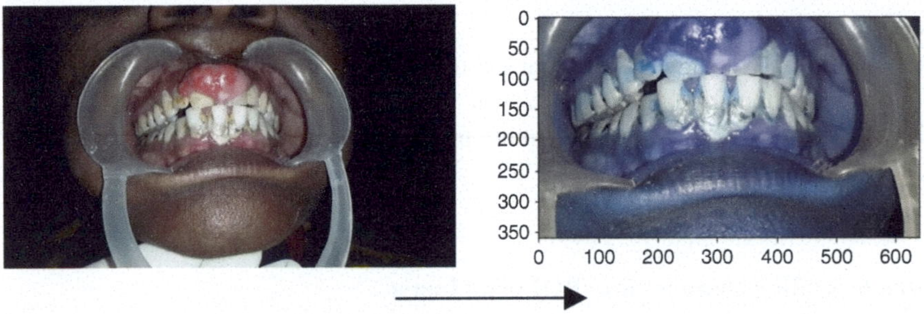

Images 2 Conversion of RGB to BGR

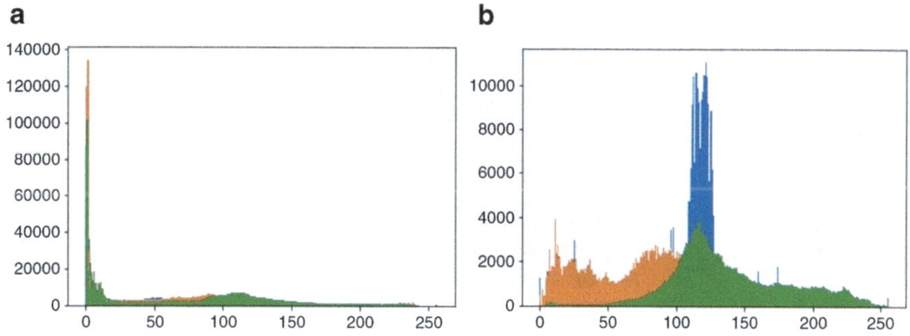

Images 3 (**a**) Histogram of RGB and (**b**) Histogram of BGR

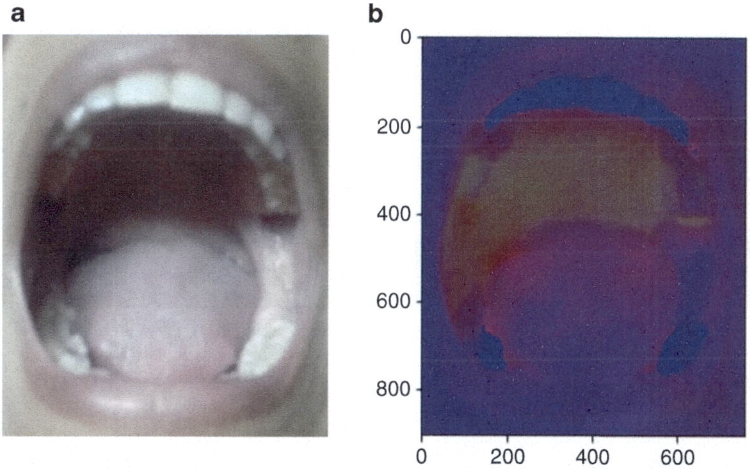

Images 4 (**a**, **b**) Image of healthy oral cavity in RGB and HSV

Step 3: RGB image is converted into HSV

In this step, four types of images are shown, as each of them includes a different parameters or includes lesions in different places of oral cavity. The main focus of this step is to detect lesions colour intensity and how they are different from a healthy region and non-healthy region. This is achieved by converting RGB into HSV, since we use opencv, RGB is converted to BGR in order to find HSV as a best approach to detect objects; in this, it is ROI of an object.

- In the first image, RGB image of healthy oral cavity and its converted HSV image are shown (Images 4).

- In the second image, RGB image of lesion on gum and the respective converted HSV image are shown (Images 5).

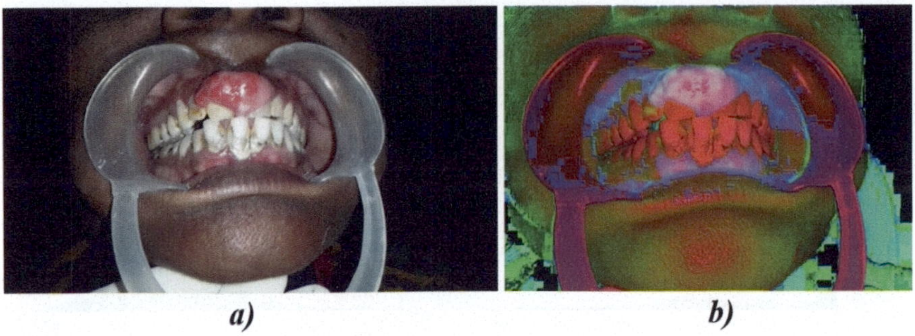

Images 5 (**a**, **b**): (f) The lesion on gum image in RGB and HSV

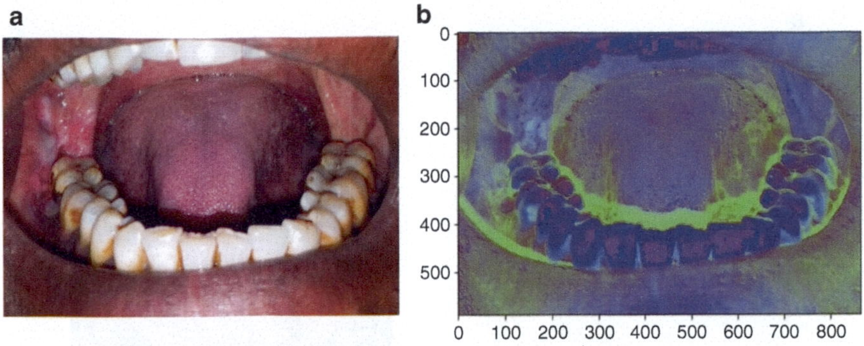

Images 6 (**a**, **b**): (g) Tissue changes in oral cavity image in RGB and HSV

In this image, we can easily identify ROI by noticing the changes in colour in lesion above the gum region.

- In the third image, RGB image of tissue changes and the respective converted HSV image are shown. In this image, only small changes of tissue are detected (Images 6).

- In the fourth image, RGB image of red lesion that occurred in soft regions of the oral cavity and the respective converted HSV image are shown. In this image, it is easy to differentiate between colour intensities of affected and non-affected regions (Images 7).

Step 4: ROI segmented into colour mask

In this step, it is focused on the main idea of the proposed work to identify the red, red white (pink) and white regions of ROI. Here track bars are used to know the range of HSV in the images. The colour mask, binary mask and values initialised to create a mask are shown in tables for respective images as shown below.

- For image named as healthy oral cavity from image 1.

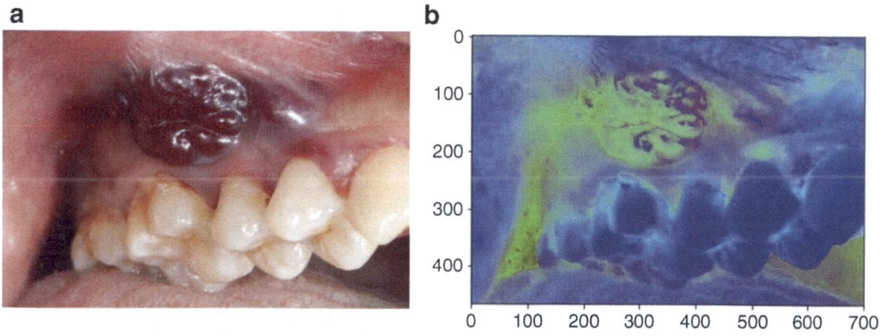

Images 7 (**a**, **b**): Images of red lesion in oral cavity in RGB and HSV

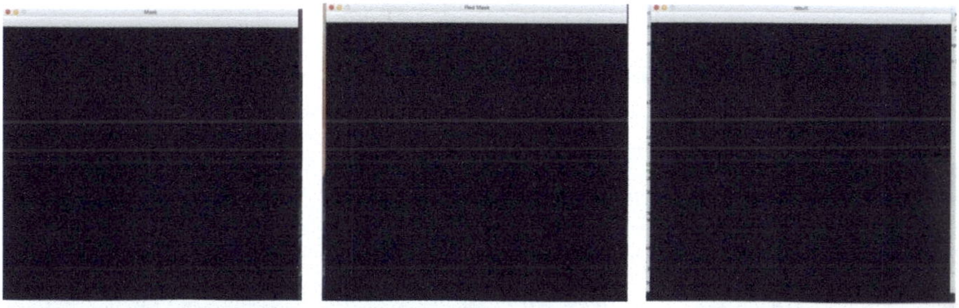

Snap shots 1 (i) Colour mask of healthy oral cavity from image 4 f

The above snapshots show no changes in intensities. Hence, no mask is determined (Tables 1 and 2).

For image named as tissue changes from image 1.

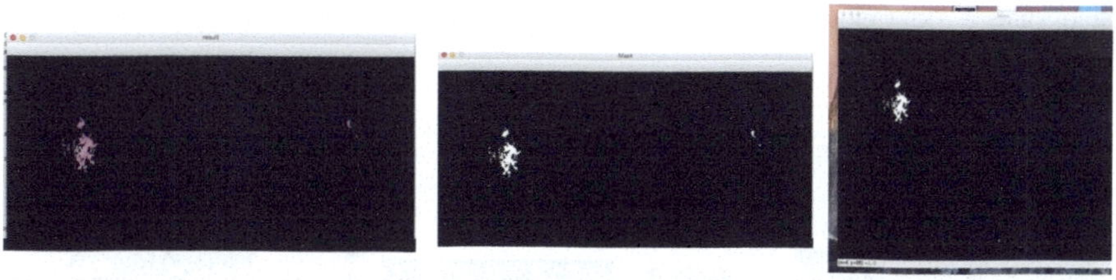

Snap shots 2 (j): Colour mask, binary mask and resultant mask of tissue changes in oral cavity from image 6

- For image named as lesion on gum from image 1 (Table 3)

Table 1 Hue saturation values for image shown in image 4

Red_mask	Hue	Saturation	Value
Lower_red	***100***	***212***	34
Upper_red	179	255	255

Table 2 Hue saturation values for image shown for image 6

Red_mask	Hue	Saturation	Value
Lower_red	***167***	***110***	186
Upper_red	179	255	255

Table 3 Hue saturation values for lesion on gum in oral cavity for image 5

Red_mask	Hue	Saturation	Value
Lower_red	***172***	***115***	201
Upper_red	179	255	255

Table 4 Hue saturation values for red lesion in oral cavity of image 7

Red_mask	Hue	Saturation	Value
Lower_red	***80***	***180***	84
Upper_red	179	255	255

- For image named as red lesion from image 1 (Table 4)

Step 5: Adding contours into ROI of an image

In this step, the contours on the mask of ROI got from previous steps are been applied on RGB images (Images 8).

- For image named as lesion on gum from image 1

Snap shots 3 (k): Colour mask, binary mask and resultant mask of lesion on gum image in oral cavity from image 5

- For image named as tissue changes from image 1 (Images 9)

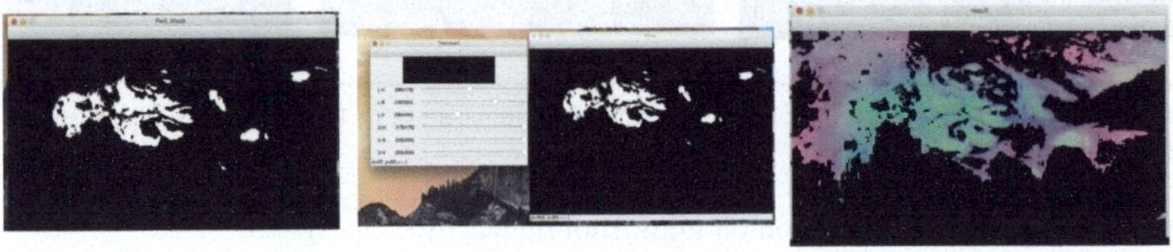

Snap shots 4 (l): Colour mask, binary mask and resultant mask of red lesion image in oral cavity of image 7

- For image named as red lesion from image 1 (Images 10)

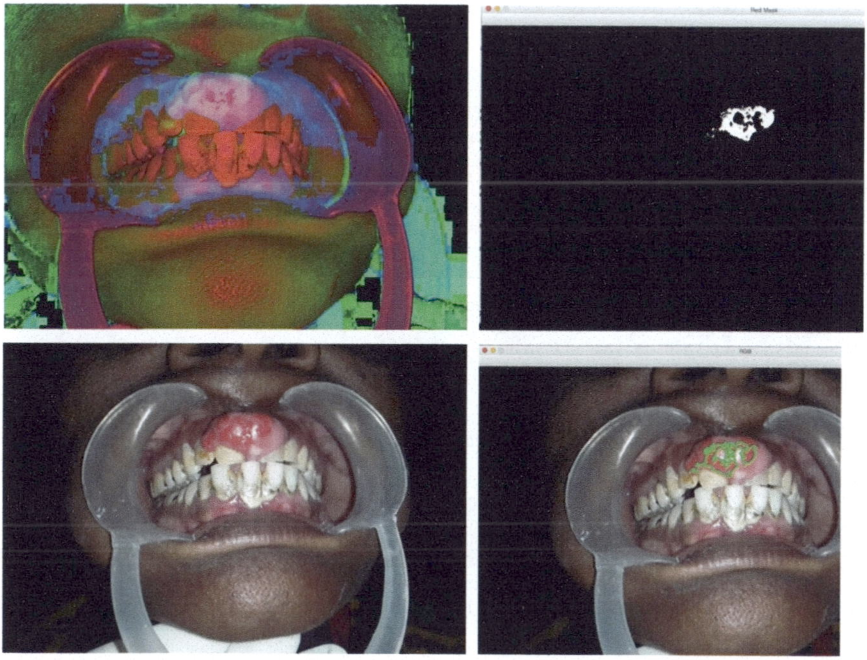

Images 8 (m) HSV and mask of ROI & (n) RGB image and contour on this RGB image

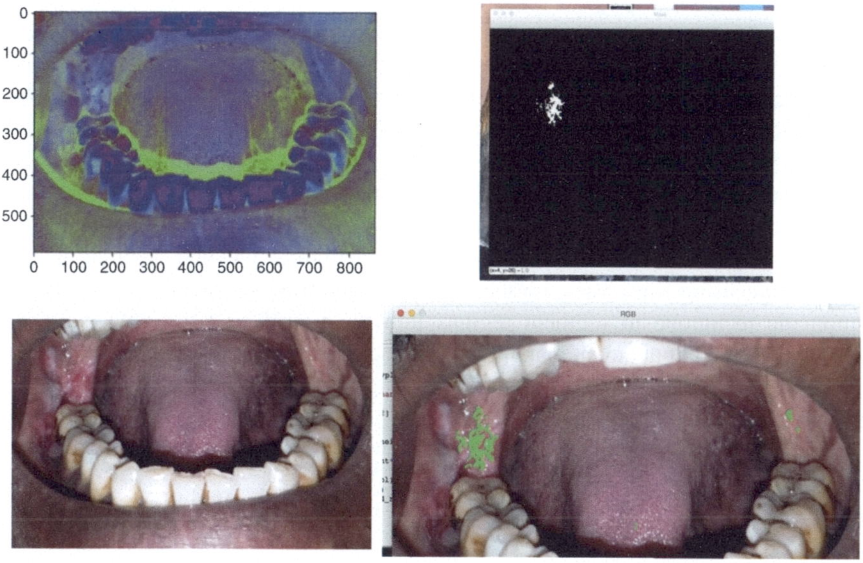

Images 9 (o) HSV and mask of ROI & (p) RGB image and contour on this RGB image

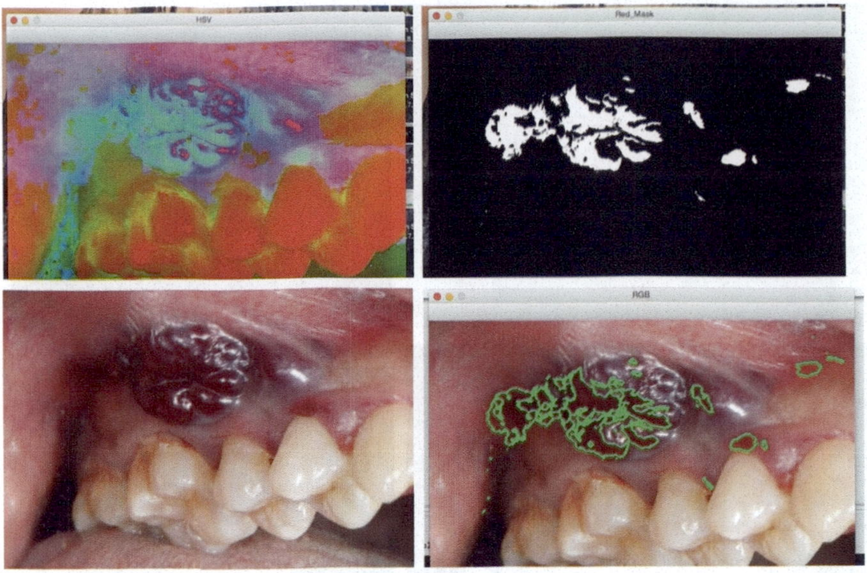

Images 10 (q) HSV and mask of ROI & (r) RGB image and contour on this RGB image

5 Conclusion and Future Work

As simple methods perform an effective solution to the problem. Implementation of these methods as basic procedure to detect such a challenging task is one of the smart methods to locate oral lesions effectively based on their colour property. The proposed method includes some noise that is removed by Gaussian blur technique. The mask, colour mask and contours applied on the image provide an easy way to detect the abnormalities in the oral cavity. Though the proposed method provides an easy method to detect abnormalities in the ROI of an image, as it implements basic techniques and manual methods such as finding the hue, saturation and value range using track bars. Further focus of this proposed work is to apply an automatic technique to segment regions into their mask and to apply contours automatically without any human intervention. Also applying this methodology can be implemented in handheld device for helping medical practitioner to detect the ROI in a promising way.

References

1. B.W. Neville, T.A. Day, Oral cancer and precancerous lesions. Cancer J. Clin. **52**, 195–215 (2002)
2. M. Awais, I. Faye, Analysis of auto-fluorescence images for automatic detection of abnormalities in Oral cavity, in *7th International Conference on Information Technology and Electrical Engineering (ICITEE)*, (Chiang Mai, Thailand, Oct. 2015)

3. C.-S. Yang, M. Ou-Yang, Portable noninvasive system for oral cancer diagnosis, in *11th IEEE SENSORS 2012 Conference*, (Taipei, Taiwan, Oct. 2012)
4. M.M. Aznaveh, H. Mirzaei, A new color BasedMethod for skin detection using RGB vector space, in *2008 IEEE Conference on Human System Interactions*, (Krakow, Poland, May 2008), pp. 25–27
5. S. Han, X. Yang, Q. Qi, Y. Pan, Y. Chen, J. Shen, H. Liao, Z. Ji, Potential screening and early diagnosis method for cancer Tongue diagnosis. Int. J. Oncol. **48**(6), 2257–2264 (2016)
6. P. Goyal, R. Yadav, S. Singh, Early detection of symptoms in tongue cancer through image processing, in *Proceedings of the International Conference on Communication and Computing Systems (ICCCS 2016)*, (2016)
7. Color Detection and Tracking from Live Stream – An Extensive Survey, International Journal of Computer Applications **168**(3) (2017)
8. S.V. Sreelatha, S. Chuppa, D. Bharati, C.R. Chowdhury, P. Shetty, Leukoplakia and erythroplakia: A clinicians perspective and histopathologist verdict – A retrospective study. Journal of Advanced Clinical & Research Insights **3**, 143–146 (2016)
9. A.F.M. Hani, L. Arshad, A.S. Malik, A. Jamil, F.Y.B. Bin, Assessment of chronic ulcers using digital imaging, in *2011 National Postgraduate Conference IEEE*, (Kuala Lumpur, Malaysia, 19–20 Sept. 2011)
10. B. Youssef, Z. Hamid, B. Abdelkhalak, Computer-aided diagnostics of facial and oral cancer, in *2015 Third World Conference on Complex Systems (WCCS)*, (2015). https://doi.org/10.1109/ICoCS.2015.7483252
11. A. Kauzman, M. Pavone, Pigmented lesions of the oral cavity: review, differential diagnosis, and case presentations. J. Can. Dent. Assoc. **70**(10), 682–683 (2004)
12. P.D. Velusamy, P. Karandharaj, Medical image processing schemes for Cancer detection: A survey, in *2014 International Conference on Green Computing Communication and Electrical Engineering (ICGCCEE) Research Gate*, (2014). https://doi.org/10.1109/ICGCCEE.2014.6922267
13. P.M.B. Ong, E.R. Punzalan, Comparative analysis of RGB and HSV color models in extracting color features of green dye solutions, in *Presented at the DLSU Research Congress 2014 De La Salle University*, (Manila, Philippines, 6–8 Mar. 2014)
14. B. Muhammad, S.A.R. Abu-Bakar, A hybrid skin color detection using HSV and YCgCr color space for face detection, in *2015 IEEE International Conference on Signal and Image Processing Applications (ICSIPA)*, (Kuala Lumpur, Malaysia, Feb. 2016), p. 25
15. M. Deswal, N. Sharma, A fast HSV image color and texture detection and image conversion algorithm. International Journal of Science and Research, 2319–7064 (2012)
16. S. Trambadia, H. Mayatra, Food detection on plate based on the HSV color model, in *2016 Online International Conference on Green Engineering and Technologies (IC-GET)*, (Nov. 2016)
17. L. Huang, G. Zhang, Y. Li, An object-based change detection approach by integrating intensity and texture differences, in *2010 2nd International Asia Conference on Informatics in Control, Automation and Robotics (CAR 2010)*, (Wuhan, China, 29 Apr. 2010)
18. D.C. Parthasarathy, Thresholding technique for color image segmentation. International Journal for Research in Applied Science & Engineering Technology (IJRASET) **3**(VI) (June 2015)
19. M. Kass, A. Witkin, D. Terzopoulos, Snakes: Active contour models. Int. J. Comput. Vis. **1**, 321–331 (1988)

Automated Pill Dispenser Application Based on IoT for Patient Medication

J. Ramkumar, C. Karthikeyan, E. Vamsidhar, and Kale Navnath Dattatraya

1 Introduction

Currently, Internet of Things (IoT) plays an important role in all real-time applications and it makes things and devices in the applications to interact through networking [1]. It provides smart home, smart environment where we are connected with the outside world and persons through mobile devices, health care, etc. When a person starts his work from morning till he goes to bed, a person comes across several IoT-based applications. Even though people are passing through several IoT applications [1], there are several scenarios where IoT plays an important role along in health care, which is very important nowadays. As there are numerous real-time applications, devices in IoT have some limitations while embedding with healthcare-based applications to work in a proper manner [2] and [3]. TJ McCue has given a summarized report on health care IoT, which will hit the market of $120 billion by 2020. Nowadays, medical devices are integrated based on networking, which prolongs some of the issues, namely security and interoperability breaches are growing more and more, which result in huge financial losses. To lower the risk among the integration of devices, consumer-based technologies or automated operational technologies need to be adopted.

Health care is one of the most important application areas related to IoT, which gives many potential applications such as health monitoring, disease diagnosis [4] and body care. The healthcare applications [5] provide identification, diagnosis,

J. Ramkumar (✉)
Department of Computer Science and Engineering, SRM Institute of Science and Technology, Chennai, India

C. Karthikeyan · E. Vamsidhar · K. N. Dattatraya
Department of Computer Science and Engineering, Koneru Lakshmaiah Education Foundation (KLEF), Guntur, Vijayawada, India

N. Gupta, S. Paiva (eds.), *IoT and ICT for Healthcare Applications*, EAI/Springer Innovations in Communication and Computing,
https://doi.org/10.1007/978-3-030-42934-8_13

treatment and regular monitoring, and these activities are visualized by the smarter devices that are embedded in IoT [6]. The main goal of healthcare applications is to provide a reduced cost, increase the user life level, easy operational customization and user experience. The above-listed objectives can be achieved by conducting smooth execution and multiple repetition of process along with the smarter devices and IoT to obtain optimality, to provide seamless and secure connectivity through a cost-effective manner among patients and healthcare organisations [7]. For seamless connectivity, efficient scheduling plays an important role based on limited resource for better usage. In healthcare applications, wireless technologies support several activities such as diagnosis of diseases, monitoring systems and medical care. In healthcare learning, several smart devices such as the gateway, the server and database play a vital role in creating data information to deliver medical services to the authorized persons [8].

In recent years, IoT plays an important role among researchers to address the issues related to healthcare applications. Across various countries in the world, health care provides new framework, applications, services and security.

1.1 Arduino

Arduino is the large manufacturer of microprocessor and microcontroller and mainly it is an open-source company based on hardware and software. Especially, the board has input and output where input reads are based on the sensor, button, etc., and output gets the results using the indication of motor or LED [9, 10]. Arduino has used several instructions set to the process based on Arduino programming language and its IDE. In these several years, Arduino has provided instruction to several applications, namely, scientific instruments, programmers, professionals, and others, who have enormous amount of knowledge access by integrating with open-source platform. In this Arduino Broad, we have several microprocessors and microcontrollers integrated, that is, basic stamp, microprocessor 8086, etc., which offer their special functionalities [11]. It provides several features as listed below:

1. It is inexpensive.
2. Cross-platform i.e. integration of multiple operating system.
3. IDE environment.
4. Open-source software and hardware.

For the above features, Arduino is less expensive when compared with micro-controller, as its modules are less costly, near \$50. Arduino works in several operating systems like windows, MAC OS, Linux, etc. Users have flexibility in using the software, as it supports Integrated Development Environment (IDE) and also provides convenience in processing the program [12]. Arduino language has extended support with C libraries to get easy experience for the programmers [13].

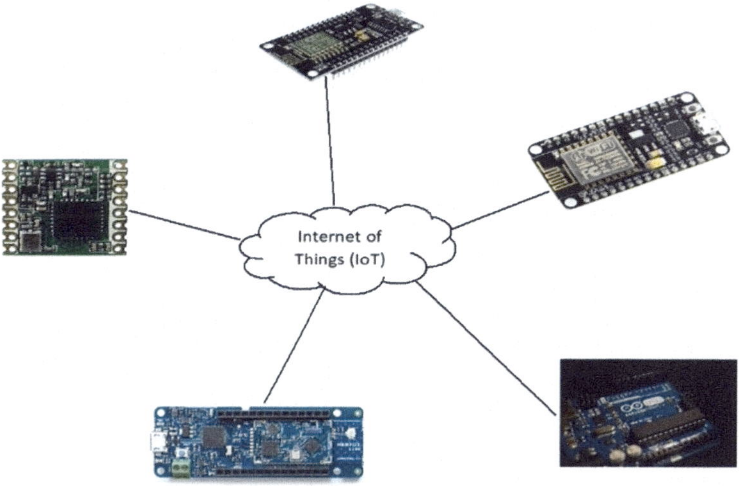

Fig. 1 IoT with microcontrollers

1.2 Arduino Uno

Arduino Uno is a microcontroller broad application, which is equipped with input and output pins and several interfacing cards based on microchips [13]. In recent days, the Arduino microcontrollers are used well in Internet of Things (IoT) technologies to connect several physical objects together [14] and it is represented in Fig. 1.

1.3 IoT Framework

IoT framework consists of physical elements, functionalities and working principles and techniques, which are represented in Fig. 2 [15]. The key elements are WLAN and WPAN, which support interoperability of the IoT gateway. It helps in streaming the video and securing communication between the IoT and its users.

2 IoT Health Care

Doctors have certain limitations to provide continuous medication to the patients in the field of medicine. To overcome the limitations, IoT enables devices into play and to integrate the concept of remote monitoring to increase the confidence of patients in terms of health and safety. Relationship between the patients and doctors

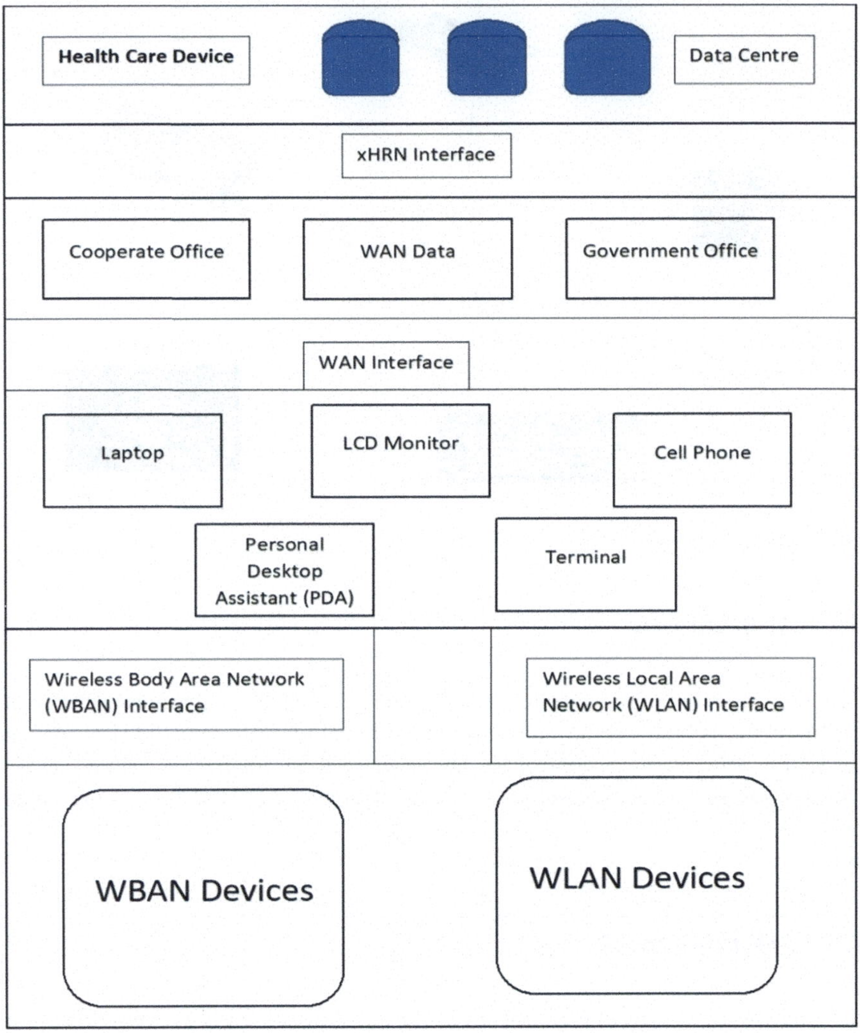

Fig. 2 IoT framework

increases and continuous assessment of health and medication is more effective and easier [16].

Generally, in the IoT health care, cloud service acts as the central database where all other components such as gateway and the medical server are integrated into the database present and it is represented in Fig. 3. Gateway is nothing but the bridge which interconnects with the Internet and another end is connected with the user terminal with hospitals which process several operations like Electronic Medical Record (EMR), real-time monitoring, fitness, etc., and it is connected with the database to store all the data [17].

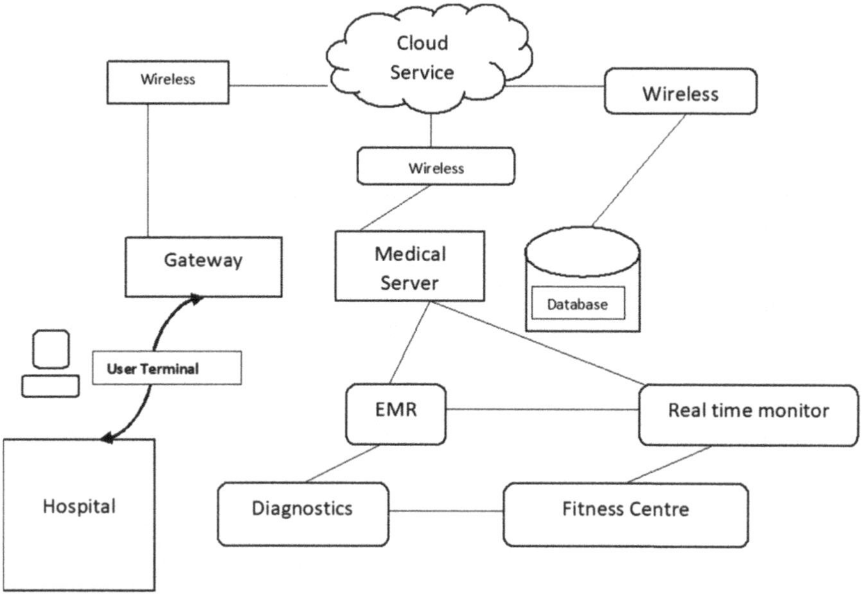

Fig. 3 IoT health care

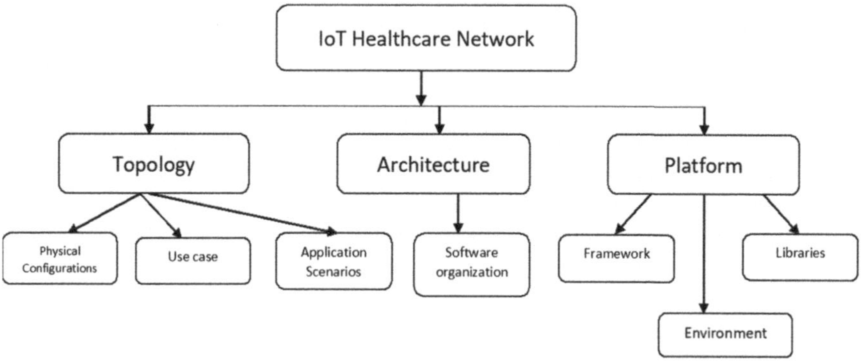

Fig. 4 IoT health care

Health care-based IoT technology plays an important role in healthcare applications and services. It supports transmission and reception of medical information to access IoT and enable the communication usage. The IoT networks' broad classification is represented in Fig. 4 and it is listed as follows:

A. *Topology Deployment*: Use case, network/element configuration, services and application environment.
B. *Framework*: Entire software system description and hierarchical reflection.
C. *Environment*: Framework, database and scenarios.

2.1 IoT Topology Deployment

IoT network arranges different elements to provide a seamless connectivity scenario in healthcare applications. In Fig. 4, coordinated heterogenous component plays a vital role which classifies three elements, such as resource providers who get the task and submit the result, data providers who will submit the work activities and get the results based on the services such as blood pressure sensor, ECG sensor, EEG sensor, and brokers who accept the services and retrieve the results to get the advertisement services [18]. Transformation occurs among the heterogenous networks and other devices such as laptop and desktop into hybrid networks.

A. Solutions of IoT Health Care:

- Coordination space integrates with the broker along with the data provider and resource provider.
- Using the various sensors present in the data, the provider will send the workload, that is, data into the coordination space to get the results for the appropriate process.
- Data which are fetched are sent to the resource provider, who will perform certain tasks based on various components such as computer, laptop, and mobile and send the results to the coordination space.
- Brokers along with the workflow manager and scheduler schedule the tasks assigned to it based on high and low priority and accept the workload and show some results to get the service advertisement.

Solutions based on IoT-based health care are diagrammatically represented in Fig. 5.

Patients' healthcare solutions can be met based on various medical devices and sensors to process certain activities by the integration of analyses and store.

Topology contains streaming of certain videos based on data served out of certain sensors, which may be the data extracted and displayed. Streaming of videos is carried out based on interconnection of the network such as Wireless interoperability of Microwave Access (WiMAX), Global System for Mobile (GSM), and Long Term

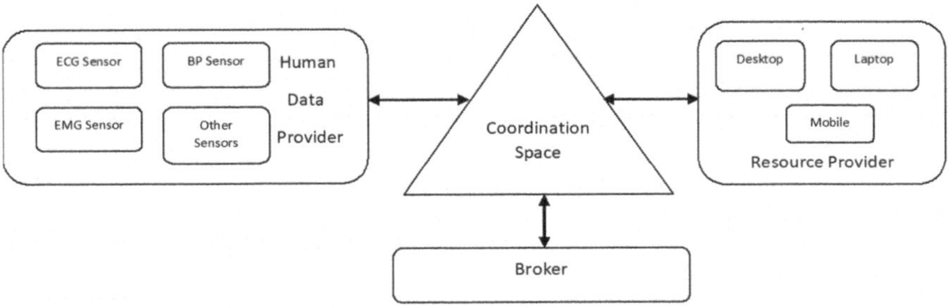

Fig. 5 Solutions based on IoT-based health care

Evolution (LTE), which processes at a higher speed. Nowadays, the medical domain is moving to intelligent processing, with IoT devices, that is, wearable sensors. It is connected by wireless through the gateway Internet with the centralized cloud, which is nothing but the IoT cloud based on health care [19]. It is processed based on investigation, storage, collection and display, and the data are extracted during this process.

2.2 Arduino-Based Health Care

Even though these activities are performed in IoT, sensor is the main component / device to provide connectivity and interactive services where the sensor fetches the data related to all changes in the activities of a person. However, the sensors help identify the variation in activities and movements in the particular scenario. IoT is embedded with the microprocessor chip which is placed in the arduino broad, which helps to integrate the relevant technology to perform all the tasks [20, 21]. If any process has to be performed, Arduino broad is self-designed to accommodate certain activities.

2.3 Android-Based IoT Health Care

In recent times, IoT is integrated into multiple users along with various components, such as computer, laptop and mobiles, as these have become more and more necessary for all people in their everyday life [22]. Mobiles are interconnected with some of the physical objects and hence they act as per the task sent by the mobile. As the number of devices is growing enormously, there is an increase in the system complexity, but the user experience should be effective at all times. For example, some air conditioner companies provide a mobile application, which will effectively perform all the activities to manage the Alternate Circuit (AC) and increase the user experience.

Here we are discussing some of the applications which are integrated with some of the physical objects other than health care, which are listed below:

A. Amazon Go.

Amazon Go application helps in integrating the supply chain management and improves effectiveness in the area of retail business. Based on IoT technology, mobile application provides a wide supply of retail management in purchasing items without any limitations. Amazon Go will take the privilege of control of the supply chain by eliminating the human factor in order to avoid abuse and fraud by the staff members.

B. Google Nest.

Google Nest is the smart home application that integrates IoT technology to make life easier and comfortable by controlling the temperature, smart appliances, etc. In the case of overheating and over cooling, the smart application will control the temperature to make the home a comfortable place. Applications will have an alarm indication for homes circulated with smoke and an alarm to make home a better place to live.

C. Fitness Band.

It is the wearable device, which is interconnected with the mobile application to keep track of the heartbeat, Blood Pressure (BP), calories, sleep and track activity during running and walking. The fitness band belongs to health care where it indicates the medication to be taken when your health goes weak. In recent days, data are collected and analysed from any IoT devices to process the required information and to identify the analysis accuracy regarding the health of the person.

Except the list of applications discussed above, a greater number of applications are available integrating with IoT Technology. Mainly in the field of health care, medications are needed for the elderly to monitor their health periodically.

3 Literature Related to Health Care in IoT

As there are several healthcare issues that are prevailing in recent days, multiple solutions to health care based on IoT are deployed by enabling various services. For health care, some of the new functionalities have to be added into the IoT framework which is discussed in sect. 1.3. New functionalities include service sharing, notification, and Internet, etc., to make the process fast, secure and by using low-power devices.

A. *Encompassing Assistance*:

It is mainly providing services to make smart home or medical IoT through which it can offer those functionalities to the elderly to make them more comfortable. For the smart homes, IoT framework integrates with Artificial Intelligence (AI) to encompass the smart home assistance and health care. Assistance will provide services mainly to older people to make them feel comfortable, convenient and safe at home. For example, if any problem arises to any elderly persons in the home, smart assistance will act like a human servant to build confidence. Smart assistance-based IoT architecture is proposed in several areas for security, communication and control [23], and it is represented in Fig. 6. Advanced technologies like Radio Frequency Identification (RFID) and Near-Field Communication (NFC) are activated, which help in processing the activities in health care. Smart assistance is provided based on IoT with cloud computing [17] to overcome the problem in interoperability, security and Quality of Service (QoS) [19].

B. *Mobile IoT*

Although several functionalities are deployed into the IoT based on health care [24], Mobile-IoT (M-IoT) provides mobility entity so that accurate sensing of a person's health can be periodically monitored. In recent days, M-IoT is

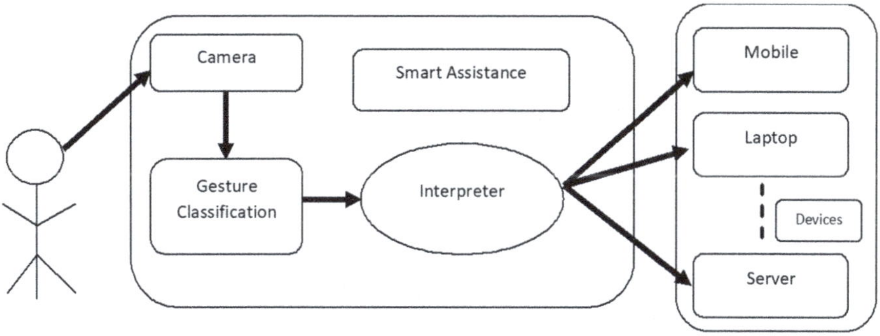

Fig. 6 Smart assistance based on IoT

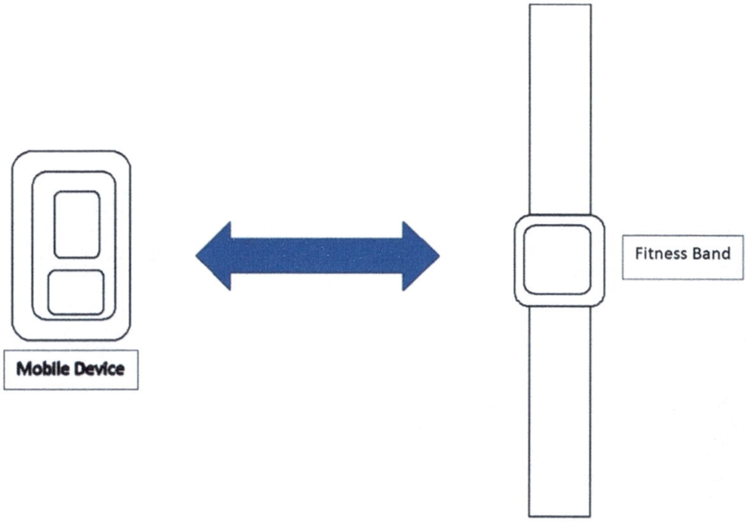

Fig. 7 Mobile-IOT

integrated with sensors, and communication technologies like 4G wireless network are applied [25]. Even though the mobile provides mobility, low-power devices are not maintained for the process of M-IoT and it is represented in Fig. 7.

C. *Drug Medication*

The patients identify the medicine based on NFC device where pharmacy-based information system helps to identify whether the medicine is suitable or not for a particular condition. It also gives the previous record of the patient health to identify the drug medication properly [26]. The medicine and medication will overcome the problems of having a single drug without proper diagnosis, prolonged medication and combining two or more medicines without proper analysis, and it is diagrammatically represented in Fig. 8.

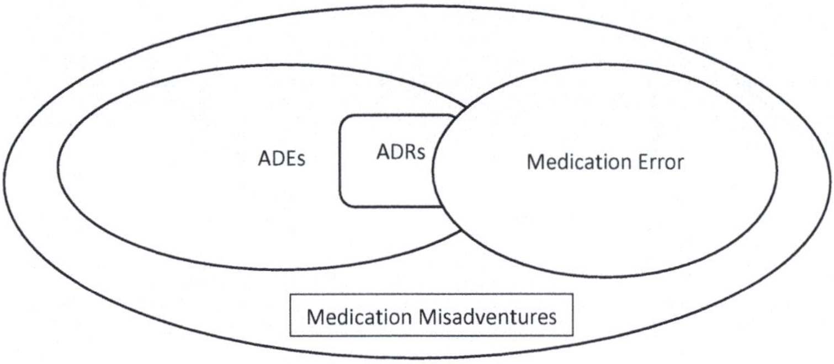

Fig. 8 Drug medication

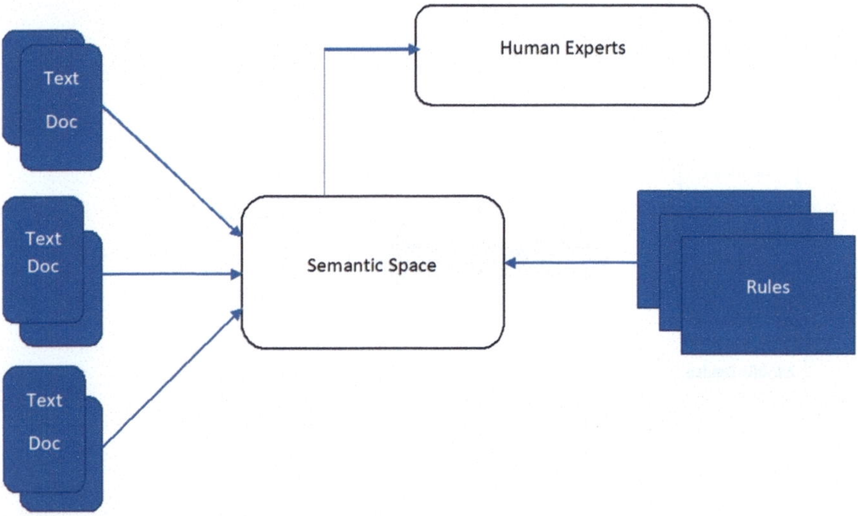

Fig. 9 Semantic medical access

D. *Semantical Medical Access*

A massive amount of medical information is in the cloud, so it cannot be classified based on healthcare applications. So categorized those data's based on web semantic and web ontology, which plays a vital role extract the specific knowledge from the datasets [27]. Data can be processed based on data collection, integration and interoperation based on IoT to provide medical service and it is represented in Fig. 9.

E. *Community Health Care*

Cooperative healthcare is monitored by establishing the local network, that is, IoT network includes municipal, urban, rural and residential network, and it is represented in Fig. 10. Healthcare information is connected to the cooperative

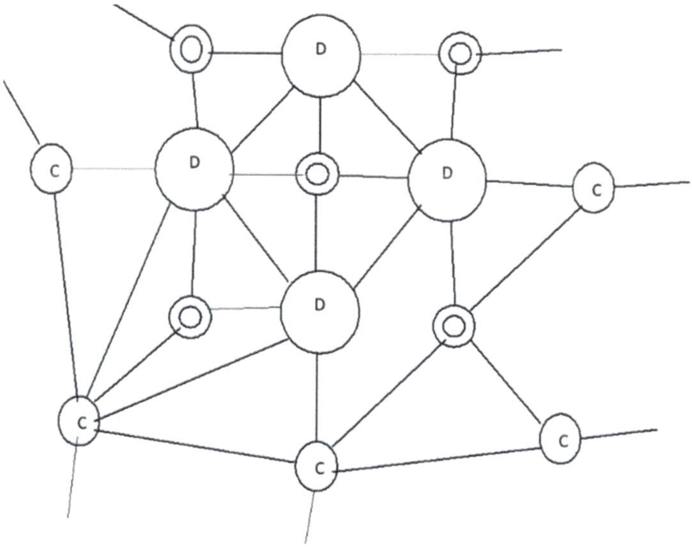

Fig. 10 Semantic medical access

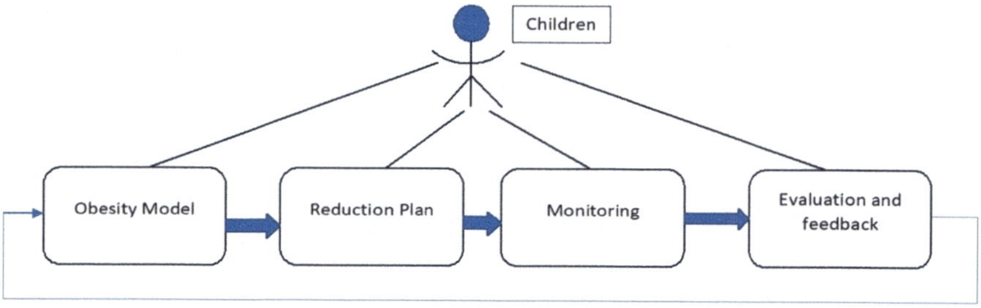

Fig. 11 Children health: example

network structure, and it collects technical requirement as a group which is energy efficient [28]. For these processes, cooperative network, authentication and authorization are needed.

F. *Health Information*

The recent scenario being so much pressure on children, especially here, there is awareness among the parents and teachers to promote children's health and needs based on emotion, or mental health, which is important. IoT-based health care among the children should be more effective to address their needs. Mainly this awareness should promote proper nutrient and effective hospitality to the children and its example is represented in Fig. 11.

G. *Wearable Devices*

Several legal sensors are available, especially in the field of health care, which will provide the services afforded by the sensors based on IoT [29]. On the other side, health care became portable, that is, wearable devices which act as a wearable product and address all the challenges with the integration of IoT and health care. The prototypes are integrated into the wearable equipment and to execute healthcare applications, mobile computing devices play a vital role. Mobile computing devices, such as smartphone and smart watches, analyse the activity performed by the sensor through the mobile application, whereas monitoring patients through a remote is possible. Wearable devices are diagrammatically represented in Fig. 7.

H. *Emergency Medication in Remote Areas*

Addressing healthcare issues is important during an emergency situation, where some post-medication is needed based on weather conditions, accident, earthquake and land collapse, etc. During those situations, IoT-based health care is needed where it will give a clear idea about how to process post-medication in the proper way.

I. *Internet-Embedded Configuration*

Internet-embedded configuration framework connects the network node directly to the patients. In this framework, the Internet plays as an intermediate, which connects the client and server to do a certain process, that is, medical equipment process [30]. Even though certain services are embedded with Internet to do certain processes, some services need some more functionalities to carry out along with the Internet. These Internet-connected processes allow intelligent and automation to monitor these processes. For some services, IoT-based health care requires mobile computing to be implemented.

J. *Context-Aware Prediction*

Third-party is in the requirement of using the generic framework [31], that is, IoT health care to implement their specific mechanisms. The context-aware healthcare system requires a context predictor over IoT network for remote monitoring.

In this chapter, we have addressed to overcome the problem of improper tablet intake where we have developed a mechanism through a mobile application that will remind the user to take the particular pills and monitor the activity of tablet intake. In order to overcome the above problem, it will help the elderly and other targeted persons to keep them under monitoring medication in case they are suffering from diseases such as Alzheimer's.

3.1 Alzheimer's Disease

It will affect the mental functions and result in memory loss. Alzheimer's disease is nothing but a brain disorder that will progress reversely, not as a normal human brain. It will destroy the brain memory slowly and some of the thinking skills will be lost and the patients will have little ability to do simple tasks. Alzheimer's disease symptoms first appeared in the mid-1960s. Based on a recent survey, 5.5 million Americans under the age of 60 and above are affected by Alzheimer's disease.

Based on the National Institute of Aging, Alzheimer's is the sixth ranked disease which causes death. Compared with heart disease and cancer, people's death due to Alzheimer's is more according to the survey. Based on this disease, mental loss happens and cognitive functionalities reduce, resulting in memory loss and reasoning, thereby affecting a person's normal life and activities. Other than Alzheimer's, the person may also have dementia and memory loss, problems in language and communication, keeping track of things, misplacing the things and changes in mood and behaviour.

Dementias is further classified into three types as follows:

A. Lewy Body Dementia

   This disease will affect when the person has extra proteins which are deposited in the brain, that is, Lewy bodies. The above activities cause problem in thinking, task movement, change in behaviour and mood.

B. Frontotemporal Disorder

   This disorder causes damage to the brain in both the frontal and temporal lobes, resulting in damage to the neurons in the brain, causing problems in normal thinking and behaviour control.

C. Vascular Dementia

   Vascular dementia is caused when stroke and other injuries related to vasculature occur, which result in thinking, memory and behaviour problems.

In some other cases, there is a combination of one or more disorders from the above lists.

3.2 Survey Related to Medication Dispenser

Basic medication dispenser mechanism, that is, controls system mechanism [32], contains several components such as keypad, Arduino broad, controller, GSM notification system and alarm. This dispenser helps the patient to take the pills at the particular time with a specific notification. Although this dispenser is effective, still it has some limitations in improper monitoring and non-automation. Pei-Hsuan Tsai et al. [33] proposed a modified medication dispenser, which is automated to schedule the activity of taking the pills and which collaborates with some other activity using the embedded system. This may have some limitations of adding new features without redefining the older dispenser system. Disadvantages nothing but it cannot be monitored by the doctor anywhere.

Some researchers are working on the GSM-based dispenser system [34], and the proposed GSM-based pills medication dispenser system can take charge of persons over the age of 60 who regularly forget to take their medicines. GSM will send the message notification to their family members, so that they help them to take their medicines periodically. With the help of the GSM, it will provide human and machine interaction based on the communication module. There are issues that arise while making continuous network connectivity.

The automatic pill dispenser is proposed [35] in order to improve medical diagnosis periodically and also to make the patient to have their medicines properly. This mechanism contains both the pill dispenser and the communication module to synchronize. Even though the automatic pill dispenser is effective, it fails in portability and does not provide evidence in monitoring the activity of the patient. If the patient is taking the pills daily for a long course of time, until now we do not have any pills dispenser to rectify these kinds of problems. To rectify the above problem, Shashank Shinde et al. proposed a smart pill dispenser [36], which will remind us to make the persons to take the to medication at the required time. The main feature of this dispenser is that it will be refilled by giving notification to the pharmacy person to fill the same as early as possible. It may also have some limitation of storage and monitoring functionalities.

Still there are some more modifications related to smart pill dispenser added to the feature of mobile computing to give notification to take the pills at the right time. Also, it refills all the tables in different sizes at different trays to facilitate some differentiation [37].

Here, we have proposed the smart pill dispenser to address the problem of improper data exchange of information between the person and mobile computing. To address the above features, EEPROM and RTC are added, which will store the notification time and will send notifications and alarm to the device wherein a proper communication is established. Notification is done based on light, sound and LED [38, 39]. No further modification can be done on the dispenser box and no monitoring. If the situation of a person or patient becomes worse based on stress and mood, during the above situation, monitoring the patient's behaviour and checking whether they have taken the pills regularly based on the integration of the third-party sensor are important [40]. Even though the third-party sensors are added, there is a limitation of not providing a proper sensor to be integrated because of compatibility issues.

Now, the modified pill dispenser [41] is proposed which will provide a proper communication module that will provide connectivity between the device and the doctor based on the periodic notification. In this process, monitoring and communication are strengthened, but a single user can work for the process and no automation is provided. A recent application of human body temperature medication is proposed [42], which will provide medication for fever and will help in the medication regime. There may be a chance of faulty sensors integrated into it as well.

As days go on, the modified pill dispenser is proposed with IoT [43] to fix a predefined time so that it will give an alarm to take medication and manually it can be stopped after taking up the pills. Major limitation observed from the above literature is that a greater volume of medications cannot be monitored at the same time.

Summary of the literature observed from the above section informs us of the need for automated pill dispenser that has to overcome all the limitations and the drawbacks listed above. Along with the automated system, storage of different pills has to be done. Dispenser based on different storages, that is, button lever, is a programme based on sensor along with mobile application. Network connectivity connects the mobile application of the device with pills dispenser and uses any advanced network technology like 3G or 4G to provide continuous connectivity.

As health care is a human-based application [44], some remedies are needed for several health applications, such as the following:

- *GPS Smart Sole*: It is a wearable device where this device is placed inside the shoes and it helps in reminding the patient [45].
- *Ingestible Sensors*: It helps to monitor the person's body and warns the patient when it identifies some indiscretions in the body [46].
- *Urosense*: This is especially meant for patient in whom a catheter is inserted into the body cavity to check the body temperature and urine temperature [47].

Based on the above, some of the applications, such as patient pill mediation, are much needed for the patient to overcome the problem of improper accessibility and poor treatment. The objective of this chapter is to have a wide discussion on the IoT and the medical dispenser based on the cloud, that is, the online database.

Based on the discussion and based on the above sections, even though the medical field has developed drastically based on the integration of new technologies, the relationship between the doctor and the patient is the same for a long period of time. Based on the relationship, there will be an improper monitoring of patients by the doctor and improper medication, at wrong time and wrong pills to the patient. As of now, there are a greater number of applications available for several scenarios like body temperature and blood pulse. To provide proper pills to the patient, a cloud is integrated into the pills dispenser where the online database is combined so that the doctor can interact with the patient very well. Here, we propose a modified pills dispenser, which will be more user-friendly as it is automated. In the proposed one, we have combined IoT and the cloud so that it will be processed with a larger number of data and to make it automated, we have applied the machine learning technique.

4 Proposed Dispenser Pills Model

The proposed DP (dispenser of pills) is so simple and easy to handle and hence, the elderly patient can operate the dispenser very easily. Medical storage is represented in a circular pattern. The storage has multiple slots where different sized medicines

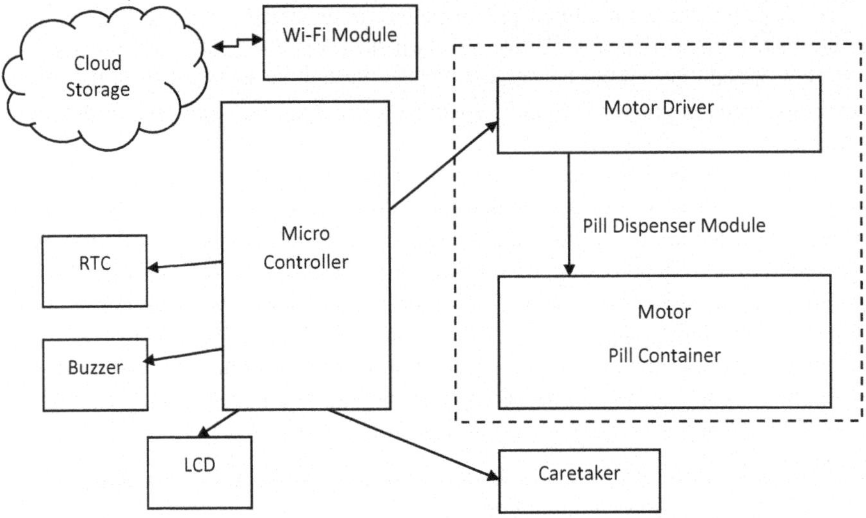

Fig. 12 Proposed dispenser pills system

can be loaded, facilitating a 360-degree rotation. Storage of medicines will be the main component which has holes at the bottom. One hole will be connected to the serving motor, which will release the appropriate medicine at a particular time. It has the indication/notification panel, which is nothing but an alarm or LED display where particular person will be reminded at the time as represented in Fig. 12.

In the working procedure, raspberry pi is used, which controls the dispenser. It acts as a server that fetches the information of the pills intake and it is directed to the online database. The information is monitored by the doctor periodically and proper guidance will be given to the patients. The proposed DP will work when there is no Internet connection, but to extract and download the information related to patient intake of pills needs an Internet connection. Here, sensors are connected to drop the pills into the container after receiving the notification. The sensor will measure the pills count and the issuing time as well. The proposed DP is controlled as follows:

1. Initially, inside the web, patients and doctor have to register and then they will be connected online.
2. After establishing the connection, the doctor would mention the dose based on the prescription given at the time when the patient came in person for the initial check-up.
3. Cloud database plays a major role where intake of pills periodically will be downloaded and updated online and it will be monitored by the doctor.
4. Cloud database is directly connected with raspberry pi.
5. Raspberry pi is used to control the pill dispenser of the storage compartment. Based on the raspberry pi, slots used in the compartment for storing different pills will rotate automatically to its position using a stepper motor.

6. Sonic sensor will check whether the hands are detected for taking pills.
7. If hands are detected, the mechanism gets activated to serve the appropriate pills and the updated information of pills taken will be updated to the raspberry pi, which is connected to the cloud database.
8. If the hands are not detected, a notification will be given based on an alarm or sound and then later, a text message will be sent.

The pill dispenser network is initiated by a server which is linked with an online database, that is, the medical information that is extracted and downloaded is stored. The network will store the information about the patient, that is, whether the patient has taken the medicines and then it is updated at regular intervals. The information which is stored in the database can be accessed by both the patient and the doctor.

Performance Evaluation:
Here, the proposed model performance is evaluated based on several parameters as listed below:

1. Data Success Rate used to determine the success rate of the proposed pill dispenser model is based on a number of medications. In the following Figs. 13, 14, 15, and 16, X-axis denotes the amount of information that varies from Most Interferers First (MIF) (0), Most Victimized First (MVF) (1), Rate Monotonic (RM) (2), Shortest Separation Difference First (SSDF) (3) and Earliest Deadline First (EDF) (4).

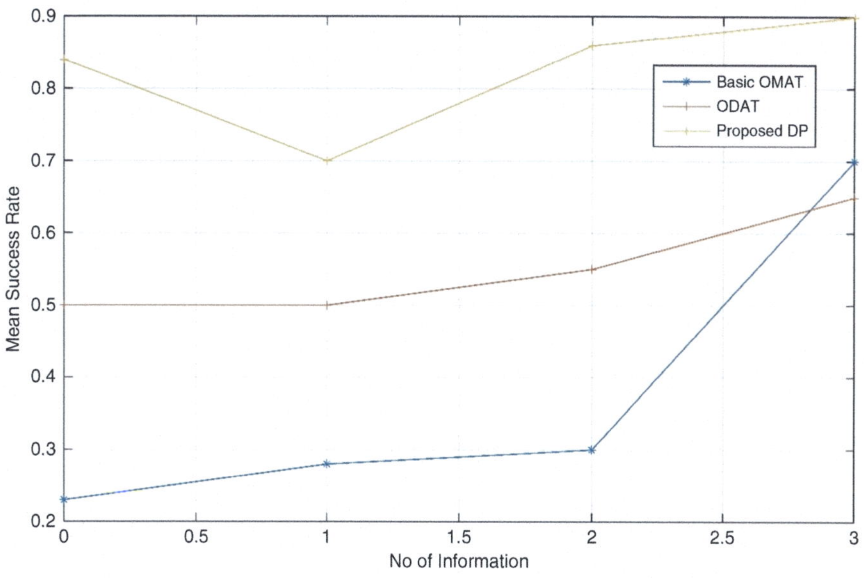

Fig. 13 Success rate based on nominal constraints (Number of Information 'n' = 5)

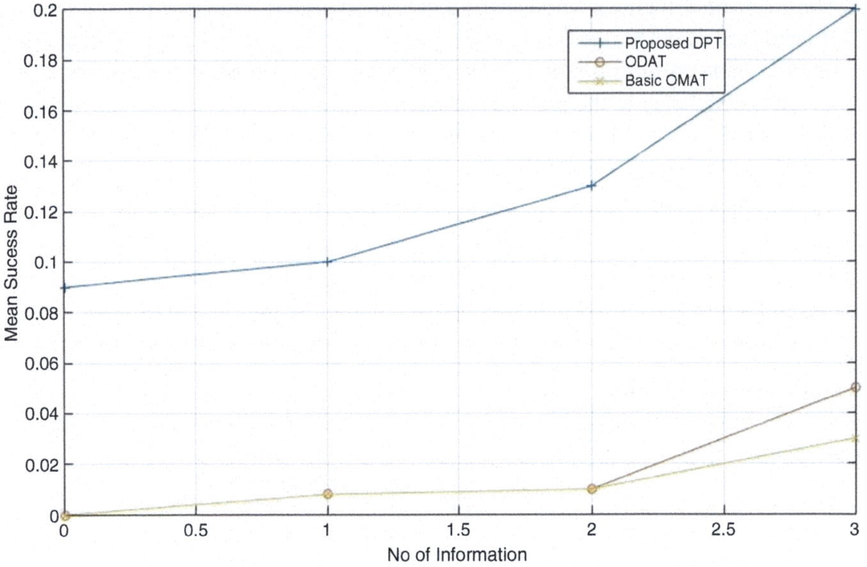

Fig. 14 Success rate based on nominal constraints (Number of Information 'n' = 10)

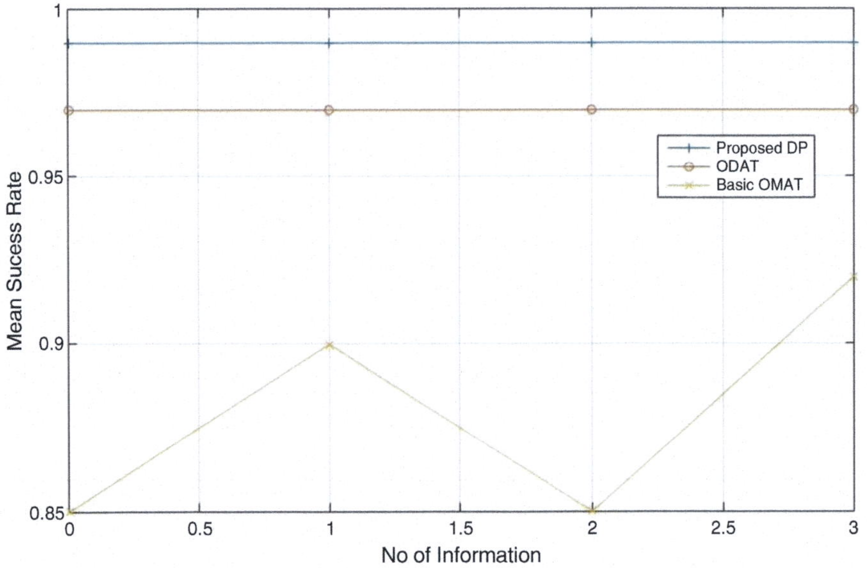

Fig. 15 Success rate based on absolute constraints (Number of Information 'n' = 5)

2. Schedule Quality Data: Percentage of dose intake given to the patient based on different ranges of normalization, which depend on the schedule generated by the proposed pill dispenser model. Based on Tables 1, 2, 3, and 4, Normalized

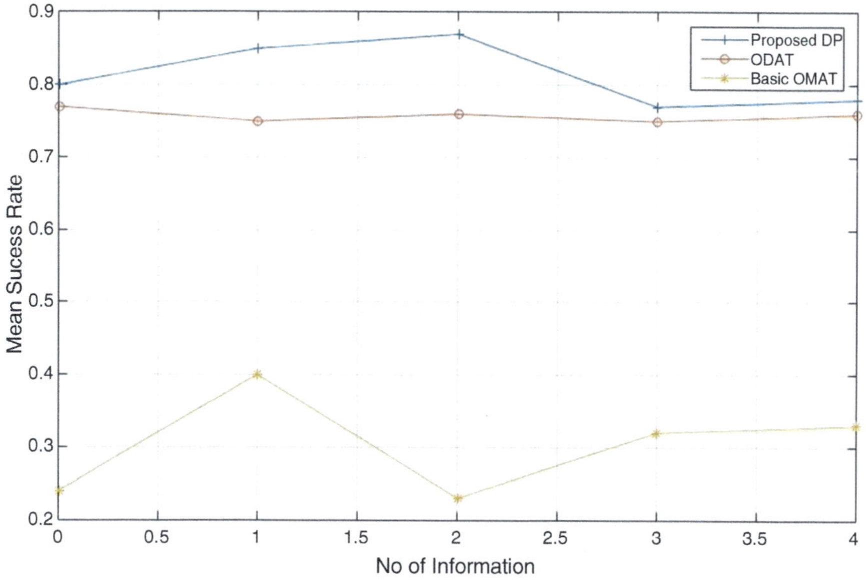

Fig. 16 Success rate based on absolute constraints (Number of Information 'n' = 10)

Table 1 NAT schedules absolute constraints, i.e. $n = 5$ with dose = 0

No of doses unfriendly	Basic OMAT	ODAT	Proposed DP
0	0.4	0.52	0.30
1	0.38	0.52	0.34
2	0.42	0.58	0.28
3	0.41	0.58	0.27
4	0.41	0.58	0.26

Table 2 NAT schedules absolute constraints, i.e. $n = 5$ with dose between 0 and 0.3

No of doses somewhat friendly	Basic OMAT	ODAT	Proposed DP
0	0.42	0.37	0.24
1	0.42	0.37	0.24
2	0.38	0.36	0.24
3	0.41	0.37	0.24
4	0.41	0.36	0.24

Allowed Tardiness (NAT) is generated by the algorithms such as Basic OMAT, ODAT and Proposed DP, which is classified into different categories, that is, Unfriendly, Some What Friendly, Friendly, very friendly to the dose which keeps varying.

Table 3 NAT schedules absolute constraints, i.e. $n = 5$ with dose between 0.3 and 0.7

No of doses friendly	Basic OMAT	ODAT	Proposed DP
0	0.10	0.37	0.24
1	0.12	0.37	0.24
2	0.	0.36	0.24
3	0.41	0.37	0.24
4	0.41	0.36	0.24

Table 4 NAT schedules absolute constraints, i.e. $n = 5$ with dose between 0.7 and 1.0

No of doses very friendly	Basic OMAT	ODAT	Proposed DP
0	0.10	0.08	0.30
1	0.12	0.08	0.26
2	0.09	0.08	0.32
3	0.08	0.07	0.32
4	0.07	0.07	0.32

Limitations Overcome by the Proposed DP Model and Its Advantages:
According to the existing literature discussed in chap. 3, the proposed automatic pill dispenser model has some advantages as listed below:

1. People of all ages forget to take their medicines regularly based on their busy schedule and living. The proposed pill dispenser model will remind you and will never make you forget to take your medicines.
2. To overcome the problem of taking up the wrong pills due to memory, the pills dispenser will be loaded with the right pills and you will know at what time the correct pills have to be dispensed.
3. A direct interaction of patient at any time is facilitated to make them feel better by using the automatic pill dispenser.

5 Summary

The modified pill dispenser is the proposed mechanism which will provide easy accessibility and functionality by the patient and will also create a good relationship and rapport between the patient and the doctor. It provides information about the activity of the pill intake and health monitoring by the patient regularly and it is sent to the doctor for medication and follow-up. This application will be useful for patients with Alzheimer's disease and suffering from memory loss, that is, not able to take the medicine periodically using the IoT device. The IoT device will provide a particular facility at any place and any time. Along with IoT, the neural network is added up to predict the patient problems related to disease based on the updated information provided by the sensor to the online database. Based on the improper diagnosis of the doctor medication, the pill dispenser is automated using artificial intelligence and provides a proper recognition of symptoms and activity of the patients, which will enable the doctor to provide appropriate medication. Based on stored information in the online Web, and information related to medicine storage and doctor's medication is linked with the pharmacy and the blood bank to provide appropriate service when necessary.

References

1. P.P. Ray, A survey on Internet of Things architectures. J. King Saud University Comput. Inform. Sci. (Elsevier) **30**(3), 291–319 (2018)
2. S.B. Baker, W. Xiang, I. Atkinson, Internet of Things for smart healthcare: technologies, challenges, and opportunities. IEEE Access **5**, 26521–26544 (2017)
3. L. Catarinucci, D. de Donno, L. Mainetti, L. Palano, L. Patrono, M.L. Stefanizz, L. Tarricone, IEEE Internet Things J. **2**(6), 515–526 (2015). Quality
4. B.E. Himes, E.R. Weitzman, Innovations in health information technologies for chronic pulmonary diseases. Respir. Res. **17**(1), 1–7 (2016)
5. P.S. Mathew, A.S. Pillai, V. Palade, Applications of IoT in Healthcare, in *Cognitive Computing for Big Data Systems Over IoT*, Lecture Notes on Data Engineering and Communications Technologies book series (LNDECT), Springer, 14, 263–288, 2017
6. A. Kiourti, R.M. Shubair, Implantable and ingestible sensors for wireless physiological monitoring: a review, In *Proc. of IEEE International Symposium on Antennas and Propagation & USNC/URSI National Radio Science Meeting*, 2017
7. A.S. Yeole, D.R. Kalbande, Use of Internet of Things (IoT) in Healthcare: a survey. In *Proc. Of ACM Symposium on Women in Research, WIR*, pp. 71–76 (2016)
8. R. Fitzpatrick, A. Fletcher, S. Gore, D. Jones, D. Spiegelhalter, D. Cox, Quality of life measures in health care. I: applications and issues in assessment. Br. Med. J. **305**, 1074–1077, (1992)
9. A.A. Galadima, Arduino as a learning tool. In *Proc. of 11th International Conference on Electronics, Computer and Computation (ICECCO)*, Oct. 2014
10. M. McRoberts, Beginning Arduino, in *Technology in Action*, (Apress publishers, 2, 2013)
11. B.N. Karthik, L. Durga Parameswari, R. Harshini, A. Akshaya, Survey on IOT & Arduino based patient health monitoring system. Int. J. Sci. Res. Comput. Sci. Eng. Inform. Technol. **3**(1), 1414–1417 (2018)
12. Y.A. Badamasi, The working principle of an Arduino," In *Proc. of 11th International Conference on Electronics, Computer and Computation (ICECCO)*, Oct. 2014
13. V. Vippalapalli, S. Ananthula, Internet of Things (IoT) based smart health care system. In *Proc. of International Conference on Signal Processing, Communication, Power and Embedded System (SCOPES)*, Oct. 2016
14. P. Gupta, D. Agrawal, J. Chhabra, P.K. Dhir, IoT based smart healthcare kit. In *Proc. of International Conference on Computational Techniques in Information and Communication Technologies (ICCTICT)*, 2016
15. J. Gubbi, R. Buyya, S. Marusic, M. Palaniswami, Internet of Things (IoT): a vision, architectural elements, and future directions. Future Gener. Comput. Syst. (Elsevier) **29**(27), 1645–1660 (2013)
16. N.P. Terry, Will the Internet of Things transform healthcare? Vand. J. Ent. Tech. **19**(2), 327–352, (2017)
17. M.S. Hossain, G. Muhammad, Cloud-assisted Industrial Internet of Things (IIoT) – Enabled framework for health monitoring. Comput. Networ. (Elsevier Publications) **101**, 192–202 (2016)
18. J. Huang, M. Yu, X. Gong, Y. Liu, Q. Duan, A novel deployment scheme for green Internet of Things. IEEE Internet Things J. **1**(2), 196–205 (2014)

19. D. Zvikhachevskiy, J.M. Sultan, K. Dimyati, Quality of service mapping over WiFi+WiMAX and Wifi+LTE networks. J. Telecommun. Electron. Comput. Eng. **5**(2), 1–9 (2013)
20. J. Granados, A.-M. Rahmani, P. Nikander, P. Liljeberg, H. Tenhunen, Towards energy-efficient HealthCare: an Internet-of-Things architecture using intelligent gateways. In *Proc. of 4th International Conference on Wireless Mobile Communication and Healthcare – Transforming Healthcare Through Innovations in Mobile and Wireless Technologies (MOBIHEALTH)*, Nov. 2014
21. H. Kemis, N. Bruce, W. Ping, T. Antonio, L.B. Gook, H.J. Lee, Healthcare monitoring application in ubiquitous sensor network: Design and implementation based on pulse sensor with Arduino. In *Proc. of 6th International Conference on New Trends in Information Science, Service Science and Data Mining (ISSDM2012)*, Oct. 2012
22. C. Doukas, I. Maglogiannis, Bringing IoT and cloud computing towards pervasive Healthcare, In *Proc. of Sixth International Conference on Innovative Mobile and Internet Services in Ubiquitous Computing*, July 2012
23. A. Gaur, B. Scotney, G. Parr, S. McClean, Smart city architecture and its applications based on IoT. In *Proc. of 5th International Symposium on Internet of Ubiquitous and Pervasive Things (IUPT 2015)*, Procedia, Elsevier Publications, pp. 1089–1094 (2015)
24. C. Doukas, I. Maglogiannis, Bringing IoT and cloud computing towards Pervasive Healthcare. In *Proc. of Sixth International Conference on Innovative Mobile and Internet Services in Ubiquitous Computing*, 2012
25. X. Chen, W. Rhee, Z. Wang, Low power sensor design for IoT and mobile healthcare applications. China Commun. IEEE **12**(5), 42–54 (2015)
26. A.J. Jara, A.F. Alcolea, M.A. Zamora, A.F. Gómez Skarmeta, M. Alsaedy, Drugs interaction checker based on IoT. *Internet of Things (IOT)*, 2010
27. J. Wan, C. Zou, S. Ullah, C.-F. Lai, M. Zhou, X. Wang, Cloud-enabled wireless body area networks for pervasive healthcare. IEEE Netw., **27**(5), 56–61 (2013)
28. M.S.D. Gupta, V. Patchava, V. Menezes, Healthcare based on IoT using Raspberry Pi. In *Proc. of International Conference on Green Computing and Internet of Things (ICGCIoT)*, 2015
29. S. Hiremath, G. Yang, K. Mankodiya, Wearable Internet of Things: concept, architectural components and promises for person-centered healthcare. In *Proc. of 4th International Conference on Wireless Mobile Communication and Healthcare – Transforming Healthcare Through Innovations in Mobile and Wireless Technologies (MOBIHEALTH)*, 2014
30. S.K. Datta, C. Bonnet, A. Gyrard, R.P.F. da Costa, K. Boudaoud, Applying Internet of Things for personalized healthcare in smart homes. In *Proc. of 24th Wireless and Optical Communication Conference (WOCC)*, 2015
31. S. Sukode, S. Gite, H. Agrawal, Context aware framework in IoT: a survey. Int. J. Adv. Trends Comput. Sci. Eng. **4**(1), 1–9 (2015)
32. S. Shinde, N. Bange, M. Kumbhar, S. Patil, Smart medication dispenser. Int. J. Adv. Res. Electron. Commun. Eng. **6**(4), 200–204 (2017)
33. P.-H. Tsai, T.-Y. Chen, C.-R. Yu, C.-S. Shih, J.W.S. Liu, Smart medication dispenser: design, architecture and implementation. IEEE Syst. J. **5**(1), 1–37 (2011)
34. B.B. Singh, S. Upadhyay, M. Sumra, GSM based automatic pill dispenser. Int. J. Eng. Sci. Comput. **7**(4), 10694–10695 (2017)
35. A. Mondragón, A. De Hoyos, A. Trejo, M. González, H. Ponce, Medi-kit: developing a solution to improve attention on medical treatment. *In Proc. of IEEE Mexican Humanitarian Technology Conference (MHTC)*, 2017
36. S. Shinde, T. Kadaskar, P. Patil, R. Barathe, A smart pill box with remind and consumption using IOT. Int. Res. J. Eng. Technol. **4**(12), 152–154 (2017)
37. M. Reddy, V. Parekh, C. Pinto, V. Pisharody, D. Ratho, Avion – the intelligent medicine box. Int. J. Sci. Eng. Res. **7**(3), 1301–1304 (2016)
38. S. Bhati, H. Soni, V. Zala, P. Vyas, Y. Sharma, Smart Medicine Reminder Box. Int. J. Sci. Technol. Eng. (IJSTE) **3**(10), 172–177 (2017)
39. J. Medina, M. Espinilla, Á.L. García-Fernández, L. Martínez, Intelligent multi-dose medication controller for fever: from wearable devices to remote dispensers. Comput. Electr. Eng. J. (Elsevier) **22**, 400–412 (2017)

40. P. Bedmuttha, N. Jain, Y. Thigale, S. Gargori, T.R. Patil, A health- IOT platform based on the biosensor and intelligent medicine box. Int. J. Comput. Sci. Mob. Comput. **6**(4), 433–438 (2017)
41. R. Huang, X. Zhao, J. Ma, The contours of a human individual model based empathetic u-pillbox system for humanistic geriatric healthcare. Comput. Syst. J. (Elsevier) **37**, 404–416 (2014)
42. S. Aakash Bharadwaj, D. Yarravarapu, S.C.K. Reddy, T. Prudhvi, K.S.P. Sandeep, O.S.D. Reddy, Enhancing Healthcare using m-Care Box (Monitoring non-compliance of medication). *International Conference on IoT in Social, Mobile, Analytics and Cloud*, IEEE 2017
43. M.R. Kinthada, S. Bodda, S.B.K. Mande, eMedicare: mHealth solution for Patient Medication Guidance and Assistance. *International conference on Signal Processing, Communication, Power and Embedded System (SCOPES)-2016*, IEEE 2016
44. V. Philip, V.K. Suman, V.G. Menon, K. Dhanya, A review on latest Internet of Things based healthcare applications. Int. J. Comput. Sci. Inform. Security (IJCSIS) **15**(1), 248–254 (2017)
45. J. Wilden, A. Chandrakar, A. Ashok, N. Prasad, IoT based wearable smart insole. In *Proc. of Global Wireless Summit (GWS)*, 2017
46. V. Sermakani, Transforming healthcare through Internet of Things, In *Proc. of Project Management Practitioner's Conference*, 2014
47. M. Pasha, *Business Development and Commercialization of Urosense, Master thesis*, (Case Western Reserve University, 2013)

The SOCIAL Platform and the Integration of Internet of Things Devices to Monitor Activities and Behaviors of Older Adults

Miguel Sousa Gomes, João Rainho, António Damasceno, and Nelson Pacheco Rocha

1 Introduction

Personal health devices, namely, the ones connected to the Internet of Things (IoT) network [1], might be used to monitor older adults in their homes. Home monitoring is useful to gather clinical data together with the identification of consistency and completeness of activities and behaviors [2, 3]. Moreover, currently, initiatives such as the Personal Connected Health Alliance (PCHA) [4] and the Integrating the Healthcare Enterprises (IHE) [5] are using interoperability standards, namely, the Fast Healthcare Interoperability Resources (FHIR) developed by Health Level Seven [6], to assure complete communication chain ranging from monitoring devices to complex infrastructures, including electronic health records.

In this context, the platform of services Social Cooperation for Integrated Assisted Living (SOCIAL) [7] was designed to allow the integration of relevant data from various data sources, including data gathered by IoT devices. Therefore, this chapter reports a research study aiming to develop IoT devices to monitor older adults' activities and behaviors, and to implement and evaluate analytical procedures

M. S. Gomes
Microio – Serviços de Eletrónica, Lda., Edifício Mirador Business Center, Aveiro, Portugal
e-mail: mgomes@microio.pt

J. Rainho · A. Damasceno
Instituto Pedro Nunes, Coimbra, Portugal
e-mail: jrainho@ipn.pt; apdamasceno@ipn.pt

N. P. Rocha (✉)
IEETA, Department of Medical Sciences, University of Aveiro, Campo Universitário de Santiago, Aveiro, Portugal
e-mail: npr@ua.pt

N. Gupta, S. Paiva (eds.), *IoT and ICT for Healthcare Applications*, EAI/Springer Innovations in Communication and Computing,
https://doi.org/10.1007/978-3-030-42934-8_14

of the SOCIAL platform designed to infer events that might characterize health conditions.

Important features related to the SOCIAL platform are the use of FHIR to guarantee information interoperability and the application of technologies such as blockchain [8] or eXtensible Access Control Markup Language (XACML) [9] to guarantee privacy, integrity, and confidentiality of personal information.

The following sections present relevant research related to this study and the methods that were applied for the specification and implementation of IoT monitoring devices and analytical procedures designed to infer significant events from the monitoring data. Moreover, the section Implementation presents the SOCIAL platform, the development of IoT devices, and analytical procedures, and how the resulting information is integrated in the SOCIAL platform. Finally, a validation is reported, and some conclusions are drawn.

2 Related Work

Digital technologies can be used for prevention, to promote the engagement between different levels of care (e.g., primary and secondary care, or health care and social care), to support paid and unpaid careers, and to improve productivity and quality by enabling new ways of working and facilitating information flow and, therefore, promoting more integrated and responsive services.

The use of technological solutions to provide information services to mediate different actors has been proposed by different authors [10–14], namely, for the management of chronic diseases [15–17]. Furthermore, there is a considerable effort to develop suitable services to manage psychosocial information [17–20], information generated by care receivers and their informal caregivers, and information resulting from automatic data collection about individuals and their environment, namely, through innovative monitoring devices [21–23].

Different groups of technologies, although focalized in specific aspects, can contribute to the care in the community, so that older adults might live safely, independently, autonomously, and comfortably, without being required to leave their own residences, but with the necessary support services to their changing needs [24], including those resulting from multiple impairments, namely, decline in cognitive or functional abilities [25].

In this respect, technologies such as mobile health (mhealth) [26] or Ambient Assisted Living (AAL) [2, 27] have been the object of relevant research.

According to the World Health Organization, mhealth can be defined as "medical and public health practice supported by mobile devices, such as mobile phones, monitoring devices, personal digital assistants, and other wireless devices" [28]. Since health care services involve multiple locations (e.g., clinics, outpatients' services, or patients' homes), they are highly mobile in nature [29]. This means mhealth might support communication and collaboration among clinicians in activities related to disease diagnosis, drug reference, medical calculations, or literature search, among others.

In turn, AAL concerns and developments are in line with the World Health Organization active ageing framework [30]. Enabling instruments such as ALL are essential, considering the fact that as people age, their quality of life (i.e., perception of the position in life in the context of the surrounding culture and value system) is largely determined by their ability to maintain autonomy (i.e., ability to control, cope with, and make personal decisions on a day-to-day basis) and independence (i.e., the ability to perform functions related to daily living with no or little help from others) [30].

Irrespective of their precise definition, technologies such as mhealth or AAL might allow the patients to receive information from a variety of media to better control their diseases (e.g., educational information or prescriptions reminders) or to provide additional clinical information resulting from home monitoring.

The advances in sensing technology, together with advances in communication technologies, such as IoT, made possible the development of mobile and wearable devices able to continuously monitor physiological parameters in out- hospital conditions (e.g., measurement of blood pressure or blood glucose) [2, 31]. Furthermore, monitoring physiological parameters together with activities and behaviors to assess, in a naturalistic and continuous way, health and cognitive status [2], might help to automate assistance, to prevent the exacerbating of disease or accidents, and to properly react to emergency situations [2, 3].

The literature reports a diversity of studies comparing usual care with the use of various devices (e.g., devices to monitor vital signals or devices to monitor behavior outcomes such as pedometers or accelerometers connected by wireless communications to measure physical activity [32]) to support home monitoring of patients with chronic conditions, including diabetes, congestive heart failure, or chronic obstructive pulmonary disease, with positive effects with moderate-to-large improvements of different outcomes [33].

Both type 1 and type 2 diabetes conditions are being the object of home monitoring [32, 34–42]. In turn, several devices are being used to remotely assess symptoms and vital signs of patients with congestive heart failure, as well as the transmission of automatic alarms [43–47]. Moreover, a broad range of devices are also being used to measure and transmit different types of information (e.g., weight, temperature, blood pressure, oxygen saturation, spirometry parameters, symptoms, medication usage, or steps in 6-min walking distance) of patients with chronic obstructive pulmonary disease [48, 49].

When dealing with diabetes, the main outcome is the control of glycemia by using glycosylated hemoglobin (HbA1c) as a proxy. However, this aim might be complemented with other health-related outcomes (e.g., health-related quality of life [34, 37, 50, 51], weight [39, 42, 50, 51], depression [51], blood pressure [34, 38, 41, 42], cholesterol level [38, 50], triglycemius level [38], fluctuation index [50]), behavior outcomes (e.g., physical activity) [31, 34, 37, 41, 42, 50, 52–54], patient self-motivation [40], patient–clinician communication [40], medication adherence [53, 54], and structural outcomes related to care coordination [32,

34]. Looking to patients with congestive heart failure, the main concerns are the impacts of home monitoring in heart failure-related hospitalizations and all-cause mortality [43], but several secondary outcomes are also considered such as self-care behavior (e.g., adherence to prescribed medication, daily weighing, or adherence to exercise recommendations [44]). In addition, mortality, admissions to hospital, or other health care utilization were primary outcomes of the studies related to chronic obstructive pulmonary disease [48]. Moreover, for these studies, secondary outcomes include, among others, health-related quality of life, patient satisfaction, physical capacity, and dyspnea [48, 49].

Other studies report different platforms able to integrate ambient monitoring and behavior recognition in order to prevent or detect dangerous spatial/temporal configurations [55, 56], to locate people [57], to detect specific situations [58, 59], to infer activities [60, 61], or to detect human behaviors and emotions [62, 63].

Additionally, considering the advances in terms of implementation of smart cities, other studies integrate data acquired by smart cities' infrastructures (e.g., air quality, pollution, noise, light conditions, temperature, or precipitation) [64] with data from monitoring devices acquiring continuous lifestyle data (e.g., location of individuals [65–68], activities [69–73], physiological parameters [69–72], social interactions [74], and crowd behaviors [75]) to support activities and participation or to promote healthy lifestyles, namely, physical activity.

However, the existence of a broad range of technological possibilities is not enough for their applicability. Mobilizing care in the community is a challenging issue and integrated solutions are required [76, 77]. This is even more relevant because the emphasis of research efforts related to the use of technological solutions to support health conditions of the individuals in their natural environment has been more oriented toward new ways of acquiring information (e.g., monitoring devices), and less on the development of new models and tools to improve information access and communication in order to build functional care provision [78]. In this respect, the SOCIAL platform [7] aims to contribute to the care integration by delivering information services to support the care and assistance provided to community-dwelling older adults. For that, among other services, the SOCIAL platform should support IoT devices designed to monitor older adults' activities and behaviors, as well as analytical procedures able to infer events that might characterize health conditions (e.g., identifying consistency and completeness of daily activities), which is the objective of the present study.

3 Methods

In the context of the present research study, the following steps were followed to identify the user requirements and to specify the applications:

- Relevant people to interview were identified in cooperation with formal care networks, city councils, borough councils, and local social intervention networks [79].
- Several interviews were conducted face-to-face by a team of five researchers from the different parties of the SOCIAL consortium. One team member was responsible for conducting the interview while the other four were responsible for observing and taking notes.
- Once all the data from the interviews were gathered, personas and scenarios were defined considering the International Classification of Functioning, Disability and Health (ICF) [80], in order to highlight users' functioning (e.g., personal factors, daily routines, and types of activities and participation) and health conditions [81].
- For the definition of personas and scenarios, the needs of the target users were considered, in order to specify the requirements of the information services to be developed [81], including situations and events that have impact (positive or negative) in personas' activities, functional requirements, data requirements, interaction requirements or business, corporate, and customer requirement (e.g., installation conditions or characteristics that should be reflected in the applications and services).
- Once the personas and scenarios were defined, a focus group was conducted with several key stakeholders (different from the ones that participated in the interviews and selected according to their relevance) to validate the results.
- Finally, a list of requirements and a set of significant use cases were defined for the specification and implementation of the relevant applications.

In terms of implementation, the approach consisted in an iterative development of a vertical solution that comprehends imperceptible IoT devices together with the integration within the SOCIAL platform and the respective services [7].

Afterwards, the implemented solution was validated by simulation techniques and real users' utilization.

4 Requirements

From January to June 2017, a total of 15 face-to-face interviews were conducted with relevant stakeholders from formal care networks with expertise on care provision.

Based on the results of the interviews, a set of personas and scenarios was created. Among the created personas, the most representative ones are a sedentary older adult, an active older adult, an institutionalized older adult, and a common citizen. In turn, among the scenarios that were created, the three most representative scenarios are an active older adult scenario, a sedentary older adult scenario, and a signalization scenario.

To validate the personas and scenarios, a focus group occurred on July 2017 and had nine participants [82].

Following the focus group, a set of significant use cases were designed, which allowed the definition of the required functions for the personal and institutional applications.

As a result of the requirements analysis, solutions within the scope of the IoT were considered the most adequate approach for behavioral and activity assessment of the older adults that, under certain circumstances, might trigger specific interventions [82].

Despite the progressive growth of the familiarity with technology within older age groups, the audience targeted by the SOCIAL platform is still characterized by little dexterity and, to some extent, little sense of ease with technology; thus, data gathering solutions should not depend on user interaction and should be as discrete as possible. In terms of financial cost, the solutions should be apt for private institutions and entities, but also suitable for single homes and thus, as economical and simple as possible, but also scalable, to enable progressive upgrades according to the specific use case requirements. Additionally, considering the wide operational scope of the SOCIAL platform, the IoT devices should feature the necessary functions to enable the compatibility, integration, and interoperability with a selected range of planned upcoming applications and services.

5 Implementation

Based on the developed personas, scenarios, and use cases, and considering the older adults' characteristics, the SOCIAL consortium defined a nonintrusive approach, with a solution that would flag alerts based only on the detection of deviations or unexpected events on routine activities. Therefore, it was decided to implement four types of IoT devices: (i) contact devices, for doors and windows; (ii) motion devices, for indoor movement detection; (iii) pressure devices, for presence detection in beds and chairs; and (iv) power devices, to detect the state of activity of electrical appliances.

The aim is to use these devices to monitor specific parameters and occurrences such as activity during specific periods of time, presence in specific rooms, or usage of specific utilities. Furthermore, data collected by the IoT devices are aggregated and transmitted to the SOCIAL platform, which detects predefined situations and promptly notifies responsible parties, based on a detailed configuration of significant events.

5.1 SOCIAL Platform

The SOCIAL platform is a services platform with a set of structural components to allow the development of innovative information services to empower community-dwelling older adults and their caregivers. Its architecture is based on a services-oriented approach and was divided into three layers: Application, Business, and Data layers.

The Application layer comprises a set of applications to fulfil the needs of the stakeholders of the SOCIAL platform, including web and mobile applications that are responsible for the information exchange between the platform and its users. Additionally, the interaction with external applications was also foreseen, thus allowing the SOCIAL platform to be interconnected with other complementary platforms.

The Business layer presents a set of interfaces that allow communication with the applications (e.g., Representational State Transfer or Advanced Message Queuing Protocol) and aggregates components such as the FHIR Gateway (i.e., validation, storage, reconstitution and search of information directly related to the care receivers), Data Analysis and Reporting (i.e., consolidation of information for secondary use), Authorization and Authentication (i.e., general mechanisms to control access to SOCIAL platform), Workflow Engine (i.e., management and configuration of workflows that must be performed according to previously defined and agreed protocols), Logging and Auditing (i.e., access and audit logging), or Complex Event Processing.

In the context of the present study, Complex Event Processing is particularly relevant since it is the unit responsible for the management of the IoT devices and the processing of the data they generate. Moreover, the Complex Event Processing unit also interacts with other services of the SOCIAL platform (e.g., FHIR Gateway or Authentication and Authorization,), so that significant events can be stored and shared [7] (Fig. 1).

In turn, the Data layer is composed of a set of databases, including: (i) information of the care receivers, subdivided into demographic, clinical, and social with distinct infrastructures, both in logical and physical terms, to safeguard the privacy of the care receivers; (ii) authentication and authorization; (iii) logging

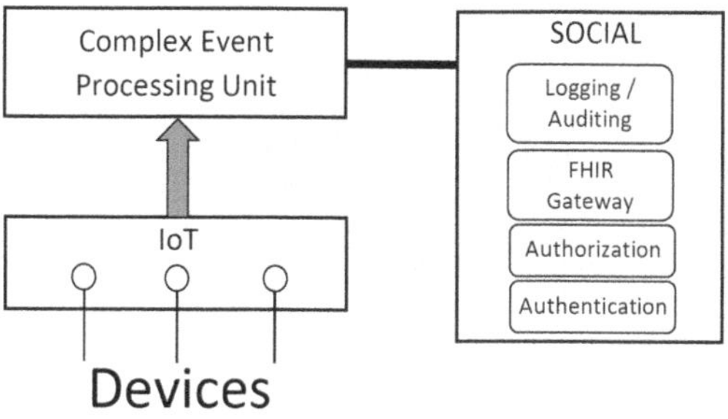

Fig. 1 General architecture

and auditing; (iv) terminology; (v) versioning; (vi) anonymized data for secondary use; (vii) and mappings between different databases. Moreover, the Data layer is also able to accommodate data from external services, including clinical systems, territory and regional repositories or foundation infrastructures (e.g., healthcare identifiers or terminology references sets).

5.2 *IoT Devices*

In terms of the technical specifications of the IoT devices, the following characteristics were considered: the devices must feature wireless communications and be battery-powered, the field communications technology must have low power requirements, and a communications gateway must be considered not only to operate as a connection hub for the devices, but also to enable the future connection of additional devices.

As depicted in Fig. 2, the different types of devices are optional in type and number per use case (e.g., A or B), according to specific requirements, and are complemented with an additional device, a Communications Gateway (GW), to bridge the connection between the monitoring devices and the SOCIAL platform (i.e., the Complex Event Processing unit).

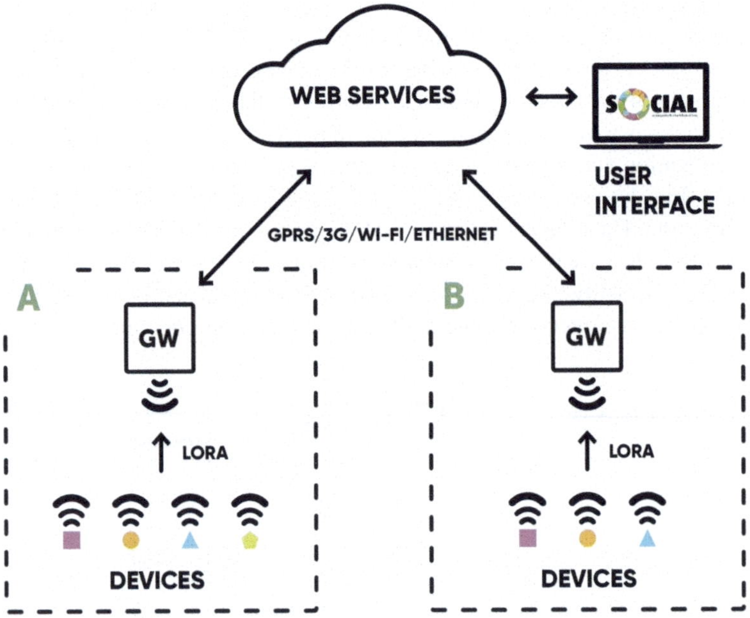

Fig. 2 Device layer architecture diagram

Long Range (LoRa) was the technology selected for the field communications between the devices and the GW. LoRa is a low-power wide-area network wireless communication technology, optimal for the predicted use cases, with battery-powered devices and a range of requirements that can vary from a few meters in smaller homes to multiple dozens of meters including obstacles in bigger institutions, which impeded the usage of technologies such as Bluetooth Low Energy (BLE) and Zigbee.

To communicate with the devices, and also to enable the connection of additional devices in the future, the developed GW features multiple communication interfaces: LoRa, Universal Serial Bus (USB), RS232, RS485, Inter-Integrated Circuit (I2C), Multi-Drop Bus (MDB), Bluetooth, or microcontroller I/Os. Moreover, a Raspberry Pi Compute Module was considered for continuous development and to support specific protocols or standards, such as the ISO/IEEE 11073 Personal Health Data, used for devices such as weighing scales, blood pressure monitoring devices, or other devices for personal use. To forward the devices' data, the GW supports Ethernet and Wi-Fi and can also feature a General Packet Radio Service/Third Generation (GPRS/3G) modem, for use cases in which an internet connection is not available.

In terms of the technical specifications of the developed devices, according to the defined approach, all devices feature a microcontroller and a LoRa transceiver for wireless communications. Furthermore, the contact, motion, and pressure devices are battery-powered with rechargeable lithium polymer (LiPo) batteries, while the power devices are powered by 230VAC as it is meant to be applied in power sockets. Looking for the specific sensing technologies, the contact devices feature a reed switch that is operated by an applied magnetic field created by a magnet, which means that the devices and their auxiliary magnets are suitable for door and window applications. In turn, the motion devices feature passive infrared (PIR) sensors, commonly used for individuals' motion detection in indoor environments, such as rooms and hallways. Moreover, the pressure devices include strain gauges whose electrical resistances vary with the applied pressures, which enable the measurement of weight variation. Finally, the power devices, to be applied on power sockets, feature a 230VAC connector for appliances, as well as current and voltage sensors to monitor the power consumption and the activity states of the connected appliances.

From the operational point of view, the approach was to implement the devices with simple operation algorithms and centralize all the intelligence and decision making at the Complex Event Processing unit, thus reducing the complexity of the devices. In general, the devices are programmed to monitor their environment with a high frequency to detect the occurrence of events when there are state transitions of their monitored parameters (i.e., motion, contact, weight, or power) and, in case of absence of transitions for a period of 15 min, to send an "I'm alive message" with the current state of the monitored parameter. On more detailed analysis, the operation algorithms of the four types of device are the following:

- Each motion device samples the environment every second and if motion is detected an event is sent to the GW, with state "1", and it enters in a sleep state for 10 min (to avoid sending a power consuming burst of messages, following eventual long-lasting movements); in the absence of movement for more than 15 min, the "I'm alive message" is sent with state "0".

- Each pressure device requires weight reset; then the device samples the strain gauge every second and if a weight variation that exceeds 10 kg is detected, an event, which is communicated with state "1"; if the detected weight decreases back to a value within the 10 kg threshold from the default value, an event is communicated with state "0"; furthermore, in the absence of transitions, for more than 15 min, the "I'm alive message" is sent with the respective occupation state "0"/"1".
- Each contact device samples the reed switch every second and, if there is a transition, an event is sent to the GW with the final state "0" (i.e., closed) or "1" (i.e., open); in the absence of occurrences for more than 15 min, the "I'm alive message" is sent with the respective state.
- Each power device samples the power consumption of the connected appliance every second and, in case the appliance is on, an event is sent to the GW with the state "1"; if the appliance is switched off, 10 min after the power off, an event is sent to the GW with the state "0" for avoiding false-positives of devices with on-off controllers; in the absence of transitions for more than 15 min, the "I'm alive message" is sent with the respective device activity state "0"/"1".

This approach enabled the standardization of both the communications protocol and the analysis algorithm on the Complex Event Processing unit, as well as the optimization of autonomy of the devices' batteries with a noncontinuous monitoring of the environment, interrupted by periods in which the microcontroller is put in sleep mode (i.e., a low-power consumption mode). The sample frequency and sleep mode were carefully defined considering the monitored parameters in real-life applications and validated in laboratory tests. Additionally, with the implementation of the "I'm alive messages", in a 15-min period, it is possible to cope with any event loss due to the nonreception of device messages, by continuously comparing the device data on the database with the newly received data.

In terms of the significant event definition, the events presented in Table 1 can be configured. Additionally, by default, the motion, pressure, and contact devices will also forward a notification if the battery is low, for recharging.

5.3 Complex Event Processing Unit

Within the services of the SOCIAL platform, the Complex Event Processing unit can both provide relevant data for decision making, with the detection of otherwise imperceptible occurrences, and a mean to trigger prompt intervention, in case of detection of emergency events. For that, it is prepared to accept details of significant events rules, based on the type of events or their occurrence (or not) within a specific time frame or limit. The unit is continuously analyzing the events received from the

Table 1 Events configuration

Type	Device(s)	Event	Parameters		#
Simple	Motion	State transition	–	Schedule (1)	01
		Absence of state transition	–	Schedule (1)	02
				Time limit (2)	03
	Pressure	State transition	–	Schedule (1)	04
		Absence of state transition	–	Schedule (1)	05
				Time limit (2)	06
	Contact	State transition	–	Schedule (1)	07
		Absence of state transition	–	Schedule (1)	08
				Time limit (2)	09
		Persistence on specific state	Open ("1")	Schedule (1)	10
				Time limit (2)	11
			Closed ("0")	Schedule (1)	12
				Time limit (2)	13
	Power	State transition	–	Schedule (1)	14
		Absence of state transition	–	Schedule (1)	15
				Time limit (2)	16
		Persistence on specific state	On ("1")	Schedule (1)	17
				Time limit (2)	18
			Off ("0")	Schedule (1)	19
				Time limit (2)	20
Combined	Pressure and Motion	Combined transition of two devices within time frame	Time frame between different devices' events (3)	Schedule (1)	21
	Pressure and Contact				22
	Contact and Motion				23

(1) Period of time the user seeks to monitor the occurrence of the event (e.g., monitor entrance door between 12H and 14H)
(2) Time limit the user seeks to monitor the occurrence of the event (e.g., monitor absence of motion state transition in hallway within a period of 24 h)
(3) Maximum time frame (in minutes) between the successions of the defined devices' events (e.g., in an institution, the detection of a bed leave followed by motion in a hallway within 5 min can indicate an unintended room exit)

IoT devices. When a significant event matches the predefined rules, a notification is forwarded to the responsible services of the SOCIAL platform.

The Complex Event Processing unit is based on a pipe and filter architecture. It is composed of filters that clean and transform the data, working all in parallel, and connectors to pass data to other components of the unit, known as pipes. The advantages of this architecture, in addition to the concurrent execution, are the simplicity and the ability to reuse the filters.

The unit can be divided into three blocks. The first corresponds to the Data Source Block that feeds the Filters Block, through a Data Stream communication

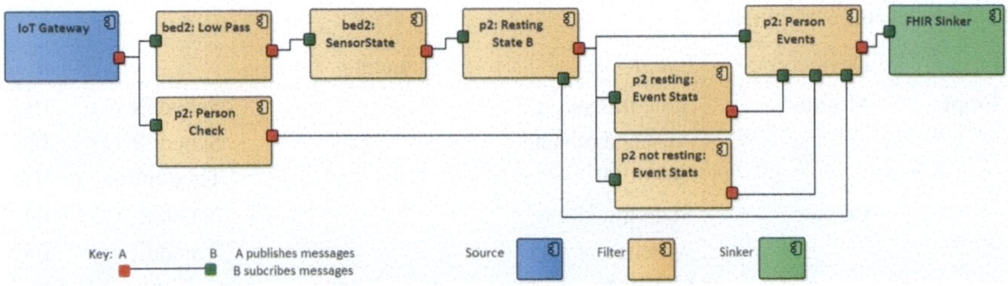

Fig. 3 Sequential plan to detect if an individual is resting or not resting, using a pressure device in bed (in the Filter Block are represented the filters necessary to detect these events and to calculate when similar events are expected)

channel. The second block contains all the filters necessary to handle the data. The operation of this block depends on the previous configuration of an Extensible Markup Language (XML) file that describes each of the filters used and the type of data that they treat. When events are determined, the processed data are passed to the Data Sink Block. There, the data can be used to alarm a significant event if it is the case.

Figure 3 shows an example of how the whole unit works sequentially to detect if an individual is resting or not resting. The detection of these events is performed by processing the data gathered by devices located in the bed.

The roles of the various filters presented in the example are the following:

- Low Pass, which is responsible to calculate the correct state of the device, based on the analysis of the last values received.
- Person Check, which is responsible for periodically checking the status of an individual to update the respective information or generate alerts.
- Sensor State, which receives the result of the state of the bed device filtered by the Low Pass filter.
- Resting State, which is based on the current resting state of the individual and the values received from the device in a given instant; this filter can detect if the individual is resting or not resting.
- Event Stats, which is based on the timestamp of each detected event; the total mean and standard deviations are calculated for each day of the week.
- Person Events, which is based on the information sent by the Person Check, Resting State, or Event Stats filters; following the detection of an event, the individuals' routine record is updated.

Figure 4 shows in detail, through a state machine, an example of how the Resting State filter works to identify if a person is resting in bed. At the start of the unit, the initial state is "Unknown", because no data have been received yet and it is impossible to know the current state of the individual. After receiving the first data from the bed device, the filter switches to the "InBed" state if the user is in bed or "NotInBed", otherwise. If the individual goes to bed, the time of the event is compared with the normal time to rest. If it is within normal time to rest, the

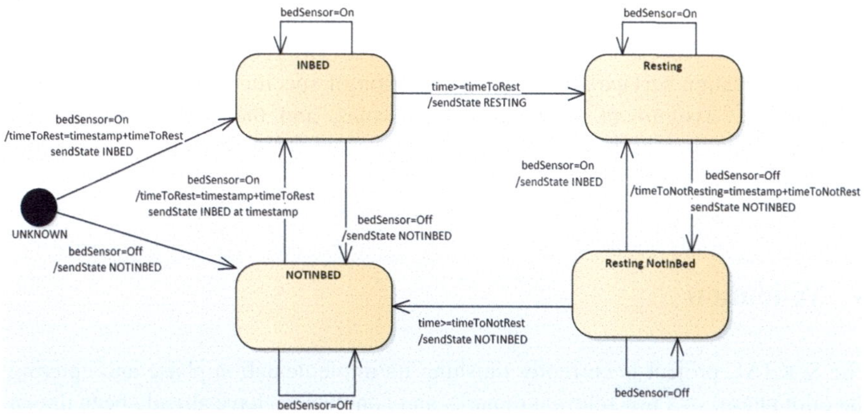

Fig. 4 State machine of Resting State filter

filter changes to the "Resting" state and remains in that state while the user is lying down. If the individual rises, the state changes to "Resting NotInBed". If this situation occurs after the expected time to wake up, it goes to the "NotInBed" state. Otherwise, it remains in that state until the expected time to wake up or return to the "Resting" state if the individual returns to the bed. If the "Resting NotInBed" state lasts more than expected, an alert is issued to the responsible services of the SOCIAL platform.

5.4 Integration of the Information in the Social Platform

In terms of the care receivers' information model defined for the SOCIAL platform, the FHIR resource identified as Person resource is the fundamental entity [83]. The Person resource might be associated to other FHIR resources (e.g., Care Plan resource or Observation resource), so that relevant information can be transmitted and stored. Particularly, with the Observation resource (i.e., measurements and simple assertions made about a care receiver, device, or another subject), it is possible to organize the information from the Complex Event Processing unit and respective IoT devices.

To store the significant events, the Complex Processing unit accesses the FHIR Gateway with a valid JSON Web Token (JWT) following the standard RFC 7519 [84]. This is one of the security measures to confine the information to undue requests. The requests are all submitted to the Authentication and Authorization service so the information can be retrieved according to permissions associated to a specific role. The JWT is used by the Authentication and Authorization service to get the role, so that rules previously defined are applied to get a decision if the requested operation can or cannot be achieved. In this respect, the XACML [9] was implemented to handle requests related to information access, according to Attribute-based Access Control (ABAC) rules [85]. Moreover, the Authentication and Authorization service allows the registration of specific users in the SOCIAL platform, the assignment of the respective roles, and the verification of their identities. Finally, a cryptographic secure blockchain [8] was adopted for logging and auditing services [86].

6 Validation

The SOCIAL project is currently finishing its implementation phase and entering the pilot phase, so a few relevant remarks and conclusions have already been drawn from the results of the controlled experimentation phases that accompanied all the implementation.

According to the iterative development procedure, the defined test strategy involved a first laboratorial phase, which included the experimentation in a controlled environment for the development of the preliminary prototypes and specific features. The laboratorial tests were followed by subsequent broader test phases, on more advanced solution versions, that comprehended stress tests and experimentation in conditions closer to those expected in real life applications, both under controlled supervisions for performance analysis and subsequent optimization of the results.

The first phase involved the experimentation on the first developed prototypes and services for overall validation of the decisions concerning the core technologies and modules, namely: (i) the experimentation of devices data gathering and processing (with a preliminary prototype temperature device, contact device, and the GW); (ii) wireless communications over LoRa; (iii) the GW operation and connection to the Complex Event Processing unit; and (iv) the data reception and brokerage on the server side.

With the satisfactory results of the first phase and the conclusion of the specification of the target personas, scenarios, and use cases, it was possible to specify the set of devices and core functionalities of the Complex Event Processing unit and proceed to their implementation. A first complete set of devices' prototypes was produced for development and experimentation of each sensing technology. This phase was characterized by successive iterations and optimizations, involving: (i) the interaction with the PIR sensor; (ii) the definition of adequate sample times and sleep times for autonomy improvement; (iii) the interaction with the strain gauge and measurement of decision weight thresholds, which lead to the implementation of the reset procedure; (iv) the application requirements of the reed switch, particularly from the application standpoint which could jeopardize its accuracy; and (v) the identification of the most adequate operation for the power devices, to avoid false-positives of devices with on-off controllers and standby power modes.

With the stability of the first batch of devices, from the hardware point of view, the SOCIAL consortium proceeded to the integration of communication mechanisms. Following the identification and implementation of the Extensible Messaging and Presence Protocol (XMPP), security login, and validation procedures and the definition of the data packets, two additional devices sets were produced for remote experimentation. This experimentation intended to bring the devices to conditions closer to the expected real-life applications, in the home of two real users. This experimentation brought up multiple improvements from application and operational standpoints, in terms of hardships experienced in device deployment and initialization, remote firmware updates, communications range, battery autonomy, connectivity, and optimizations of communication protocols.

In parallel, the Complex Event Processing unit was validated using a simulator to produce events, in order to solve the lack of data. The simulator was capable of injecting data for different types of devices similar to those that will be installed in real houses. For that, it was configured with multiple house setups in terms of types of devices and their locations, multiple end-user expected behaviors, and data simulations for a specific number of days. Additionally, it was possible to have settings for different days of the week and to insert some randomness based on temporal windows as well as abnormal events on specific days. This last point is extremely important in order to validate the Complex Event Processing unit [87].

The goal was to verify if the unit correctly identifies a set of end-user behaviors (e.g., in bed, not in bed, resting, or not resting). The simulations were executed for a period of 10 days and the results show that the expected events were detected.

To give even more realism to the simulation, it was possible to add noise to the data generated by the devices. Some filters have been developed to help in processing and cleaning up the received data and, at the end of the whole process, to compare if the data generated by the simulator corresponds to those initially predicted. When injecting noise, it was verified that the unit detected all the expected events with 10% noise, but it was not reliable with 40% noise [87].

Moreover, the Complex Event Processing was exposed to stress tests, for validating its performance critical operation circumstances (e.g., low communications signal, operation close to the defined thresholds, or high number of successive events).

Finally, two sets of devices were installed in two environments closer to the expected real-life applications: a household and a test facility resembling an institutional application. These tests are still underway and being accompanied using a web platform for real-time notification of the significant events.

7 Conclusion

The scientific literature reports a considerable number of studies developing and applying a broad range of devices to monitor health conditions. However, the integration of information from these monitoring devices to provide coherent data analysis is a central concern and current solutions are considered to lack interoperability and obstruct the establishment of a remote patient monitoring solution market [88].

In this respect, interoperability is assumed as being an essential requirement as well as the guarantee of the privacy, integrity, and confidentiality of personal information.

Therefore, the authors have proposed the development of IoT devices to monitor older adults' activities and behaviors that are integrated in a services platform, the SOCIAL platform. This platform provides the detection of predefined situations and a set of common services, including interoperability, authentication, authorization, logging, and auditing mechanisms to allow an interoperable and secure access to the generated information. This can contribute for innovative applications be able to reach their potential in terms of impact they can have on care provision, considering a diversity of settings and care providers.

The SOCIAL consortium is now producing a final batch of IoT devices for the pilot phase, which aims to evaluate the solution with target users in real-life environments with minimal technical intervention.

Acknowledgments This work was supported by Sistema de Incentivos à Investigação e Desenvolvimento Tecnológico (SI I&DT) of the Programa Portugal 2020, through Programa Operacional Competitividade e Internacionalização and/or Programa Operacional do Centro do FEDER – Fundo Europeu de Desenvolvimento Regional, under Social Cooperation for Integrated Assisted Living (SOCIAL), project number 017861.

References

1. B.M. Lee, J. Ouyang, Intelligent healthcare service by using collaborations between IoT personal health devices. Int. J. Biosci. Biotechnol. **6**(1), 155–164 (2014)
2. P. Rashidi, A. Mihailidis, A survey on ambient-assisted living tools for older adults. IEEE J. Biomed. Health Inform. **17**(3), 579–590 (2012)
3. I. Awolusi, E. Marks, M. Hallowell, Wearable technology for personalized construction safety monitoring and trending: Review of applicable devices. Autom. Constr. **85**, 96–106 (2018)
4. Personal Connected Health Alliance, Continua design guidelines (2019), https://www.pchalliance.org/continua-design-guidelines
5. IHE International, Mobile access to health documents (MHD) (2019), https://wiki.ihe.net/index.php/Mobile_access_to_Health_Documents_(MHD)
6. D. Atkins, T. Cullen, The future of health information technology: Implications for research. Med. Care **51**(3), S1–S3 (2013)
7. M. Sousa et al., SOCIAL platform, in *World Conference on Information Systems and Technologies*, (Springer, Cham, 2018)
8. L. Aniello et al., A prototype evaluation of a tamper-resistant high performance blockchain-based transaction log for a distributed database, in *13th European Dependable Computing Conference (EDCC)*, (IEEE, Piscataway, 2017)
9. OASIS, XACML (2013) [Online]. Available: http://docs.oasis-open.org/xacml/3.0/xacml-3.0-core-spec-os-en.html. Visited on 24 Apr 2019
10. A.R. Mori et al., Holistic health: Predicting our data future (from inter-operability among systems to co-operability among people). Int. J. Med. Inform. **82**(4), e14–e28 (2013)

11. C. Bossen et al., CareCoor: Augmenting the coordination of cooperative home care work. Int. J. Med. Inform. **82**(5), e189–e199 (2013)
12. N. Gusew et al., eHealth vision towards cooperative patient care – domain fields and architectural challenges of regional health care networks. Stud. Health Technol. Inform. **160**(Pt 1), 386–390 (2010)
13. M. Hägglund et al., Bridging the gap: A virtual health record for integrated home care. Int. J. Integr. Care **7**(2), e26 (2007)
14. M. Hägglund, R. Chen, S. Koch, Modeling shared care plans using CONTsys and open EHR to support shared homecare of the elderly. J. Am. Med. Inform. Assoc. **18**(1), 66–69 (2010)
15. S.-H. Li et al., Design and implementation of a telecare information platform. J. Med. Syst. **36**(3), 1629–1650 (2012)
16. M. Haritou et al., A technology platform for a novel home care delivery service to patients with dementia. J. Med. Imaging Health Inform. **2**(1), 49–55 (2012)
17. D. Capurro et al., Effectiveness of eHealth interventions and information needs in palliative care: A systematic literature review. J. Med. Internet Res. **16**(3), e72 (2014)
18. M. Rigby et al., Social care informatics as an essential part of holistic health care: A call for action. Int. J. Med. Inform. **80**(8), 544–554 (2011)
19. R.E. Glasgow et al., Patient-reported measures of psychosocial issues and health behavior should be added to electronic health records. Health Aff. **31**(3), 497–504 (2012)
20. W.J. Spitzer, K.W. Davidson, Future trends in health and health care: Implications for social work practice in an aging society. Soc. Work Health Care **52**(10), 959–986 (2013)
21. M.J. Rantz et al., Developing a comprehensive electronic health record to enhance nursing care coordination, use of technology, and research. J. Gerontol. Nurs. **36**(1), 13–17 (2010)
22. B. Reeder et al., Framing the evidence for health smart homes and home-based consumer health technologies as a public health intervention for independent aging: A systematic review. Int. J. Med. Inform. **82**(7), 565–579 (2013)
23. M. Memon et al., Ambient assisted living healthcare frameworks, platforms, standards, and quality attributes. Sensors **14**(3), 4312–4341 (2014)
24. L.A. Pastalan, *Aging in Place: The Role of Housing and Social Supports* (Routledge, London, 2013)
25. K. Connelly, M. Mokhtari, T.H. Falk, Approaches to understanding the impact of technologies for aging in place: A mini-review. Gerontology **60**(3), 282–288 (2014)
26. M.N.K. Boulos et al., Mobile medical and health apps: State of the art, concerns, regulatory control and certification. Online J. Public Health Inform. **5**(3), 229 (2014)
27. A. Queirós et al., Usability, accessibility and ambient-assisted living: A systematic literature review. Univ. Access Inf. Soc. **14**(1), 57–66 (2015)
28. World Health Organization, *mHealth: New Horizons for Health Through Mobile Technologies* (World Health Organization, Geneva, 2011)
29. A.S.M. Mosa, I. Yoo, L. Sheets, A systematic review of healthcare applications for smartphones. BMC Med. Inform. Decis. Mak. **12**(1), 67 (2012)
30. World Health Organization, *Active Ageing: A Policy Framework. A Contribution of the World Health Organization to the Second United Nations World Assembly on Ageing* (World Health Organization, Geneva, 2002)
31. C. Arambepola et al., The impact of automated brief messages promoting lifestyle changes delivered via mobile devices to people with type 2 diabetes: A systematic literature review and meta-analysis of controlled trials. J. Med. Internet Res. **18**(4), e86 (2016)
32. M. Jaana, G. Paré, Home telemonitoring of patients with diabetes: A systematic assessment of observed effects. J. Eval. Clin. Pract. **13**(2), 242–253 (2007)
33. A. Queirós et al., Technologies for ageing in place to support home monitoring of patients with chronic diseases, in *BIOSTEC 2017: Proceedings of the 10th International Joint Conference on Biomedical Engineering Systems and Technologies, Volume 5: HEALTHINF*, (SciTePress, Setúbal, 2017), pp. 66–76
34. F. Verhoeven et al., The contribution of teleconsultation and videoconferencing to diabetes care: A systematic literature review. J. Med. Internet Res. **9**(5), e37 (2007)

35. J. Baron, H. McBain, S. Newman, The impact of mobile monitoring technologies on glycosylated hemoglobin in diabetes: A systematic review. J. Diabetes Sci. Technol. **6**(5), 1185–1196 (2012)
36. O. El-Gayar et al., A systematic review of IT for diabetes self-management: Are we there yet? Int. J. Med. Inform. **82**(8), 637–652 (2013)
37. M. van Vugt et al., Use of behavioral change techniques in web-based self-management programs for type 2 diabetes patients: Systematic review. J. Med. Internet Res. **15**(12), e279 (2013)
38. C.K.L. Or, D. Tao, Does the use of consumer health information technology improve outcomes in the patient self-management of diabetes? A meta-analysis and narrative review of randomized controlled trials. Int. J. Med. Inform. **83**(5), 320–329 (2014)
39. Z. Huang et al., Management of endocrine disease: Effects of telecare intervention on glycemic control in type 2 diabetes: A systematic review and meta-analysis of randomized controlled trials. Eur. J. Endocrinol. **172**(3), R93–R101 (2015)
40. H.D. Tildesley, M.D. Po, S.A. Ross, Internet blood glucose monitoring systems provide lasting glycemic benefit in type 1 and 2 diabetes: A systematic review. Med. Clin. **99**(1), 17–33 (2015)
41. H. Riazi et al., Managing diabetes mellitus using information technology: A systematic review. J. Diabetes Metab. Disord. **14**(1), 49 (2015)
42. L.F. Garabedian, D. Ross-Degnan, J.F. Wharam, Mobile phone and smartphone technologies for diabetes care and self-management. Curr. Diab. Rep. **15**(12), 109 (2015)
43. A. Conway, S.C. Inglis, R.A. Clark, Effective technologies for noninvasive remote monitoring in heart failure. Telemed. e-Health **20**(6), 531–538 (2014)
44. Y. Ciere, M. Cartwright, S.P. Newman, A systematic review of the mediating role of knowledge, self-efficacy and self-care behaviour in telehealth patients with heart failure. J. Telemed. Telecare **18**(7), 384–391 (2012)
45. S.I. Chaudhry et al., Telemonitoring for patients with chronic heart failure: A systematic review. J. Card. Fail. **13**(1), 56–62 (2007)
46. S. Dang, S. Dimmick, G. Kelkar, Evaluating the evidence base for the use of home telehealth remote monitoring in elderly with heart failure. Telemed. e-Health **15**(8), 783–796 (2009)
47. N. Nakamura, T. Koga, H. Iseki, A meta-analysis of remote patient monitoring for chronic heart failure patients. J. Telemed. Telecare **20**(1), 11–17 (2014)
48. S. Lundell et al., Telehealthcare in COPD: A systematic review and meta-analysis on physical outcomes and dyspnea. Respir. Med. **109**(1), 11–26 (2015)
49. C. Pedone, D. Lelli, Systematic review of telemonitoring in COPD: An update. Adv. Respir. Med. **83**(6), 476–484 (2015)
50. A. Ramadas et al., Web-based interventions for the management of type 2 diabetes mellitus: A systematic review of recent evidence. Int. J. Med. Inform. **80**(6), 389–405 (2011)
51. K. Pal et al., Computer-based diabetes self-management interventions for adults with type 2 diabetes mellitus. Cochrane Database Syst. Rev. **3**, CD008776 (2013)
52. C.L. Jackson et al., A systematic review of interactive computer-assisted technology in diabetes care: Interactive information technology in diabetes care. J. Gen. Intern. Med. **21**(2), 105–110 (2006)
53. M. Cassimatis, D.J. Kavanagh, Effects of type 2 diabetes behavioural telehealth interventions on glycaemic control and adherence: A systematic review. J. Telemed. Telecare **18**(8), 447–450 (2012)
54. C.W. Hunt, Technology and diabetes self-management: An integrative review. World J. Diabetes **6**(2), 225 (2015)
55. D. Cacciagrano, F. Corradini, R. Culmone, Resourcehome: An rfid-based architecture and a flexible model for ambient intelligence, in *2010 Fifth International Conference on Systems*, (IEEE, Piscataway, 2010), pp. 6–11
56. A. Almeida et al., An approach to dynamic knowledge extension and semantic reasoning in highly-mutable environments, in *3rd Symposium of Ubiquitous Computing and Ambient Intelligence 2008*, (Springer, Berlin, Heidelberg, 2009)

57. A. Fortier et al., Dealing with variability in context-aware mobile software. J. Syst. Softw. **83**(6), 915–936 (2010)
58. R. Hartmann et al., Camera-based system for tracking and position estimation of humans, in *2010 Conference on Design and Architectures for Signal and Image Processing (DASIP)*, (IEEE, Piscataway, 2010)
59. P. Siciliano et al., A networked multisensor system for ambient assisted living application, in *2009 3rd International Workshop on Advances in Sensors and Interfaces*, (IEEE, Piscataway, 2009)
60. N. Bicocchi, M. Mamei, F. Zambonelli, Detecting activities from body-worn accelerometers via instance-based algorithms. Pervasive Mob. Comput. **6**(4), 482–495 (2010)
61. N. Mirarmandehi, H.R. Rabiee, An asynchronous dynamic bayesian network for activity recognition in an ambient intelligent environment, in *5th International Conference on Pervasive Computing and Applications*, (IEEE, Piscataway, 2010)
62. A. Abbasi, A. Rehman, A. Hussain, N.V. Afzulpurkar, Towards context-adaptive affective computing, in *ECTI-CON2010: The 2010 ECTI International Conference on Electrical Engineering/Electronics, Computer, Telecommunications and Information Technology*, (IEEE, Piscataway, 2010)
63. E. Leon et al., Affect-aware behaviour modelling and control inside an intelligent environment. Pervasive Mob. Comput. **6**(5), 559–574 (2010)
64. F. Casino et al., Healthy routes in the smart city: A context-aware mobile recommender. IEEE Softw. **34**(6), 42–47 (2017)
65. G. Villarrubia et al., Wireless sensor networks to monitoring elderly people in rural areas, in *Ambient Intelligence-Software and Applications*, (Springer, Cham, 2014), pp. 171–181
66. Z. Liouane et al., A genetic-based localization algorithm for elderly people in smart cities, in *Proceedings of the 14th ACM International Symposium on Mobility Management and Wireless Access*, (ACM, New York, 2016)
67. F. Chuang, Construction and value study of IT-based smart senior citizens' communities, in *2014 Sixth International Conference on Measuring Technology and Mechatronics Automation*, (IEEE, Los Alamitos, 2014)
68. C.A. Gomes et al., VIAS | Viseu InterAge stories: Developing an app to foster social inclusion and healthy lifestyles, in *2017 International Symposium on Computers in Education (SIIE)*, (IEEE, Piscataway, 2017)
69. C. Zschippig, T. Kluss, Gardening in ambient assisted living. Urban For. Urban Green. **15**, 186–189 (2016)
70. G. Trencher, A. Karvonen, Stretching "smart": Advancing health and well-being through the smart city agenda. Local Environ. **24**, 610–627 (2019)
71. N. Bryant et al., IoT and smart city services to support independence and wellbeing of older people, in *2017 25th International Conference on Software, Telecommunications and Computer Networks (SoftCOM)*, (IEEE, Piscataway, 2017)
72. R. Mulero et al., An AAL system based on IoT technologies and linked open data for elderly monitoring in smart cities, in *2017 2nd International Multidisciplinary Conference on Computer and Energy Science (SpliTech)*, (IEEE, Piscataway, 2017)
73. D. López-de-Ipiña et al., Towards ambient assisted cities and citizens, in *2013 27th International Conference on Advanced Information Networking and Applications Workshops*, (IEEE, Piscataway, 2013)
74. A. Kötteritzsch, M. Koch, S. Wallrafen, Expand your comfort zone! smart urban objects to promote safety in public spaces for older adults, in *Proceedings of the 2016 ACM International Joint Conference on Pervasive and Ubiquitous Computing: Adjunct*, (ACM, New York, 2016)
75. A. Stibe, K. Larson, Persuasive cities for sustainable wellbeing: Quantified communities, in *International Conference on Mobile Web and Information Systems*, (Springer, Cham, 2016)
76. S. Santana et al., The Domiciliary Support Service in Portugal and the change of paradigm in care provision. Int. J. Integr. Care **7**(1), e01 (2007)
77. V.T. Cruz et al., Web-based cognitive training: Patient adherence and intensity of treatment in an outpatient memory clinic. J. Med. Internet Res. **16**(5), e122 (2014)

78. N.P. Rocha et al., Information persistence services designed to support home care. JMIR Med. Inform. **3**(1), e15 (2015)
79. S. Social, *Redes Locais de Intervenção Social (RLIS)* (Segurança Social, Lisbon, 2016)
80. World Health Organization, *International Classification of Functioning, Disability and Health* (WHO, Geneve, 2001)
81. A. Queirós et al., ICF inspired personas to improve development for usability and accessibility in ambient assisted living. Procedia Comput. Sci. **27**, 409–418 (2014)
82. A.I. Martins et al., Applications to help local authorities to support community-dwelling older adults, in *International Conference on Information Technology & Systems*, (Springer, Cham, 2019)
83. N.P. Rocha et al., The social platform: Profiling FHIR to support community-dwelling older adults. J. Med. Syst. **43**(4), 86 (2019)
84. IETF, *JSON-Based Open Standard (RFC 7519)* (Internet Engineering Task Force, Washington, DC, 2015)
85. M. Rosa, J.P. Barraca, N.P. Rocha, Access control for social care platforms using fast healthcare interoperability resources, in *World Conference on Information Systems and Technologies*, (Springer, Cham, 2019)
86. M. Rosa, J.P. Barraca, N.P. Rocha, Logging integrity with blockchain structures, in *World Conference on Information Systems and Technologies*, (Springer, Cham, 2019)
87. J. Rainho et al., Testing a significant event detector in lab environment, in HCist – the Conference on Health and Social Care Information Systems and Technologies, 2019
88. A. Dias et al., Interoperability in pervasive health: Is it tackled as a priority?, in Healthinf, 2018

Intelligent Wearable IOT Continuous Monitoring System for Elderly Based on Deep Learning Algorithm

Ahmad Hoirul Basori

1 Introduction

IoT is the acronym for Internet of Things, and it is a new technology that enables people to interact wirelessly with devices, a sensor that attached into physical object/environment [1]. On the other hand, the elderly and people keep increasing over time. Therefore, IoT has come up with technology for controlling the home equipment such as a lamp, water pump, ac or heater through the internet connection. Some researcher focused on providing the patient with wearable IoT equipment that can be used for rehabilitation [2]. The wearable device also planted or attached into the skin to analyse perspiration to conduct analysis on human physiological signal [3]. Nowadays, most of the elderly stayed in the home unattended due to their relative has activities such as working, shopping or any other activities that need to be done outside their home. Therefore, it becomes an obstacle for both parties: the elderly and their family members to do activities separately.

Therefore, this research proposed a unique solution by providing great devices and apps for the elderly and their relatives to continuously observe the health of the elder or patient without blocking out their activities. The structure of this chapter can be described as follows: The first section focuses on the introduction of the project that can explain the general idea of the project, followed by related works that discuss the theory, tools and algorithm that are used in the similar works. Section 3 concentrates on methodology and material of the project and then Sect. 4 on results and discussion. The last section depicts a conclusion of the whole project and future work of the research.

A. H. Basori (✉)
Faculty of Computing and Information Technology Rabigh, King Abdulaziz University, Makkah, Kingdom of Saudi Arabia
e-mail: abasori@kau.edu.sa

N. Gupta, S. Paiva (eds.), *IoT and ICT for Healthcare Applications*, EAI/Springer Innovations in Communication and Computing,
https://doi.org/10.1007/978-3-030-42934-8_15

2 Related Works

The technology of IoT that relatively new has been exposed excessively in most of the field recently. The IoT is applied in medical, business, education or even industry [1–4]. The trend of self-monitoring for health also increased rapidly because of the era where everybody busy with their activities. While elderly people number also increased dramatically by 2020 [2], this condition has triggered the researcher to produce an innovative solution that can be attached to their clothes and did not suppress their daily activities [3]. The real-time and portable device that is capable of watching the heart based on digital signal processor (DSP) has been introduced almost two decades ago. Therefore, the appearance of IoT has attracted society attention easily due to the internet connection, and hardware technology has become cheaper and easy to be used in general [5]. Wearable devices and IoT have a lot of variation depending on their purpose of use: skin instrument, gas radar, cardio/pulse sensor, a soil sensor, etc. [6–10]. The sensor for human gait analysis also widely used to measure body activities or abnormality of the human body [11, 12]. While everything seems connected, security and automation can be considered an important factor in the development of IoT device and system [13, 14]. Al-Makhadmeh, Z., and Tolba A. (2019) focused on collecting and analysing the data set of the patient before and after the heart attack. They used higher order Boltzmann deep belief neural network to train the network about their past disease and come up with a solution to reduce heart disease death rate [15].

Meanwhile, some researchers also intensively studied IoT technology to come up with a wearable device for heart monitoring [16, 17]. Furthermore, the others concentrated on the acceptance model of an individual to wear an IoT device. They said that people tend to use a wearable device if they have confidence in devices [18].

3 Research Method and Material

This chapter discusses the material and research method used to develop the current project. The material used for development can be detailed out as follows:

- IoT Devices: Arduino Board or Raspberry Pi.
- Pulse sensor.
- Heat sensor.
- Cable/wired.
- Battery.
- Bluetooth transmitter/USB data cable.
- PC/laptop.
- Smartphone (Android based).

Figure 1 shows the system architecture that consists of smart cloth, database (cloud) and then mobile apps. The smart cloth has IoT devices that can capture the heart beat (HB), heart beat inconsistency (HBI) and human body temperature (HBT). Afterwards, data will continuously sent to NoSQL database (Google Firebase). The data in the cloud database will be pulled out by mobile apps that continuously displayed the heart rate of the elderly. The mobile apps will notify the relative that hold information for their elder. In case of an emergency, apps will alert the mobile phone and send the emergency assistance request to the health authority or nearby family members/neighbourhood to provide immediate assistance.

Figure 2 shows that the IoT device is attached to the cloth of the elderly. At the final stage, it will be located in a special pocket that will not suppress any movement

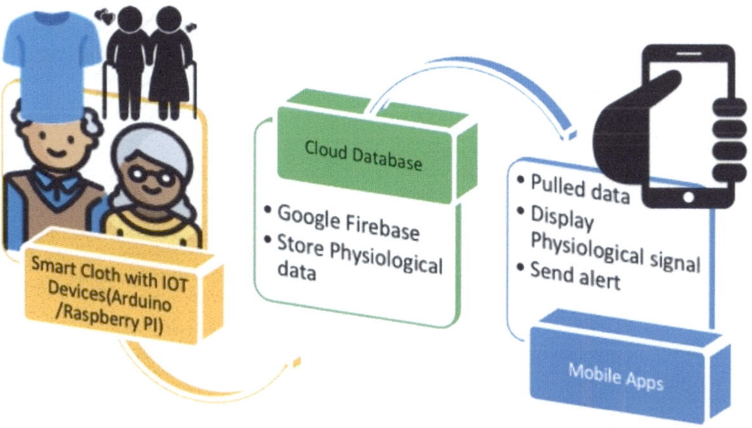

Fig. 1 System architecture

Fig. 2 The prototype of wearable smart cloth

of the elderly. Furthermore, the detail of the business process of our proposed system is depicted as the use case diagram in Fig. 3. It has shown that user actor can observe the heart rate monitoring, body temperature and request for immediate health assistance. While system actor is responsible for sending data to the cloud database (Google firebase), if a certain condition is met (HB increase tremendously, or immediate body temperature drops or rise), it will send a notification to the guardian/relative as emergency information.

Table 1 is a sample of use case specification for heart monitoring that consists of a flow of use case, pre- and postcondition, as well as an exception condition.

In addition to the use case diagram, we generate the activity diagram for two conditions for heart beat (HB) and body temperature (BT). Figure 4 shows that if the heartbeat is either lower than 60 or more than 100, it will send notification and request for immediate health assistance.

While Fig. 5 shows if body temperature more or less than 37.5 as standard human body temperature, it will trigger alert to the relative/guardian of the elderly.

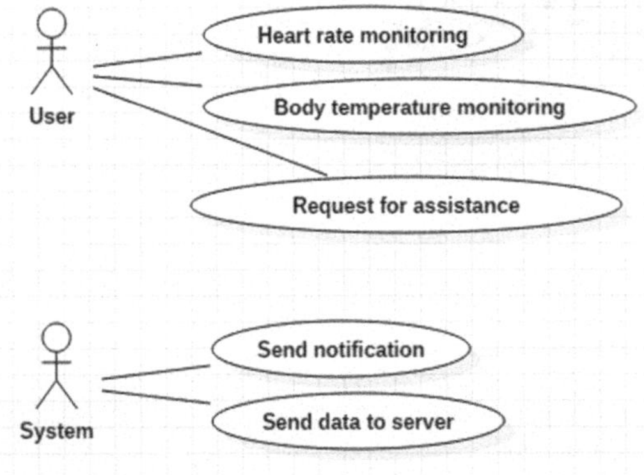

Fig. 3 Use case diagram for WIOTE system

Table 1 Use case specification for heart rate monitoring

Use case name:	Heart rate monitoring	
Scenario:	Display the HB data	
Description:	Load the HB data sent by IoT devices to the cloud database	
Actors:	User	
Related use cases:		
Preconditions:	HB data captured by IoT devices and sent to the database	
Postconditions:	Display the chart of HB	
The flow of activities:	User	System
Exception conditions:	Load HB data and display in the chart	The chart displayed successfully
	Data empty Data pull error due to network connection	

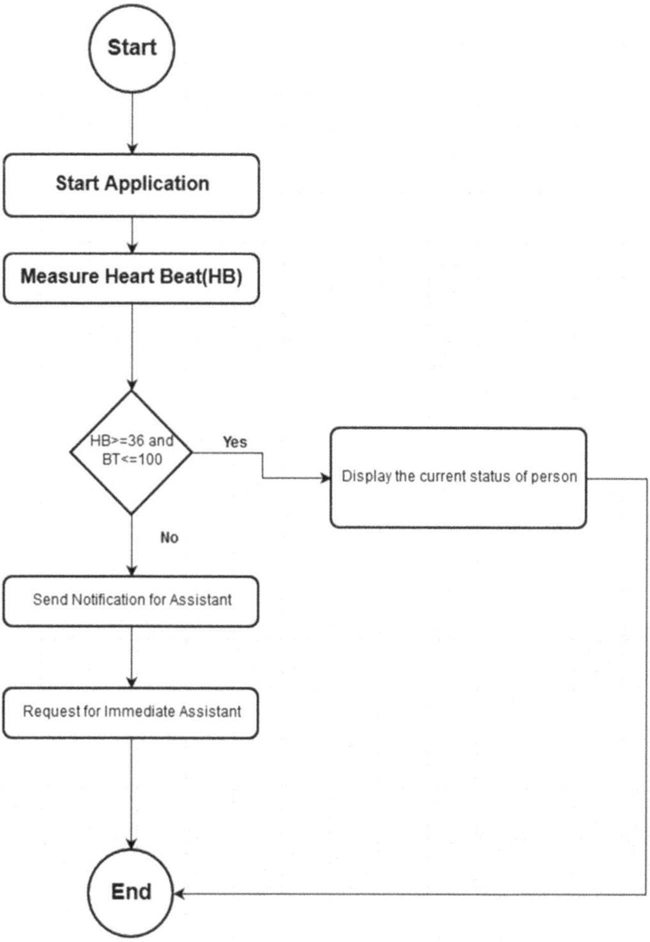

Fig. 4 Activity diagram for heart beat case

4 Design and Prototyping

The implementation of the projects involved three main components, as mentioned previously: IoT devices, cloud database and mobile apps. The IoT device used for this project is Arduino UNO. However, it can be replaced easily by raspberry Pi3 (refer to Figs. 6 and 7).

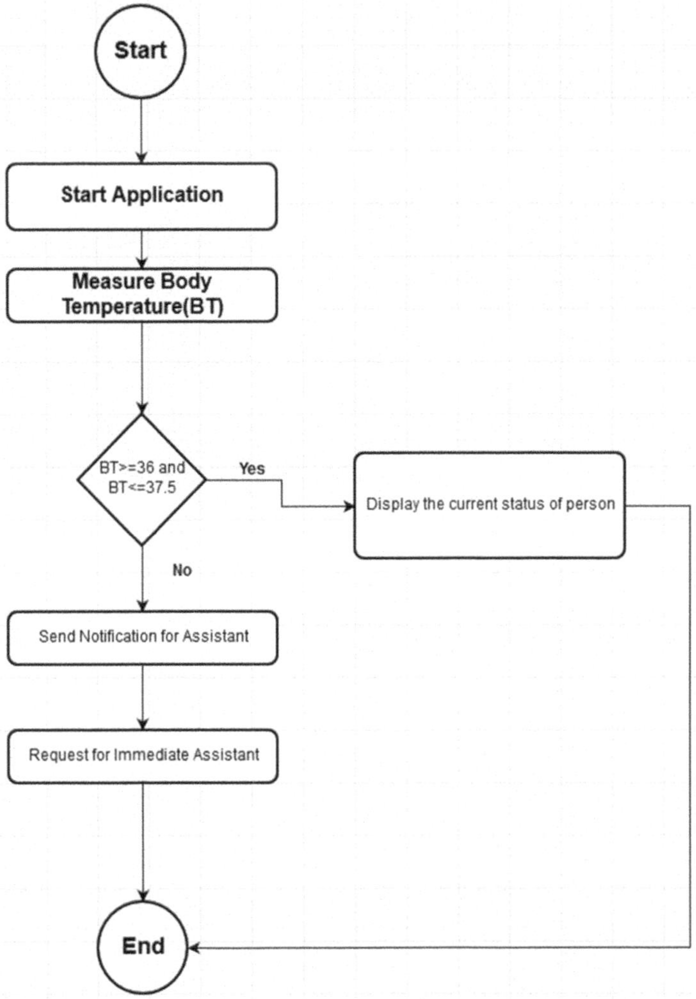

Fig. 5 Activity diagram for body temperature case

The data will be sent wirelessly through Bluetooth transmitter due to the IoT devices attached to the elderly cloth. Arduino board will keep sending the HB and BT to the cloud server and store once it synchronises with the server. Figure 8 shows the fragment of code for reading data from the pulse sensor, which initiates BPM (beat per minute). The initial test uses Arduino UNO R3 with pulse sensor from

Fig. 6 Arduino Uno, Image courtesy of Arduino [19]

Fig. 7 Raspberry Pi 3, Image courtesy of Raspberry [20]

pulsesensor.com (refer to Fig. 8), while body temperature is measured using the LM35 sensor (refer to Fig. 9). The network communication is established by using Bluetooth HC-5 module.

In Fig. 2, we have shown the initial design of the wearable IoT; the cloth will have a pocket to keep the mainboard; meanwhile, the pulse and temperature sensor is attached to the skin of the body with tape. We aim to achieve a 96 kHz sampling

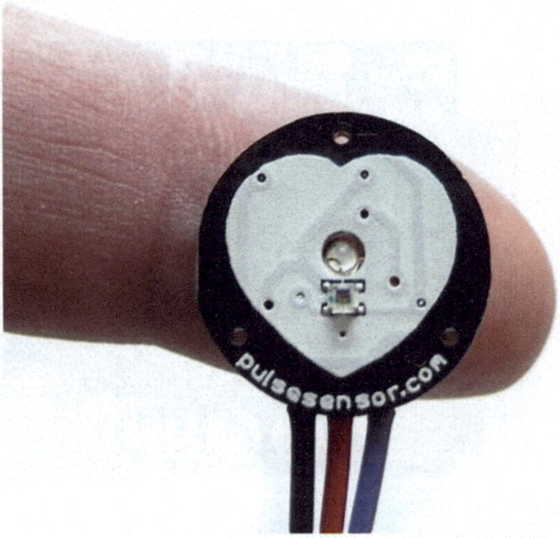

Fig. 8 Pulse sensor from pulsesensor.com. (Image courtesy of https://pulsesensor.com/)

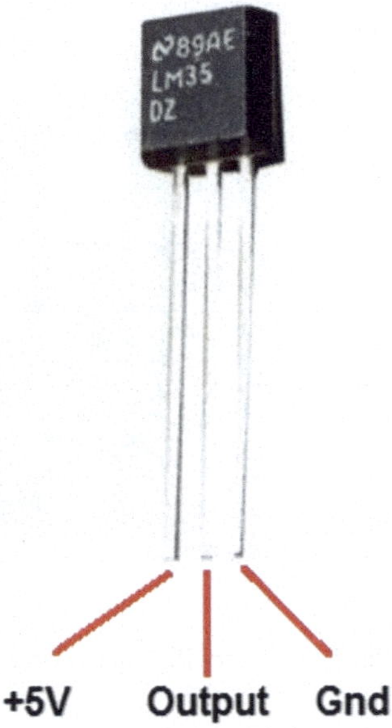

Fig. 9 Temperature sensor from pulsesensor.com. (Image courtesy of http://arduinolearning.com/code/lm35-temperature-sensor.php/)

rate with transfer speed of 115,200 bits/s (refer to Fig. 10). However, during testing, we still only achieve around 56 kHz. The cause might have come from Bluetooth transmitter limitation, while in the future we intend to use Wi-Fi card to amplify the transfer and sampling rate, as well as the coverage of the area.

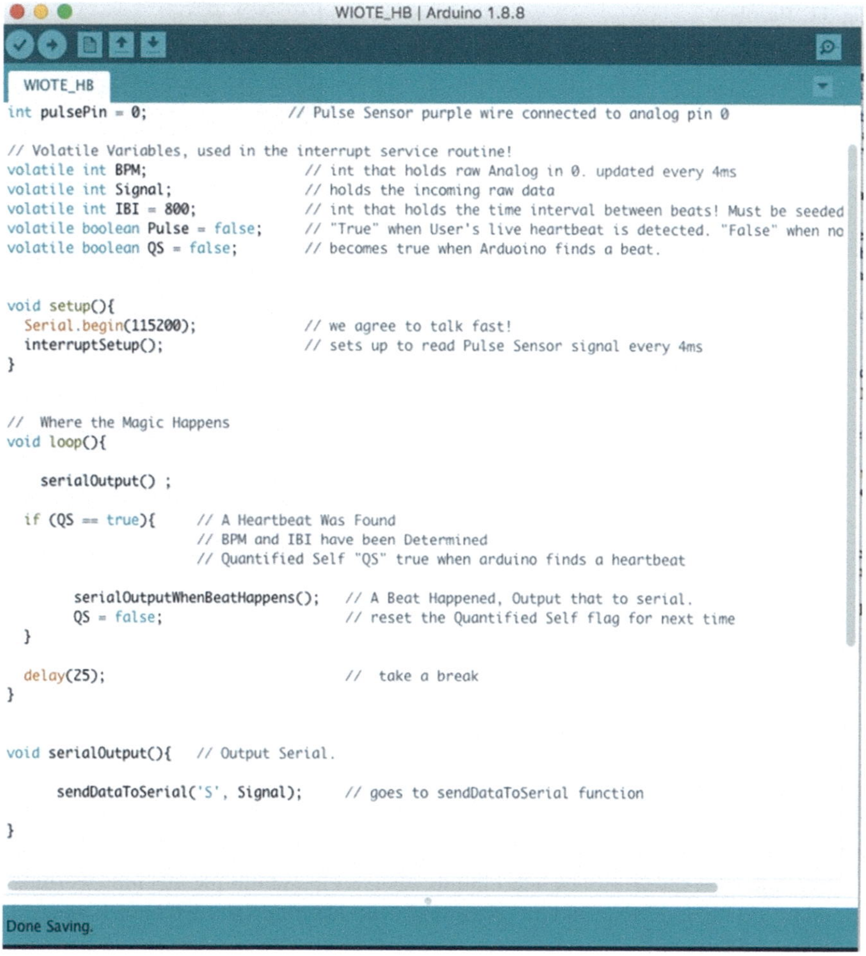

WIOTE_HB | Arduino 1.8.8

WIOTE_HB

```
int pulsePin = 0;                 // Pulse Sensor purple wire connected to analog pin 0

// Volatile Variables, used in the interrupt service routine!
volatile int BPM;                   // int that holds raw Analog in 0. updated every 4ms
volatile int Signal;                // holds the incoming raw data
volatile int IBI = 800;             // int that holds the time interval between beats! Must be seeded
volatile boolean Pulse = false;     // "True" when User's live heartbeat is detected. "False" when no
volatile boolean QS = false;        // becomes true when Arduoino finds a beat.


void setup(){
  Serial.begin(115200);             // we agree to talk fast!
  interruptSetup();                 // sets up to read Pulse Sensor signal every 4ms
}


//  Where the Magic Happens
void loop(){

    serialOutput() ;

  if (QS == true){     // A Heartbeat Was Found
                       // BPM and IBI have been Determined
                       // Quantified Self "QS" true when arduino finds a heartbeat

        serialOutputWhenBeatHappens();    // A Beat Happened, Output that to serial.
        QS = false;                       // reset the Quantified Self flag for next time
  }

  delay(25);                              //  take a break
}


void serialOutput(){   // Output Serial.

      sendDataToSerial('S', Signal);      // goes to sendDataToSerial function

}
```

Done Saving.

Fig. 10 Arduino code for reading data from the pulse sensor

Besides, the mobile apps will keep pulling the data from the cloud database and displayed the graph of HB, as shown in Fig. 11. In the case of irregularity emerged either in a heartbeat or in a body temperature, it will distress an alert towards the guardian of the elderly and request for immediate health assistance.

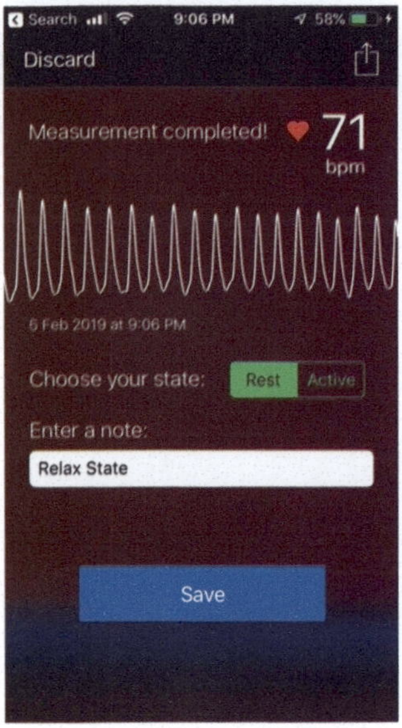

Fig. 11 Heartbeat monitoring through mobile apps

5 Deep Learning Analysis for Elderly

The initial prototype has successfully read the pulse data of elderly heart, record it into a database and then displayed in the mobile apps. To give deep observation towards proposed elderly monitoring system, we use a standard data set from Janosi et al. [21], Aha et al. [22] and Gennari et al. [23], which contains more detailed heart failure description [21–23]. The data have been filtered and only select the elderly data with age 60+. The heart failure analysis uses multiple machine learning techniques such as KNN, random forest and deep learning. The KNN or K-nearest neighbour relied on the distance function such as:

$$D_{\text{Euclidean}} = \sqrt{\sum_{i=1}^{k} (x_i - y_i)^2} \tag{1}$$

$$D_{\text{Manhanttan}} = \sum_{i=1}^{k} |x_i - y_i| \tag{2}$$

$$D_{\text{Minkowski}} = \left(\sum_{i=1}^{k} (|x_i - y_i|)^q\right)^{1/q} \tag{3}$$

Random Forest Algorithm

1. Find and select "**k**" from features set "**m**", **k<<m**
2. Between "**k**" features, calculate the "**d**" **note** based on best split point characteristic
3. Divide the node into **child nodes** by **best-split approach**
4. Repeat **1-3** phases till "**l**" number of nodes has been reached
5. Generate the forest by reiterating process number **1-4** for "**n**" to produce trees with "**n**" **numbers**.

Fig. 12 Random forest algorithm

If the *k* value selection has been done, then the prediction of KNN might be computed through Eq. 4:

$$Y = \frac{1}{K}\sum_{i=1}^{k} y_i \tag{4}$$

Besides, the random forest algorithm is depicted by using the steps in Fig. 12.

While, the cost function *C* of deep learning respect to any weight (ω) or bias inside the network(b), for backpropagation, we need to consider two assumptions with quadratic function as depicted in Eq. 5.

$$C = \frac{1}{2n}\sum_{x} \left\| y(x) - a^{L}(X) \right\|^{2} \tag{5}$$

where *n*, characterize the total number of training, the sigma $\sum$ is the summation of discrete training, while $x;y = y(x)$ signify the output;L represent layers number and $a^L = a^L(x)$ denote activation vector if *x* used as an input.

The statistic of the data set describes 57.8% male and 42.2% female as portrayed in Fig.13.

There are four main types of pain in the data set: type 1: typical angina, type 2: atypical angina, type 3: non-anginal pain and type 4: asymptomatic, refer to Fig. 14. Figure 14 represents the blood pressure in the resting/relax position by a boxplot chart, refer to Fig. 14, the value hanging around 120–150.

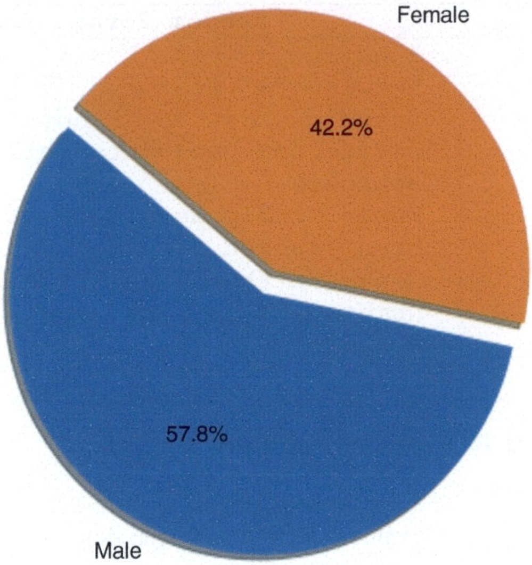

Fig. 13 Gender statistic

Figure 15 illustrates the general view of blood pressure condition of the patients, along with age as the second parameter. This figure reveals the correlation between age and resting blood pressure level.

Figure 16 provides a general graphic of age, blood pressure, cholesterol level and maximum heart rate of the patient.

The heatmap graphic, which is shown by Fig. 17, describes that the pain types reach quite high value of 0.48, followed by a high correlation between the major vessel and the old peak of depression situation test. Afterwards, the deep learning test along with random forest and KNN has shown that deep learning neural network is little bit superior compared to the random forest and superior compared to KNN with 75% accuracy (refer to Fig. 18).

Then, we also provide ROC analysis and confusion matrix (refer to Figs. 19 and 20) for the deep learning neural network. It seems the deep learning neural network and random forest are comparable with each other, and both are superior to KNN. It can be predicted with accuracy 76%. We still be working towards greater recognition rate in the future.

6 Conclusion and Future Works

Continuous monitoring is essential for the elderly when they have stayed alone in their house without an assistant. The proposed system has enabled the new concept of smart health monitoring that allows both patient and their family to keep mobile

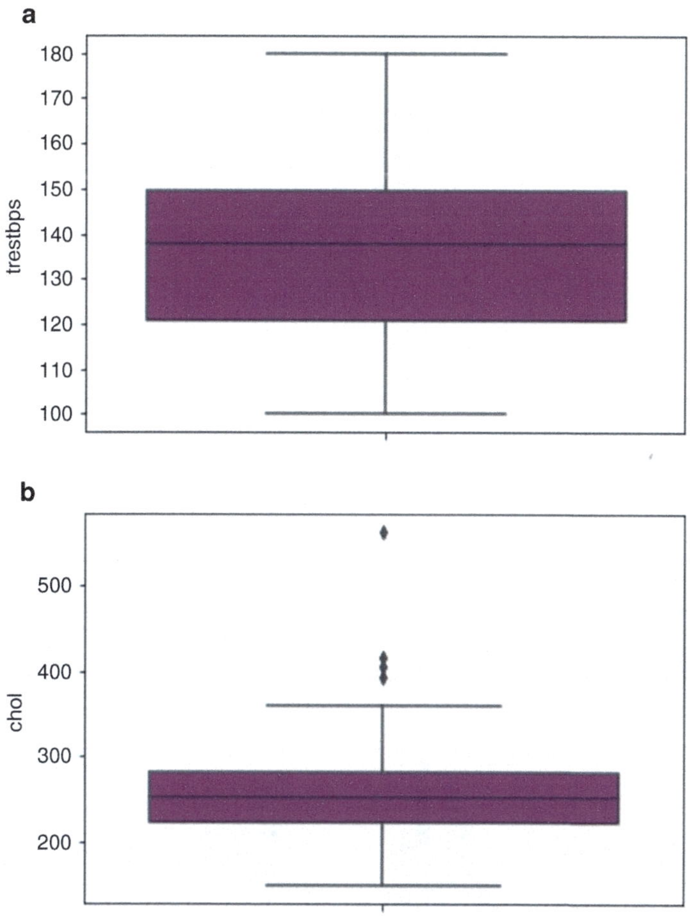

Fig. 14 (**a**) Blood pressure range – resting position, while (**b**) shows the cholesterol level in mg/dl

and live in confidence. Elderly may do their regular activities without lying on the bed, and their family members are also able to do their job and keep watching them remotely. The proposed system consists of IoT devices that are attached to the elderly cloth and the sensor with the skin. Initially, the apps can calculate and provide suggestion intelligently by analysing the data that sent through IoT sensor and compare with the standard physical signal. Currently, the proposed system still has limited data for complex analysis; however as mentioned, in Sect. 5, we did an extra exploration towards existing data set that is related to heart disease with deep

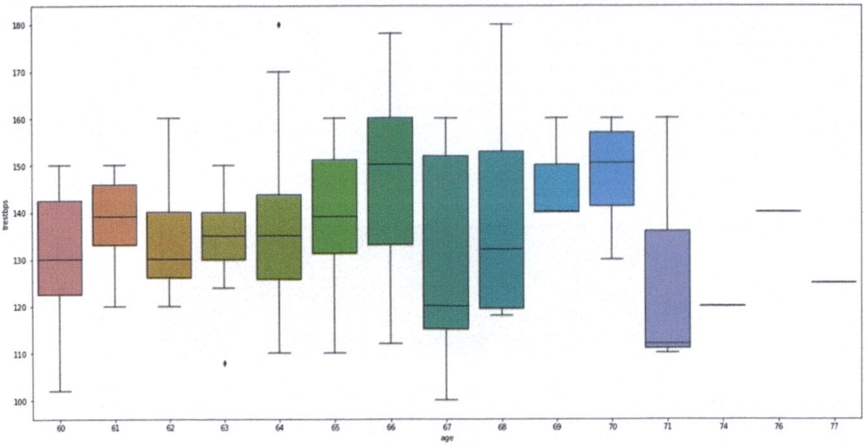

Fig. 15 Age and blood pressure correlation

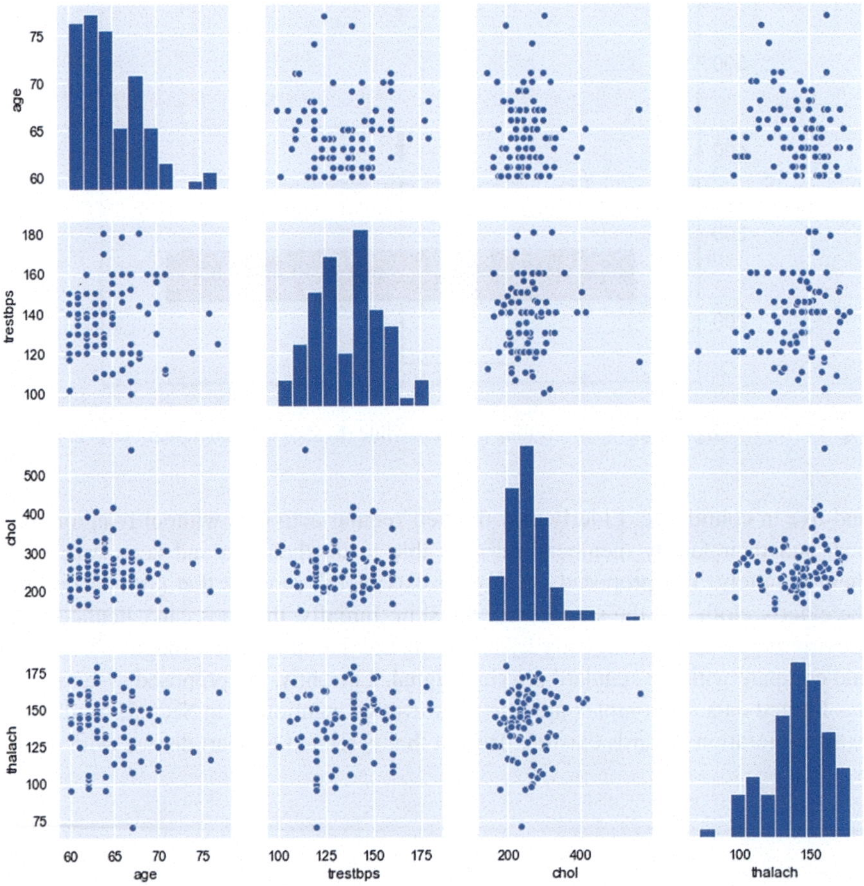

Fig. 16 The graphic of age, blood pressure, cholesterol level and maximum heart rate

Fig. 17 The graphic of age, blood pressure, cholesterol level and maximum heart rate

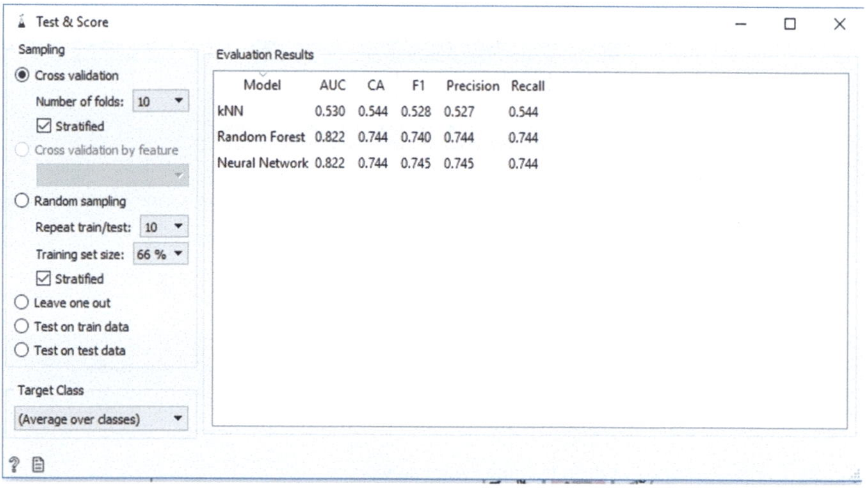

Fig. 18 Evaluation result of deep learning analysis testing

learning analysis. The data set is preselected only for elderly with age more than 60. Also, the deep learning analysis has proven to be useful for giving analysis towards the heart failure/disease with 76% accuracy. This percentage might be improved further in the future works with a more suitable data set. The decision

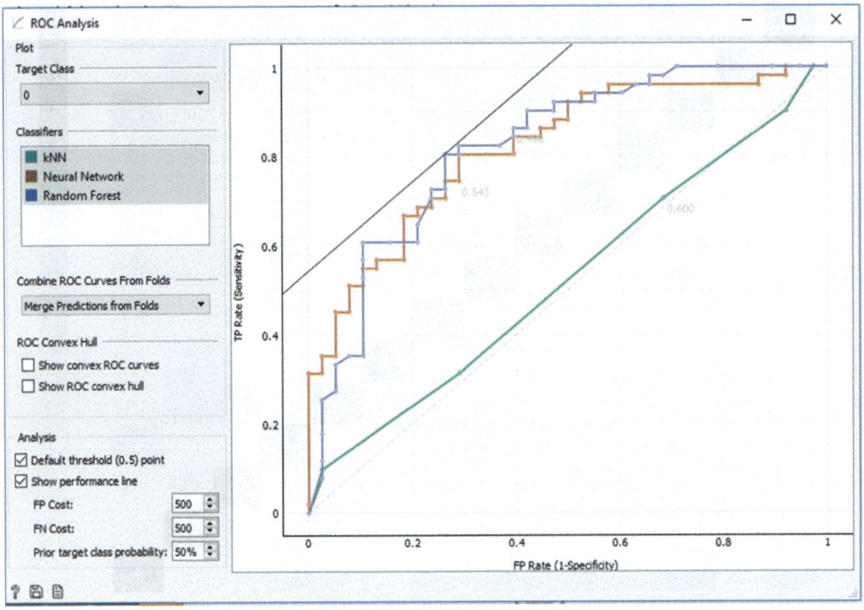

Fig. 19 ROC analysis of three classifiers

Fig. 20 Confusion matrix of deep learning neural network

of deep learning analysis will notify or not notify the relative according to the elderly condition. The future improvement will focus towards a more advanced wearable sensor that can observe oxygen rate inside blood and sweat and even with a CCTV/portable drone to provide video data and analysis.

Acknowledgement This work was supported by the Deanship of Scientific Research (DSR), King Abdulaziz University, Jeddah Saudi Arabia. The authors, therefore, gratefully acknowledge the DSR technical and financial support.

References

1. LeHong H, Velosa A. Hype cycle for the Internet of Things, 2014 [Internet]. Stamford (CT): Gartner Inc.; 2014 [cited at 2017 Jan 25]. Available from: https://www.gartner.com/doc/2804217/hype-cycle-internet-things
2. Business Wire, Finding success in the new IoT ecosystem: market to reach $3.04 trillion and 30 billion connected "Things" in 2020, IDC says [Internet]. San Francisco (CA): Business Wire; 2014 [cited at 2017 Jan25]
3. S. Patel, H. Park, P. Bonato, L. Chan, M. Rodgers, A review of wearable sensors and systems with application in rehabilitation. J. Neuroeng. Rehabil. **9**(1), 21 (2012)
4. W. Gao, S. Emaminejad, H.Y. Nyein, S. Challa, K. Chen, A. Peck, et al., Fully integrated wearable sensor arrays for multiplexed in situ perspiration analysis. Nature **529**(7587), 509–514 (2016)
5. E. Jovanov, P. Gelabert, B. Wheelock, R. Adhami, P. Smith, Real time portable heart monitoring using low power DSP. *Proceedings of International Conference on Signal Processing Applications and Technology (ICSPAT)*; 2000 Oct 16–19; Dallas, TX, pp. 16–9
6. D. Evans, *The Internet of Things: How the Next Evolution of the Internet is Changing Everything [Internet].* San Jose (CA): Cisco Internet Business Solutions Group; 2011 [cited at 2017 Jan 25]
7. S. Xu, Y. Zhang, L. Jia, K.E. Mathewson, K.I. Jang, J. Kim, et al., Soft microfluidic assemblies of sensors, circuits, and radios for the skin. Science **344**(6179), 70–74 (2014)
8. T. Martin, E. Jovanov, D. Raskovic, Issues in wearable computing for medical monitoring applications: a case study of a wearable ECG monitoring device. *Proceedings of the 4th International Symposium on Wearable Computers*; 2000 Oct 16–17; Atlanta, GA, pp. 43–9
9. U. Anliker, J.A. Ward, P. Lukowicz, G. Troster, F. Dolveck, M. Baer, et al., AMON: a wearble multiparameter medical monitoring and alert system. IEEE Trans. Inf. Technol. Biomed. **8**(4), 415–427 (2004)
10. Gas sensor developer kit [Internet]. Newark (CA): SpecSensors; c2015 [cited at 2017 Jan 25]. Available from: https://www.spec-sensors.com/product-category/gassensorgassensor-developer-kits/
11. P.H. Veltink, H.B. Boom, 3D movement analysis using accelerometry theoretical concepts, in *Neuroprosthetics: From Basic Research to Clinical Applications*, ed. by A. Pedotti, M. Ferrarin, J. Quintern, R. Reiner, (Springer, Berlin, 1996), pp. 317–326
12. M.J. Mathie, A.C. Coster, N.H. Lovell, B.G. Celler, Detection of daily physical activities using a triaxial accelerometer. Med. Biol. Eng. Comput. **41**(3), 296–301 (2003)
13. J. Wilson, Infonetics survey unveils businesses' plans for mobile security, Internet of Things, wearables [Internet]. London: HIS Inc.; 2015 [cited at 2017 Jan 25]. Available from: https://technology.ihs.com/527159/infoneticssurvey-unveils-businesses-plans-for-mobile-securityinternet-of-things-wearables
14. S. Neubert, *Automation Requires Process Information Technologies* [Internet]. Rostock: Center for Life Science Automation (celisca); c2016 [cited at 2017 Jan 25]
15. Z. Al-Makhadmeh, A. Tolba, Utilizing IoT wearable medical device for heart disease prediction using higher order Boltzmann model: a classification approach. Measurement (Elsevier) **147**, 106815 (2019)
16. V. Pardeshi, S. Sagar, S. Murmurwar, P. Hage (2017) Health monitoring systems using IoT and Raspberry Pi – a review. International Conference on Innovative Mechanisms for Industry Applications (ICIMIA 2017), IEEE

17. P. Sasidharan, T. Rajalakshmi, U. Snekhalatha, Wearable cardiorespiratory monitoring device for heart attack prediction. International Conference on Communication and Signal Processing, April 4–6, 2019, India, IEEE (2019)
18. H. Yildirim, A.M.T. Ali-Eldin, A model for predicting user intention to use wearable IoT devices at the workplace. J. King Saud University Comput. Inform. Sci. (2018). https://doi.org/10.1016/j.jksuci.2018.03.001
19. Arduino, Arduino Uno, 2018., https://www.arduino.cc/en/main/boards. Last Accessed: December 2018
20. Raspberry, Raspberry PI3, 2018., https://www.raspberrypi.org/products/raspberry-pi-3-model-b-plus/. Last Accessed: December 2018
21. D.R. Janosi, A. Steinbrunn, W. Pfisterer, M. Schmid, S. Guppy, K. Lee, V. Froelicher, International application of a new probability algorithm for the diagnosis of coronary artery disease. Am. J. Cardiol. **64**, 304–310 (1989)
22. D.W. Aha, D. Kibler, Instance-based prediction of heart-disease presence with the Cleveland database (1980)
23. J.H. Gennari, P. Langley, D. Fisher, Models of incremental concept formation. Artif. Intell. **40**, 1161 (1989)

Index

N. Gupta, S. Paiva (eds.), *IoT and ICT for Healthcare Applications*, EAI/Springer Innovations in Communication and Computing,
https://doi.org/10.1007/978-3-030-42934-8

W

The manufacturer's authorised representative in the EU is Springer Nature Customer Service Centre GmbH, Europaplatz 3, 69115 Heidelberg, Germany. If you have any concerns regarding our products, please contact ProductSafety@springernature.com

Printed and bound by CPI Group (UK) Ltd, Croydon, CR0 4YY
30/09/2026
02225503-0007